A THERANOSTIC AND PRECISION MEDICINE APPROACH FOR FEMALE-SPECIFIC CANCERS

A THERANOSTIC AND PRECISION MEDICINE APPROACH FOR FEMALE-SPECIFIC CANCERS

Edited by

RAMA RAO MALLA

Cancer Biology Lab, Department of Biochemistry and Bioinformatics, Institute of Science, GITAM (Deemed to be University), Visakhapatnam, Andhra Pradesh, India

GANJI PURNACHANDRA NAGARAJU

Department of Hematology and Medical Oncology, Winship Cancer Institute, Emory University, Atlanta, GA, United States

ELSEVIER

ACADEMIC PRESS

An imprint of Elsevier

Academic Press is an imprint of Elsevier
125 London Wall, London EC2Y 5AS, United Kingdom
525 B Street, Suite 1650, San Diego, CA 92101, United States
50 Hampshire Street, 5th Floor, Cambridge, MA 02139, United States
The Boulevard, Langford Lane, Kidlington, Oxford OX5 1GB, United Kingdom

Notices
Knowledge and best practice in this field are constantly changing. As new research and experience broaden our
understanding, changes in research methods, professional practices, or medical treatment may become necessary.

Practitioners and researchers must always rely on their own experience and knowledge in evaluating and using any
information, methods, compounds, or experiments described herein. In using such information or methods they
should be mindful of their own safety and the safety of others, including parties for whom they have a
professional responsibility.

To the fullest extent of the law, neither the Publisher nor the authors, contributors, or editors, assume any liability
for any injury and/or damage to persons or property as a matter of products liability, negligence or otherwise, or from
any use or operation of any methods, products, instructions, or ideas contained in the material herein.

Library of Congress Cataloging-in-Publication Data
A catalog record for this book is available from the Library of Congress

British Library Cataloguing-in-Publication Data
A catalogue record for this book is available from the British Library

ISBN: 978-0-12-822009-2

For information on all Academic Press publications
visit our website at https://www.elsevier.com/books-and-journals

Publisher: Stacy Masucci
Acquisitions Editor: Rafael Teixeira
Editorial Project Manager: Billie Jean Fernandez
Production Project Manager: Punithavathy Govindaradjane
Cover Designer: Miles Hitchen

Typeset by SPi Global, India

Contents

Contributors

Phaniendra Alugoju
Department of Biochemistry and Molecular Biology, School of Life Sciences, Pondicherry University, Puducherry, India

Dinakara Rao Ampasala
Centre for Bioinformatics, School of Life Sciences, Pondicherry University, Puducherry, India

Neelakantan Arumugam
Department of Biotechnology, School of Life Sciences, Pondicherry University, Puducherry, India

Dariya Begum
Department of Biosciences and Biotechnology, Banasthali University, Banasthali, Rajasthan, India

L.V.K.S. Bhaskar
Department of Zoology, Guru Ghasidas Vishwavidyalaya, Bilaspur, Chhattisgarh, India

Ishita Bhattacharyya
Centre for Bioinformatics, School of Life Sciences, Pondicherry University, Puducherry, India

Narayan P. Burte
Department of Pharmacology, Viswabharathi Medical College, Kurnool, Andhra Pradesh, India

Nyshadham S.N. Chaitanya
Department of Animal Biology, School of Life Sciences, University of Hyderabad, Gachibowli, Hyderabad, India

V. Dixit
Department of Botany, Guru Ghasidas Vishwavidyalaya, Bilaspur, Chhattisgarh, India

Bhavya Kavitha Dwarapureddi
Department of Environmental Science, GITAM Institute of Science, GITAM (Deemed to be University), Visakhapatnam, Andhra Pradesh, India

Mohan Krishna Ghanta
Department of Pharmacology, SRMC & RI, Sri Ramachandra Institute of Higher Education and Research, Chennai, Tamil Nadu, India

Manoj Kumar Gupta
Department of Biotechnology and Bioinformatics, Yogi Vemana University, Kadapa, Andhra Pradesh, India

Santosh C. Gursale
Department of Pharmacology, BKL Walawalkar Rural Medical College, Sawarde, Ratnagiri, Maharashtra, India

Pavan Kumar Kancharla
Department of Biotechnology, School of Life Sciences, Pondicherry University, Puducherry, India

Manoj Kumar Karnena
Department of Environmental Science, GITAM Institute of Science, GITAM (Deemed to be University), Visakhapatnam, Andhra Pradesh, India

V.K.D. Krishna Swamy
Department of Biochemistry and Molecular Biology, School of Life Sciences, Pondicherry University, Puducherry, India

Rama Rao Malla
Cancer Biology Lab, Department of Biochemistry and Bioinformatics, Institute of Science, GITAM (Deemed to be University), Visakhapatnam, Andhra Pradesh, India

Neha Merchant
Department of Hematology and Medical Oncology, Winship Cancer Institute, Emory University, Atlanta, GA, United States

Mathavan Muthaiyan
Centre for Bioinformatics, School of Life Sciences, Pondicherry University, Puducherry, India

Ganji Purnachandra Nagaraju
Department of Hematology and Medical Oncology, Winship Cancer Institute, Emory University, Atlanta, GA, United States

Leimarembi Devi Naorem
Centre for Bioinformatics, School of Life Sciences, Pondicherry University, Puducherry, India

Jayshree Nellore
Department of Biotechnology, School of Bio and Chemical Engineering, Sathyabama Institute of Science and Technology, Chennai, Tamil Nadu, India

Kiranmayi Patnala
Department of Biotechnology, Institute of Science, GITAM (Deemed to be University), Visakhapatnam, Andhra Pradesh, India

Sujatha Peela
Department of Biotechnology, Dr. BR Ambedkar University, Srikakulam, Andhra Pradesh, India

P.S. Pradeep
Centre for Laboratory Animal Technology and Research, Sathyabama Institute of Science and Technology, Chennai, Tamil Nadu, India

Samrat Rakshit
Department of Zoology, Guru Ghasidas Vishwavidyalaya, Bilaspur, Chhattisgarh, India

A. Ram Sailesh
Department of Environmental Science, Institute of Science, GITAM (Deemed to be University), Visakhapatnam, India

Vadde Ramakrishna
Department of Biotechnology and Bioinformatics, Yogi Vemana University, Kadapa, Andhra Pradesh, India

K. Santhiya
Department of Biochemistry, Biotechnology and Bioinformatics, Avinashilingam Institute for Home Science and Higher Education for Women, Coimbatore, Tamil Nadu, India

S. Saxena
Department of Medical Laboratory Sciences, Lovely Professional University, Phagwara, India

S. Shinde
Department of Biotechnology, Guru Ghasidas Vishwavidyalaya, Bilaspur, Chhattisgarh, India

D. Shukla
Department of Biotechnology, Guru Ghasidas Vishwavidyalaya, Bilaspur, Chhattisgarh, India

Sreedevi Muttathuveliyil Sivadasan
Department of Biotechnology, School of Life Sciences, Pondicherry University, Puducherry, India

J. Sivaprabha
Department of Biochemistry, Biotechnology and Bioinformatics, Avinashilingam Institute for Home Science and Higher Education for Women, Coimbatore, Tamil Nadu, India

Nagarjuna Sivaraj
Department of Biochemistry and Bioinformatics, GITAM Deemed to be University, Visakhapatnam, Andhra Pradesh, India

D. Sivaraman
Centre for Laboratory Animal Technology and Research, Sathyabama Institute of Science and Technology, Chennai, Tamil Nadu, India

N. Srinivas
Department of Environmental Science, Institute of Science, GITAM (Deemed to be University), Visakhapatnam, India

S. Sumathi
Department of Biochemistry, Biotechnology and Bioinformatics, Avinashilingam Institute for Home Science and Higher Education for Women, Coimbatore, Tamil Nadu, India

K. Suresh Kumar
Department of Environmental Science, Institute of Science, GITAM (Deemed to be University), Visakhapatnam, India

A.K. Tiwari
Department of Zoology, Bhanwar Singh Porte Government Science College, Pendra, India

Saritha Vara
Department of Environmental Science, GITAM Institute of Science, GITAM (Deemed to be University), Visakhapatnam, Andhra Pradesh, India

Amouda Venkatesan
Centre for Bioinformatics, School of Life Sciences, Pondicherry University, Puducherry, India

K. Vijaya Rachel
Department of Biochemistry and Bioinformatics, GITAM Deemed to be University, Visakhapatnam, Andhra Pradesh, India

N.K. Vishvakarma
Department of Biotechnology, Guru Ghasidas Vishwavidyalaya, Bilaspur, Chhattisgarh, India

Soumya Vishwas
Department of Biotechnology, Institute of Science, GITAM (Deemed to be University), Visakhapatnam, Andhra Pradesh, India

About the editors

Professor Rama Rao Malla is a faculty member of the Department of Biochemistry and Bioinformatics, Institute of Science, GITAM (Deemed to be University), Visakhapatnam, Andhra Pradesh, India. Dr. Malla obtained his MSc and PhD in Biochemistry from Andhra University, Visakhapatnam, Andhra Pradesh, India. Dr. Malla did postdoctoral research at the University of Illinois, College of Medicine, United States. His research focus is to explore tetraspanin CD151 as a therapeutic target and its miR candidates or exosomal CD151 as early diagnostic markers for TNBC. Dr. Malla also focuses on exploring the role of CD151 in tumor microenvironment and drug resistance mechanisms. He has published more than 85 research articles in peer-reviewed international journals and presented more than 100 abstracts in national and international conferences. Dr. Malla is author and coauthor for more than 20 book chapters published by international publishers. He was an editorial board advisor and member of various international journals. He has received one research and one academic excellence award.

Dr. Ganji Purnachandra Nagaraju is a faculty member in the Department of Hematology and Medical Oncology at Emory University School of Medicine. Dr. Nagaraju obtained his MSc and his PhD, both in Biotechnology, from Sri Venkateswara University in Tirupati, Andhra Pradesh, India. Dr. Nagaraju received his DSc from Berhampur University in Berhampur, Odisha, India. Dr. Nagaraju's research focuses on translational projects related to gastrointestinal malignancies. He has published more than 90 research papers in highly reputed international journals and has presented more than 50 abstracts at various national and international conferences. Dr. Nagaraju is author and editor of several published books including *Role of Tyrosine Kinases in Gastrointestinal Malignancies*, *Role of Transcription Factors in Gastrointestinal Malignancies*, *Breaking Tolerance to Pancreatic Cancer Unresponsiveness to Chemotherapy*, *Theranostic*

Approach for Pancreatic Cancer, and *Exploring Pancreatic Metabolism and Malignancy*. He serves as editorial board member of several internationally recognized academic journals. Dr. Nagaraju is an associate member of the Discovery and Developmental Therapeutics Research Program at Winship Cancer Institute. Dr. Nagaraju has received several international awards including FAACC. He also holds memberships with the Association of Scientists of Indian Origin in America (ASIOA), the Society for Integrative and Comparative Biology (SICB), the Science Advisory Board, the RNA Society, the American Association for Clinical Chemistry (AACC), and the American Association of Cancer Research (AACR).

Preface

In 2002, Funkhouser introduced the concept of theranostics, a strategy that combines diagnostic and therapeutic methods through the simultaneous probing, imaging, and targeted delivery of cytotoxic compounds to cancer tissues. This approach aims at improving cancer diagnosis and prognosis to inform personalized therapeutic regimens tailored to each patient's specific needs. These new approaches could finally accomplish better treatment results for patients with female-specific cancers (FSCs), resulting in a precision medicine approach allowing early detection, management, and targeted tumor therapy. In this book, I compile this information thoroughly and accurately by exploring the new theranostic (i.e., diagnostic, therapeutic, and precision medicine) strategies currently being developed for FSCs.

FSCs include breast cancer, cervical cancer, ovarian cancer, and endometrial cancer. They are very lethal and are projected to become more malignant in the near future. This heavy burden is due to the lack of effective early detection methods and to the emergence of chemoradioresistance. Attempts at improving the outcome of FSCs by incorporating cytotoxic agents such as chemo drugs have been so far disappointing. These results indicate that the main challenge remains in the primary resistance of FSC cells to chemotherapy in the majority of patients. Therefore, improvement in the outcomes of FSCs is dependent on the introduction of new agents that can modulate the intrinsic and acquired mechanisms of resistance.

The increased understanding of the genetic, epigenetic, and molecular pathways dysregulated in FSCs has revealed the complexity of the mechanisms implicated in tumor development. These include alterations in the expression of key oncogenic or tumor-suppressive miRNAs, modifications in methylation patterns, the upregulation of key oncogenic kinases, and so on. This knowledge will allow the development of novel biomarkers that aid in the early diagnosis and management of these deadly diseases. It will also pave the way for the design of innovative therapeutic compounds targeted against specific signaling pathways upregulated during cancer progression.

In this book, we precisely focus on the subject matter with broader range of treatment options. Further, we compile this information thoroughly and accurately by exploring the new theranostic (i.e., diagnostic, therapeutic, and precision medicine) strategies currently being developed for FSCs. Further, this book provides an understanding of the roadblocks of chemotherapy in patients with newly diagnosed and metastatic FSCs.

It also provides an in-depth understanding of the treatment options available currently as well as prospective options. Finally, it explores how these various advances can be integrated into a precision and personalized medicine approach that can eventually enhance patient care.

Rama Rao Malla

Cancer Biology Lab, Department of Biochemistry and Bioinformatics, Institute of Science, GITAM (Deemed to be University), Visakhapatnam, Andhra Pradesh, India

Ganji Purnachandra Nagaraju

Department of Hematology and Medical Oncology, Winship Cancer Institute, Emory University, Atlanta, GA, United States

CHAPTER 1

Role of selected phytochemicals on gynecological cancers

Dariya Begum[a], Neha Merchant[b], and Ganji Purnachandra Nagaraju[b]
[a]Department of Biosciences and Biotechnology, Banasthali University, Banasthali, Rajasthan, India
[b]Department of Hematology and Medical Oncology, Winship Cancer Institute, Emory University, Atlanta, GA, United States

Abstract

Gynecological malignancies constitute cervical, uterine, and ovarian cancers with cervical cancer being rated as the second-most predominant malignancy in women globally. The therapeutic outcome always depends on the stage of cancer correlated with the disease level and its metastasis. The conventional therapeutic options include resection through surgery, chemotherapy, and radiotherapy. These treatment options cause tumor recurrence or adverse toxic effects. Furthermore, they develop dysregulated oncogenes and tumor suppressors. This inhibits apoptosis and enhances metastasis. This disillusioned clinical outcome is associated with poor prognosis. Thus, the conventional therapies require novel medications to prevent toxic effects and sensitize tumor cells toward the chemoradiotherapies. Phytochemicals include bioactive compounds derived naturally from plants. These bioactive compounds are found to antagonize the dysregulated gene and enhance the efficacy of conventional therapies when used in combination. In this chapter we examine the role of selected phytochemicals, including resveratrol, genistein, and curcumin, that are now widely used for cancer therapy alone and in combination with conventional therapies. We also discuss the formulation of these bioactive compounds and novel nano-formulations to improve the bioavailability, stability, and pharmacokinetics of the drug used.

Keywords: Gynecological cancer, Prognosis, Uterine cancer, Ovarian cancer, Cervical cancer, Phytochemicals, Resveratrol, Genistein, Curcumin.

Abbreviations

AgNPs	silver nanoparticles
AHR	aryl hydrocarbon receptor
ATR	serine/threonine–protein kinase
COX-2	cyclooxygenase-2
CXCR4	(C-X-C) chemokine receptor type 4
DHA	dihydroartemisinin
DHS	dihydroxystilbene
EGCG	epigallocatechin gallate
EGFR	epidermal growth factor receptor
EMT	epithelial-mesenchymal transition
ER α, β	estrogen receptor alpha, beta
ER	endoplasmic reticulum
ERK	extracellular signal-regulated kinase

G.P. Nagaraju, R.R. Malla (eds.)
A Theranostic and Precision Medicine Approach for Female-Specific Cancers
ISBN 978-0-12-822009-2
https://doi.org/10.1016/B978-0-12-822009-2.00001-7

1

FOXO3	Forkhead box transcription factors
GAL-3	galectin-3
HER	human epidermal growth factor receptor
HIF-1α	hypoxia inducible factor-1 alpha
HNPG	5-hydroxy-4′-nitro-7-propionyloxy-genistein
HPV	human papillomavirus
IL	interleukin
MK	midkine
MMP	matrix metalloproteinases
mTOR	mammalian target of rapamycin
NF-κB	nuclear factor kappa B
NQO1	NAD(P)H:quinone oxidoreductase 1
OCSLCs	ovarian cancer stem-like cells
P-gp	permeability glycolprotein
PIK3	phosphatidylinositol-4,5-bisphosphate 3-kinase
PTEN	phosphatase and tensin homolog
STAT3	signal transducer and activator of transcription 3
TGFA	transforming growth factor alpha
TPGS	tocopheryl polyethylene glycol
VEGF	vascular endothelial growth factor

1. Introduction

Gynecological cancers are a group of malignant neoplasms of the female reproductive system that constitute the fourth most common cancers recorded in women [1]. The lack of awareness about these cancers, differentiating pathology, and shortage of screening opportunities are the chief causes for delayed diagnosis; these cancers are usually detected only at advanced stages. Among the gynecological cancers, endometrial, uterus, cervical, and ovarian cancers are the most commonly diagnosed that unfavorably affect the prognosis and clinical outcome of the patient [1]. The incidence and mortality rates of these types of cancers vary [2]. As per estimations from previous studies, 109,000 women were diagnosed with gynecological cancers in 2019; 33,100 deaths were recorded in the same year [3]. The common risk factors associated with genital system cancers are obesity, advanced age, use of fertility drugs, hormone replacement therapies, immunosuppression, and smoking. Infections including human papillomavirus (HPV) and chlamydia are also risk factors for cervical cancers [4]. Additionally, hereditary and family history of colorectal, breast, and ovarian cancers are also associated with the occurrence of disease [1]. Hysterectomy and tubal ligation may reduce the risk for uterine and ovarian cancers [5]. Thus, it is very crucial that women should be aware of the symptoms and signs that should be followed with regular screening and prevention strategies [1]. Moreover, therapy of cancer at its advanced stage and recurrence is always a challenge. Several attempts have been made by researchers to target the signaling pathways to control progression of cancer cells and to improve overall survival and prognosis of the patient.

The conventional therapies including chemo-radio-, immune-, and hormone therapy are included as therapeutic options prescribed based on the stage and type of malignancy. The chemo drugs that are frequently combined with conventional therapies are paclitaxel, doxorubicin, and cisplatin. However, these drugs are associated with toxic side effects and recurrent disease that develops from multidrug resistance. This resistance results from dysregulation of signaling pathways. Thus, there is a need to develop sensitive and personalized therapeutic strategies to reach higher therapeutic standards. Phytochemicals are naturally available molecules obtained from plants that constitute various bioactive properties. It was found that these compounds play a crucial role in antagonizing the dysregulated signaling pathways that cause malignancy. They can be easily obtained from healthy diets and can potentiate the conventional therapeutic strategies and enhance their efficacy when used together. Additionally, these phytochemicals incur no side effects and can target multiple signaling pathways responsible for cancer progression and proliferation [6]. Thus, these phytochemicals possess anticancer properties and can treat various cancers.

In this chapter we focus on gynecological cancers and the role of specific phytochemicals acting on signaling pathways to control tumor progression and induce apoptosis.

2. Uterine cancer

The uterus is a hollow organ located at the pelvic region that functions to support the development of a fetus. Anatomically, it has an inner layer, outer later, and bottom layer called the endometrium, myometrium, and cervix, respectively. Uterine cancer is the abnormal growth of cells in the uterine muscle. It occurs at two different parts of the uterus, based on which it is divided into two categories: commonly occurring endometrial cancer that forms the inner lining of the uterus and rarely occurring uterine sarcoma that forms the other tissue of the uterus. The uterine cancer is considered as one of the gynaecological tumors to have worst prognosis, and therapy includes resection of uterus. However, tumor recurrence requires effective therapeutic drugs. Pazopanib and trabectedin are most widely used for therapy, but are less effective as the tumor cells develop resistance resulted from the dysregulated signaling pathways. For instance, the mammalian target of rapamycin (mTOR) signaling pathway was found in elevated levels in women with gynecological cancer [7]. Similarly, endometrioid carcinoma results from the dysregulated DNA mismatch repair genes and mutations in phosphatase and tensin homolog (PTEN) and Kras [8–10]. Additionally, mutation in TP53, aberrantly acting HER2, and inactive expression of E-cadherin also results in endometrial cancer [11]. Moreover, uterus serous carcinoma results from mutated genes of p53, PIK3CA, HER2, and PPP2R1A, overexpression of cyclin D1/E, and downregulated expression of E-cadherin and p16 [12, 13]. Thus, the downregulated tumor suppressor genes and apoptotic genes, actively acting survival genes, and their signaling pathways develops chemoresistance. Therefore, detecting dysregulated genes and targeting them with active phytochemicals reflects a better therapeutic strategy for promoting chemosensitivity and better survival.

2.1 Resveratrol

The natural polyphenolic bioactive compound resveratrol, also known as 3,5,4′-trihydroxystilbene, can be found in grapes, peanuts, berries, and in high concentration in red wine [14]. It is widely popular for its antioxidant and anticancerous properties. As an anti-inflammatory, it regulates activation of certain proinflammatory proteins like nuclear factor kappa B (NF-κB) [15] and enzymes like cyclooxygenase-2 (COX-2) [16]. Resveratrol was determined to induce autophagy through activation of the AMPK-dependent signaling pathway. However, previously it was determined that autophagy also induces cancer progression [17]. Thus, Fukuda et al. [18] suggested that resveratrol along with chloroquine, an autophagy inhibitor, effectively induces apoptosis in Ishikawa endometrial cancer cells. This proved to be a therapeutic strategy for inducing apoptosis. However, the molecular mechanism behind resveratrol inducing autophagy and apoptosis remains a question. Adrenomedullin was found to exist in the cytoplasm of epithelial cells and stromal compartments [19]. Furthermore, adrenomedullin is upregulated at the time of endometrial repair and promotes angiogenesis by inducing endothelial cell proliferation and tube formation. Moreover, previously it was determined that the mRNA and peptide level of adrenomedullin is found in high levels under hypoxic conditions in various cancers by the activation of hypoxia inducible factor-1 alpha (HIF-1α) [20–23]. Evans et al. [24] further determined that this protein induces vascular endothelial growth factor (VEGF) secretion from primary cells of endometrial cancer. They also investigated for the first time that resveratrol and epigallocatechin gallate (EGCG) inhibit the secretion of adrenomedullin, thus reducing cancer incidence. The Wnt signaling pathway induces the accumulation of β-catenin that translocates to the nucleus. It later activates the transcription of c-myc and cyclin D, which promote tumorigenesis in various cancers [19]. Resveratrol was found to inhibit the Wnt signaling pathway in gastric and colorectal cancers. In uterine cancer cells, Sexton et al. [25] determined that resveratrol induces apoptosis when taken in high doses through COX-2 inhibition, which was tested in a group of uterine cancer cell lines, and regulates cancer cell proliferation. It was found to inhibit the expression of β-catenin and c-myc in a dose-dependent manner in uterine sarcoma cells. Thus, it inhibits cell proliferation and induces apoptosis via Wnt signaling pathway deactivation [26]. Later, Chen et al. [27] also determined that resveratrol inhibits β-catenin to control uterine fibroids. Uterine fibroids are the most common neoplasm of the uterus [28, 29] resulting from the aggressive production of the extracellular matrix (ECM) that includes collagens, fibronectin, and proteoglycans [30, 31]. Additionally, α-SMA, COL1A1, and β-catenin are found to be upregulated in ECM, developing into uterine fibroids. Resveratrol decreased their expression to mRNA and protein level in vitro and in vivo. However, the molecular mechanism involved is not yet explored and is yet to be determined [27]. Thus, resveratrol can be used as a complementary medicine for the therapy of cancer, however, further research is still essential for determining the efficacy of this compound.

2.2 Curcumin

Curcumin is an active phytocompound derived from the therapeutic Indian spice, *Curcuma longa*. It is used against various cancers like uterine leiomyosarcoma. It was previously demonstrated that curcumin inhibits tumor cell growth via downregulating the expression of tumor-promoting genes like mTOR [32, 33]. Wong et al. [34] determined that this bioactive compound downregulates phosphorylation of mTOR (Ser2448) together with its effectors p70S6 and S6 in uterine leiomyosarcoma cells. Later, they also proved the efficacy of intraperitoneal curcumin, which was administered in vivo in inhibiting the tumor growth in uterine leiomyosarcoma cells. They controlled the growth via inhibition of mTOR and inducing apoptosis. Low curcumin inhibits mTOR but showed no effect with Akt. Therefore, use of peritoneal curcumin enhanced the bioavailability and efficacy of curcumin in inhibiting Akt as well as mTOR. Androgen receptors are found to induce the activation of Wnt signaling pathways and are known to develop resistance. In another work, Feng et al. [35] revealed that curcumin in human endometrial carcinoma cells induce apoptosis via downregulating AR and β-catenin expression through the Wnt signaling pathway. Furthermore, curcumin also downregulates the expression of matrix metalloproteinase-2 (MMP-2) to its mRNA and protein level in a dose-dependent manner. Additionally, E-cadherin showed an inverse relationship with MMP-2 that is upregulated with exposure to high concentration of curcumin [36]. It was even evidenced from past studies that curcumin in endometrial cancer cells controlled the migration of tumor cells through the downregulation of MMP2 and MMP9-mediated extracellular signal-regulated kinase (ERK) pathway [37]. Slit-2, a secretory protein, identified as a molecule for axon guidance, is detected in various cancer cells. It plays a crucial role in inhibiting tumor cells and metastasis, inducing apoptosis, and cell cycle arrest. In support of this, Sirohi et al. [38] determined that curcumin induces the expression of Slit-2, to inhibit migration of tumor cells mediated via downregulation of CXCR4, MMP2, MMP9, and SDF-1. Thus, curcumin selectively affects aggregation and metastasis of endometrial tumor cells. However, the bioavailability and stability of curcumin is always a limitation. Researchers, therefore, have modified the structure of curcumin to produce analogues. For instance, the curcumin analogue, HO-3867, as suggested from earlier studies, mediates inhibition of signal transducer and activator of transcription 3 (STAT3) phosphorylation and induces apoptosis in various cancer cells [39, 40]. Tierney et al. [41] investigated that expression of pSTAT3 Ser727 is responsible for endometrial tumor growth and survival. They also suggested that HO-3867 was found to target pSTAT3 Ser727 to inhibit Ishikawa cell growth and proliferation via downregulation of active CDK5 and ERK1/2. Additionally, the upregulation of tumor suppressor protein p53, cleavage of caspase-3 and -7, and decrease in Bcl-2 and Bcl-xL was determined. It also induced cell cycle arrest at G2/M phases. This suggested that HO-3867 would potentially serve as a promising therapeutic strategy for cancer treatment. Furthermore, the development of nanotechnology enhanced the efficacy of phytochemical drugs. For instance, curcumin for the first time was successfully encapsulated in mixed micelles using PEG (15)-hydroxystearate and D-α-tocopheryl PEG (TPGS). The nano-designed micelles of curcumin were determined to have efficient anticancer property, enhanced intracellular uptake, and induced apoptosis. This resulted from downregulation of surviving expression in endometrial cancer cell lines. Additionally, it also regulated permeability glycolprotein (P-gp) efflux, preventing tumor cells from developing resistance and downregulating the expression of tumor necrosis factor-α (TNF-α), interleukin (IL)-6, and IL-10 [42]. Similarly, the liposome encapsulated with curcumin was evidenced from the work of Xu et al. [43]. They determined that the liposome encapsulated with curcumin effectively suppressed the activation of NF-κB in vitro and in vivo against endometrial cancer cells. Additionally, it was also found to downregulate the expression of MMP-9. This indicates the therapeutic potentiality of curcumin against uterine cancer and its efficacy when encapsulated in nanoparticles. The clinical trials recorded for curcumin are illustrated in Table 1. However, further exploration of the phytochemical's molecular mechanism and evaluation of its safety and efficacy is essential.

2.3 Genistein

Genistein is a naturally available isoflavone extracted from soya-based products. It is widely known for its estrogenic effects and is encouraged for hormonal therapies as an alternative option [44, 45]. Moreover, it was elucidated from previous studies that genistein has antioxidant and antimetastatic properties. It was found to regulate the expression of various genes responsible for cell growth, angiogenesis, and apoptosis in vitro and in vivo. The antiproliferative efficacy of genistein was tested for clinical trials against various cancers including colorectal, prostate, and bladder cancers. It was also found to modulate the proliferation of uterine sarcoma. For instance, Hu et al. [46] determined that using Chinese herbal products including genistein, coumestrol, and daidzein along with tamoxifen controlled the risk of occurrence of endometrial cancer in Taiwanese female patients. Tamoxifen is administered for estrogen receptor–positive breast cancer in order to reduce recurrence; however, its continuous usage develops the risk for endometrial cancer. Therefore, consuming phytoestrogen Chinese herbal products along with tamoxifen reduces the risk for endometrial cancer. Similarly, genistein was also compared

Table 1 The clinical trial record of phytochemicals tested for gynecological cancers.

Gynecological cancer	Phytochemical	Intervention drugs	Title	Phase	ClinicalTrials. gov identifier
Cervical cancer Uterine cancer Endometrial cancer	Curcumin (dietary supplement)	Pembrolizumab, radiation, aspirin, lansoprazole, cyclophosphamide, vitamin D	Patient administered with immunomodulatory cocktail	Phase 2	NCT03192059
Endometrial cancer	Curcumin (dietary supplement)	Curcuphyt	Effect of curcumin in tumor inhibition and inducing inflammation in endometrial cancer patients	Phase 2	NCT02017353
Endometrial cancer	Genistein	Genistein	Study of efficacy of genistein in preventing endometrial cancer in healthy postmenopausal women	Phase 1	NCT00099008
Cervical intraepithelial neoplasia	Curcumin	Curcumin	Effect of curcumin in treating squamous cervical intraepithelial neoplasia	Early Phase 1	NCT02554344
Neoplasms/ cervical lesions	Curcumin	Curcumin	Topical application of curcumin for precancerous cervical lesions	Phase 2	NCT02944578

Source: www.clinicaltrials.gov.

to estrogen given immediately or in later stages to ovariectomy rats for the risk of developing endometrial cancer. It was determined that the rat that was given estrogen in the later stage of ovariectomy developed an enhanced estrogen signaling cascade. This resulted in vigorous cell proliferation due to the activation of Ki67 and VEGF-A, which increased the risk of endometrial cancer. However, no such proliferation was detected in the rat that was given genistein, regardless of the time it was given. Thus, genistein renders low risk for the occurrence of endometrial cancer as compared to estrogen [47]. Moreover, the estrogenicity of genistein was also found to be similar to that of estradiol when compared in the expression profile of Ishikawa cells exposed to both genistein and estradiol prepared by Naciff et al. [48]. They determined that genistein also upregulates and downregulates the expression of multiple genes in a time- and dose-dependent manner involved in biological functions. These include FOS, EGFR1, FGFR2, SOX4, PTEN, and TGFA, which play a crucial role in the human endometrium. It is also found to induce DNA fragmentation and enhance apoptosis via upregulating the expression of pro-apoptotic proteins including BAX and BAD and activation of caspase-3 [49]. Furthermore, it increased the activation of p27, p53, and p21 and reduced the expression of β-catenin to control the growth of endometrial cancer cell progression [49]. Moreover, it induced apoptosis in endometrial cancer cells that developed resistance against doxorubicin. It was previously determined that taking genistein in high doses suppresses tyrosine kinase and DNA topoisomerases with the release of TGF-β and induces apoptosis [50]. Later, Di et al. [51] investigated the biological mechanisms involved with the activity of genistein to inhibit uterine leiomyoma via the TGF-β signaling pathway and the downstream genes actin A and Smad3. Their results suggest that downregulation of actin A and Smad3 control growth of uterine leiomyoma effectively when exposed to high doses of genistein. Even though genistein is widely known for its antiproliferative activity against tumors, its usage is limited due to its poor absorption in the gastrointestinal tract. Therefore, the structural alteration would potentiate the effect of the phytochemical. Bai et al. [52] determined the effect of 5-hydroxy-4′-nitro-7-propionyloxy-genistein (HNPG), an analogue of genistein. It was found to induce cell cycle arrest at G1 phase in Ji endometrial cell in vitro. Furthermore, it also downregulates the expression of cyclin D1, MMP-2, MMP-7, MMP-9, C-Myc, and β-catenin, inactivating the Wnt/β-catenin signaling pathway. Subsequently, this enhanced the stability and half-life of the drug. The clinical trials performed to determine the efficacy of genistein are illustrated in Table 1. Future investigations are essential to enhance the efficacy and stability of genistein to improvise the therapeutic strategies for the benefit of cancer patients.

3. Cervical cancer

Cervical cancer occurs in the cervical cells present toward the bottom part of the uterus that connects with the vagina. As estimated by the World Health Organization (WHO), cervical cancer is the fourth most recurrent cancer in females [53]. Cervical cancer in its preinvasive stage shows no symptoms, but the abnormal cells behave aberrantly and invade the adjacent tissues. The symptoms of cervical cancer include abnormal vaginal bleeding in between menstrual cycles and heavier bleeding than usual. Risk factors include persistent HPV infection, which subsequently leads to cancer. Additionally, smoking and using contraceptive pills are also associated with the risk of cancer occurrence [54]. Squamous cell carcinoma and adenocarcinoma are the two main types of cancers of the cervix, which is the determining factor for deciding the therapy and prognosis of the patient [55]. Surgery is always a primary choice for treatment, but tumor relapse and metastasis results in poor prognosis of the patient [56]. Chemotherapy would be the option to avert recurrence in the postoperative stage. The chemodrugs commonly prescribed include bevacizumab, pembrolizumab, and gemcitabine, with cisplatin as a combinational drug. However, development of resistance and adverse side effects always affects the quality of therapy. Therefore, there is a serious need for the exploration of therapeutic strategies that are less toxic to healthy cells and more consistent.

3.1 Resveratrol

Resveratrol is a phytoalexin, which is a naturally available bioactive compound obtained from berries, peanuts, and grapes [57]. It was demonstrated that phytoalexin exhibits multiple biological properties including antiproliferative, antimetastatic, antioxidative, and cardioprotective [58, 59]. Investigations from earlier research studies determined that phytoalexin has anticarcinogenic and apoptotic properties against various cancers. In cervical cancer, resveratrol induces radio-sensitization and apoptotic effect that are inhibitory to various signaling pathways associated with tumorigenesis [60]. For instance, STAT3, Notch, and Wnt signaling pathways play a crucial role in promoting cervical cancer and are found to be regulated by the activity of resveratrol. It was suggested to induce apoptosis and inhibit progression in HeLa and SiHa cell lines. This was further accompanied by the suppression of STAT3/JAK3, Notch, and Wnt signaling pathways [61]. The activity of STAT3 in cervical cancer was analyzed and it was determined that the three proteins, including suppressor of cytokine signaling 3 (SOCS3), SHP2, and protein inhibitor of activated STAT protein 3 (PIAS3), negatively regulate the STAT3 signaling pathway. These proteins are found to be downregulated in cervical cancer. It was determined that resveratrol inhibits the STAT3 signaling pathway to control progression in cervical cancer, elucidating changes in these three proteins [62]. Moreover, it was investigated that the expression of PIAS3 and SOCS3 are reduced in cervical cancer cells [62]. When treated with resveratrol, PIAS3 was upregulated threefold greater than SOCS3, which showed moderate enhancement. This eventually resulted in inhibiting JAK 1/2 enzyme activity and significantly suppressed the STAT3 signaling pathway in vivo [62]. Thus, triggering PIAS3 could serve as a therapeutic prognostic factor for determining the efficacy of resveratrol. Furthermore, Li et al. [63] determined that resveratrol activates the mitochondrial apoptotic signaling pathway. It upregulates caspase-3 and caspase-9 expression and downregulates antiapoptotic proteins including Bcl-2. It was found to induce cell cycle arrest at G2 phase via downregulating cyclin D1. The same group suggested that resveratrol potentiates the upregulation of tumor suppressor p53 in HeLa cells. Moreover, HPV infections are responsible for approximately 55% of cervical cancers. This results from the activation of E6 and E7 genes that inactivate various tumor suppressor proteins like p53. These genes are deactivated via ubiquitination and degraded by E3 ubiquitin ligase (E6-associated protein) [64]. Moreover, E3 ubiquitin ligase proteins like Mdm2 are activated when the environmental contaminants like halogenated hydrocarbons bind to the aryl hydrocarbon receptor (AHR) [65]. Thus, researchers suspect that this AHR activation may cause p53 ubiquitination and promote cervical cancer, however, its connection to cervical cancer proliferation is unclear. Flores et al. [66] determined that resveratrol along with α-naphthoflavone induced apoptosis and inhibited proliferation. Together they increased the half-life of p53 through the activation of E2F4/5 and induced G1/S phase cell cycle arrest. E2F4/5 activation induces suppression of tumors as well as induces cell cycle arrest at G1/S phase. Later, they suggested that AHR functions as per the cell type and cell transformation state. They showed that knocking of the AHR gene did not show any kind of decrease in cell proliferation or promote apoptosis. Thus, both α-naphthoflavone and resveratrol inhibit cell proliferation and induce apoptosis in HeLA, showing no effect on AHR. Similarly, Mukherjee et al. [67] used a combinational phytochemical drug TriCurin, which includes curcumin, EGCG, and resveratrol, against HPV E6 and E7 expression. They determined that TriCurin administered as a subcutaneous injection exhibited no side effects as detected in mice. It was found to downregulate the expression of HPV18 E6 and NF-κB with the upregulation of p53 and activation of caspase-3. Moreover, the combo drug used provided high stability and decreased cervical cancer tumor growth (80%–90%). Thus, this novel combo phytochemical drug can be administered as a promising therapeutic drug against HPV-related cancers. Previously, it was determined that HPV infection causes the dysregulation of p53, inducing apoptosis. Additionally, TRAIL, another protein, promotes apoptosis, but in a p53 independent way [68–70]. Thus, TRAIL can be encouraged as a therapeutic strategy to control cervical cancer apoptosis. Nakamura et al. [71] further determined that resveratrol inhibits the expression of survivin via phosphorylating its transcription factor STAT3. Inhibiting STAT3 enhances TRAIL expression and induces apoptosis. They even performed knockdown of survivin with siRNA that induced cell cycle arrest at G2/M phase and upregulated the expression of E-cadherin in the cervical cancer cell lines. Moreover, this treatment led to enhancement of cisplatin-induced apoptosis. Thus, this would be an ideal strategy to enhance the sensitivity of drug and induce apoptosis. Similarly, pterostilbene, otherwise called 3′,5′ dimethoxy-resveratrol, also exerts anticancer effects like resveratrol. It was found to induce apoptosis via disrupting the mitochondrial membrane and inhibiting the mTOR/PI3K/Akt pathway [72]. Phytoalexin is an effective anticancer drug, however, its bioavailability is always a limitation and it is not encouraged for further clinical trials. Recently, a few analogues were synthesized to enhance the bioavailability, stability, and efficacy against cancer progression. For instance, N-(4-methoxyphenyl)-3-5-dimethoxybenzamide [73] and (E)-8-acetoxy-2-[2-(3,4-diacetoxyphenyl)ethenyl]-quinazoline [74] are found to induce cell cycle arrest at G2/M phase via upregulating Chk1/2-cdc25 and p53-p21 tumor suppressor protein in ATM/ATR dependent way. Additionally, researchers have also synthesized nanoparticles encapsulated with resveratrol to improve the efficacy of the drug. More recently, the green synthesized gold nanoparticle that is capped with resveratrol was combined together with doxorubicin to enhance its efficacy in inhibiting cervical carcinoma [75]. This novel drug nano-vehicle efficiently improved the efficacy of the drug without causing cytotoxicity to the healthy cells. Thus, developing nanocomplexes is essential as an advanced application for cancer diagnosis and therapy.

3.2 Curcumin

Curcumin is a natural phenolic compound obtained from the roots of *Curcuma longa*. It was previously determined that curcumin promotes tumor suppressor proteins, including p53, and deactivates various signaling pathways, including PI3K/Akt, Ras, and β-catenin that promote carcinogenesis. Moreover, it was also found to promote the expression of Nrf2, a transcription factor that targets NAD(P)H: quinone oxidoreductase 1 (NQO1) [76–79]. NQO1 plays a crucial role in inducing stability of p53. Therefore, curcumin was found to promote the complex formation NQO1-p53 and also prevent negative regulators like E6AP to bind at its promoter region [80]. Furthermore, curcumin induces its antiproliferative effect via downregulating various pathways including β-catenin, FOXO3 (Forkhead box transcription factors) and COX-2 and reduces the expression of cyclin D1, Akt, and HIF-1α. It was later determined by Ghasemi et al. [81] that curcumin efficiently inhibits NF-κB activation and the Wnt/β-catenin signaling pathway. This inhibits progression and invasion of cervical cancer cells. Moreover, the deactivated NF-κB inhibits cyclin D1, MMP-9, pro-MMP2, and COX-2 expression [82]. Additionally, they also determined that curcumin along with 5-fluoro uracil induces cell cycle arrest at G2/M phase followed by sub-G1 apoptosis. As mentioned earlier, HPV contributes to high risk of cervical cancer due to the overexpressed viral E6 and E7 oncoproteins. These viral oncoproteins control pirin expression. Pirin plays a crucial role in developing sensors for oxidative stress during epithelial-mesenchymal transition (EMT) and migration of cells. Curcumin was found to downregulate the expression of the pirin gene and thereby decrease EMT. Moreover, the group also found that knockdown of pirin with siRNA significantly increased the expression of E-cadherin and decreased the expression of N-cadherin. Similar results were detected when cell lines and pirin siRNA were exposed to curcumin [83]. Among the traditional therapies including surgery and chemo-radio therapy, photodynamic therapy is also found to be advantageous in cervical cancer therapy as it improves the functionality of the organ and reduces the tumor instead of completely resecting it. He et al. [84] combined curcumin with photodynamic therapy to control invasiveness via targeting the Notch signaling pathway, as this pathway and its downstream genes VEGF and NF-κB were associated with the carcinogenesis of cervical cancer. Moreover, combining DAPT together with curcumin and PDT potentiated inhibition via blocking the Notch pathway and inhibiting the expression of NF-κB and VEGF. DAPT is a γ-secretase complex inhibitor that controls the Notch pathway. Thus, DAPT indirectly blocks the Notch pathway.

It is clear that the yellow bioactive compound is the principal curcuminoid of turmeric. Additionally, it has other related compounds including bisdemethoxycurcumin and demethoxycurcumin [85]. Both these compounds are found to induce apoptosis via mitochondrial dysfunction in many cancers. In cervical cancer, these compounds are found to inhibit metastasis via supressing MMP-2 and MMP-9 signaling pathways. These compounds were also found to regulate the expression of proteins involved in promoting metastasis, which includes N-cadherin, Ras, SNAIL, vimentin, β-catenin, ERK1/2, and NF-κB that are downregulated significantly. Meanwhile, PERK1/2 and E-cadherin are found to be upregulated [86, 87]. Thus, these compounds can be used as an antimetastatic agents for cervical cancer therapy. At present, researchers are exploring the structural alteration of curcumin to potentiate the kinetic stability and enhance the efficacy of the compound. Chaudhary et al. [88] performed in silico analysis for a novel series of 11 curcumin derivatives substituted with pyrazoline at the aryl rings. They performed molecular docking and simulation studies and determined that the designed analogues interact efficiently with IKκβ protein. The 4-bromo-4′-chloro analogue was found to interact more efficiently with three H bonds (docking score of -11.534 kcal/mol) compared to the parent compound (docking score of -7.12 kcal/mol). The cocrystallized structure of this lead molecule revealed the highest inhibitory effect via directly connecting at the ATP binding pocket. This determined increased cytotoxic effect compared to the parent compound and paclitaxel. The apoptotic effect evaluated in vitro was about 70.5% greater than curcumin with only 19.9% inhibition of tumor cells. The high rate of apoptosis was determined from the increase in caspase-3 cleavage. However, further preclinical and clinical studies are needed to determine curcumin's clinical use as an anticancer drug. Furthermore, curcumin was found to be quite effective in combinational therapies. For instance, with paclitaxel, curcumin sensitizes the cells and promotes apoptosis via upregulating the expression of p53. Additionally, the apoptotic protein caspase (-3, -7, -8 and -9) and PARP cleavage with the release of cytochrome c were enhanced [89, 90]. Curcumin was found to reverse MDR due to paclitaxel. NF-κB, Akt [89] and their downstream proteins COX-2 and cyclin D1 [91] are found to be activated when treated with paclitaxel, but curcumin inhibited NF-κB and Akt. Together curcumin and paclitaxel were found to inhibit antiapoptotic proteins such as XIAP, survivin and cIAP1 via NF-κB, Akt inhibition [90, 92]. The advanced nanotechnologies are now developing encapsulated curcumin to enhance its bioavailability and efficacy and reduce its toxicity to healthy cells. More recently, a green synthesized silver nanoparticle (AgNP) was designed using the derivative curcumin ST06–AgNPs against cervical cancer in vitro [93]. However, majority of the data did not correlate with the research. Later, Li et al. [94] compared the delivery of two nanocarriers of curcumin and cisplatin. They were characterized according to their particle size, efficiency of drug encapsulation, drug release, and zeta potential. The two nanocarriers included a lipid polymer hybrid nanoparticle and a polymeric nanoparticle. The in vitro and in vivo results revealed that the nanocarrier of lipid polymer hybrid with both curcumin and cisplatin showed higher cytotoxicity as compared to the other nanocarriers. Thus, the lipid polymer nanocarrier would serve as an effective drug carrier to target the tumor site. Microbubble-mediated delivery of curcumin enhanced its efficacy against cervical cancer. The use of a microbubble improves the stability, bioavailability, and the uptake of drugs that have poor water solubility for which the use of phytochemicals is limited. Moreover, microbubbles are used along with ultrasound to enhance the uptake of drugs and inhibit tumor progression. Upadhyay et al. [95] developed a curcumin-loaded protein microbubble. They synthesized it using bovine serum albumin as shell material and perfluorobutane as core gas for in vitro delivery of curcumin. Release of curcumin through the microbubble was fourfold higher with enhanced uptake (~250 times) by tumor cells. Cell viability was decreased (~71%), as detected from the 3-(4,5-dimethylthiazol-2-yl)-2,5-diphenyltetrazolium bromide assay. The clinical trials performed so far for determining the efficacy of curcumin against cervical cancer are illustrated in Table 1. Future effective drug systems are still warranted to enhance the delivery and efficacy of the drug.

3.3 Genistein

Previous epidemiological studies recognized genistein as a phytoestrogen isoflavone obtained from soybean-enriched diets. It plays a protective role against various cancers like cervical cancer. More like an anticarcinogenic agent, it is accounted for its antiproliferative chemo-preventive properties that control the cell cycle and induce apoptosis. It was suggested by Hussain et al. [96] that treatment with genistein promotes chromatin fragmentation and condensation of the nucleus, thus causing apoptosis. This results in increased accumulation of apoptotic bodies in a time-dependent manner. They further determined from their flow cytometric analysis that genistein induces cell cycle arrest at G2/M phase. MMPs that degrade the ECM and promote migration of cells from the tumor mass were also found to be downregulated by the activity of genistein. Additionally, TIMPs, MMPs inhibitors, which are downregulated in cervical cancer cells, are found to be upregulated when exposed to genistein. Endoplasmic reticulum (ER) stress under certain pathophysiological conditions induces apoptosis. They respond by producing GRP78 as released from the ER stress sensors. Genistein was found to induce apoptosis via ER stress. It significantly increases GRP78 expression and CHOP proteins, as they mediate ER stress to induce apoptosis [97]. Aberrant epigenetic modifications are some of the risk factors that cause cervical cancer. These result from the activation of DNA methyltransferase and histone deacetylases. Genistein was found to reverse epigenetic modifications through deactivating the expression of DNA methyltransferase and histone deacetylase [98]. This improved the expression of tumor suppressor genes like p21, E-cadherin, DAPK1, and RARβ. These genes are actually found to be methylated in cervical cancer as the methylating enzymes bind at their promoter site. Furthermore, isoflavone compound together with polyphenol is suggested to have more efficient anticancer properties. Dhandayuthapani et al. [99] found that resveratrol together with genistein or alone depolarizes mitochondria and dysfunctions it to activate caspase (caspase-3 and caspase-9) enzymes. Caspase-3 sequentially cleaved cellular and nuclear components leading to DNA fragmentation and inducing apoptosis. Additionally, it was also found that both phytochemicals inhibit the expression of HDM2 and promote p53 expression. HDM2 gene was previously found overexpressed in various cancers and is correlated with poor prognosis. It targets p53 and causes proteasomal degradation of this tumor suppressor protein, thereby inhibiting apoptosis and inducing cell proliferation. As determined from earlier studies, the PI3K/Akt pathway promotes tumorigenesis [100, 101] and mTOR is the downstream of this pathway that promotes growth and survival of tumor cells [102]. mTOR exists as TORC1 and TORC2 complexes. These are activated by the binding protein 4EBP1, translational initiation factor eIF4E, and ribosomal kinase p70S6, whose upregulation is detected in various cancers. Sahin et al. [103] determined that genistein sensitizes HeLa cells to cisplatin that showed decrease in mTOR phosphorylation, 4E-BP1, and p70S6 expression, which are upregulated due to cisplatin upon initial treatment. Furthermore, cisplatin that was previously determined to upregulate the expression of NF-κB [104] was also found to be downregulated when HeLa cells were treated with genistein together. Similarly, Liu et al. [105] found that genistein when combined together with cisplatin inhibits Erk1/2 phosphorylation. They also reduced Bcl2 levels with increased expression of p53 and caspase-3 that promote apoptosis in cervical cancer cell lines. Researchers are working to improve the bioavailability and pharmacokinetics of genistein via developing analogues. Xiong et al. [106] designed, synthesized, and evaluated eight genistein analogues. They determined that these analogues were highly antiproliferative and showed improved cytotoxicity. They also evaluated that the OH group present at C-5 and C-7 is responsible for the cytotoxicity. The 1-alkyl-1H-pyrazol-4-yl and pyridine-3-yl are found to be appropriate bioisosteres for 4′-hydroxyphenyl moiety present in genistein. Similarly, a 7-difluoromethyl-5,4′-dimethoxygenistein analogue of genistein was investigated for its effects. It was found to downregulate the expression of c-myc to its protein and mRNA level. This further upregulated apoptotic protein including Bax, caspase-9, and cytochrome-c and down-regulated Bcl-2 [107]. Recently, the application of nanotechnology has improved genistein's anticancer properties and enhanced stability and loading of the drug at the tumor site. For instance, Cai et al. [108] formulated a novel nanoparticle with folic acid conjugated with genistein loaded in a chitosan nanoparticle. They reported that FRs-α are highly exposed to blood circulation in tumor cells rather than in normal cells. These are attractive to folic acid and can be easily targeted to a folic acid-conjugated nanoparticle. They compared the chitosan-loaded nanoparticle with the folic acid nanoparticle, and found that the latter showed efficient apoptosis. Similarly, a biodegradable d-α-tocopheryl polyethylene glycol 1000 succinate TPGS-b-PCL (poly (ε-caprolactone)) nanoparticle was formulated to encapsulate genistein against cervical cancer cell lines. Genistein is found to inhibit tumor progression more efficiently in vitro and in vivo with enhanced cytotoxicity [109]. The preceding reports reflect advanced application of genistein against cervical cancer inhibition. However, further clinical and preclinical studies are essential for determining the efficacy of the drug.

4. Ovarian cancer

Ovaries are the reproductive glands present in pairs on either side of the uterus that form the major source for estrogen and progesterone secretion. As estimated by the American Cancer Society, ovarian cancer stands fifth amongst cancer-related deaths in women. There are around 21,750 newly diagnosed cases with approximately 13,940 ovarian cancer-related deaths in women [5]. This large variation is due to the face that four out of five women with ovarian cancer are only detected at the advanced stages. There are three main types of ovarian cancers: epithelial (90%), germ cell (3%), and sex cord-stromal (2%) cancers [110]. The major risk factors associated with ovarian cancer are family history of breast and ovarian cancer, menopausal hormone therapy, and obesity [111–113]. Additionally, mutations in genes including BRCA1 and BRCA2 are highly susceptible and are the major cause of about 40% of ovarian cancer cases [114]. Resection of the organ followed by chemotherapy is always the primary choice for ovarian cancer therapy. Chemotherapy is associated with serious toxicities and multidrug resistance. Therefore, use of phytochemicals as discussed earlier reduces adverse side effects and sensitizes the tumor cells.

4.1 Resveratrol

As previously discussed, resveratrol potentiates anticancer properties in various cancer cells like ovarian cancer cells. It was determined that resveratrol inhibits progression and induces apoptosis of the ovarian cancer cell lines including A2780 and SKOV3 in time- and dose-dependent manners via reactive oxygen species generation. Moreover, it decreases Notch1 signaling in a ROS-dependent way and promotes PTEN and downregulates Akt phosphorylation in human ovarian cancer cells [115]. Ovarian cancer progression is also promoted due to aerobic glycolysis, which plays a crucial role in promoting malignancy, which was found to be inhibited by resveratrol. It acts via decreasing the uptake of glucose and induces apoptosis interrupting Akt. This regulates GLUT1 plasma membrane trafficking in ovarian cancer cell lines. Moreover, it was also found to induce apoptosis via promoting the levels of cleaved caspase-3 via mitochondrial pathways. In addition to glycolysis inhibition, resveratrol also activates the p-AMPK pathway and downregulates mTOR. As determined, the activated AMPK always activates the TSC1/2 complex, which inhibits AMPK downstream mTOR protein, which plays a major role in proliferation and survival of tumor cells [116]. Resveratrol was also found to behave efficiently as a chemo drug. For instance, resveratrol acts like oxaliplatin, a chemo drug that was found to promote immunogenic cell death in addition to apoptosis of cancer cells. Zhang et al. [117] determined that resveratrol play a potential role in inducing immunogenic cell death in vivo and in vitro. Moreover, the cue signaling during the immunogenic cell death caused by the chemo drug was also confirmed to be similar when exposed to resveratrol. The cue signaling includes secretion of HMGB1, cell surface exposure of calreticulin, and release of ATP. They also found mature dendritic cells and active cytotoxic T cells in the microenvironment of the tumor. This further showed reduced secretion of TGF-β and upregulation of IFN-γ and IL12p7. Altogether resveratrol could render a novel therapeutic strategy with the combination of PD-L1 and PD-1 antibodies (immune checkpoint inhibitors) against ovarian cancer. To determine the effect of resveratrol on ovarian cancer cell lines Ferraresi et al. [118] performed a comparison with IL-6 via profiling the microRNome and transcriptomes. They determined that mRNA and miRNA contrastingly altered the expression when exposed to resveratrol and IL-6. Moreover, the epigenetic modifications in miRNA caused alteration in the gene expression. Out of six miRNAs, hsa-miR486-3p (targets ULK2) and hsa-miR-21-5p (targets ATG10) are upregulated by the expression of IL-6. These miRNAs target STAT3 in order to promote progression and metastasis. miRNA activated via IL-6 was also found to inhibit tumor suppressor ARH-1 in ovarian cancer cell lines, which regulates autophagy and inhibits motility of cells. miR-21 is found to be abnormally expressed in ovarian cancer cells that develop chemoresistance and escape from apoptosis. Resveratrol on the other hand was found to downregulate the expression of STAT3 activated by IL-6 and induced the synthesis of ARH-1, which regulates autophagy positively. ARH-1 upregulated via resveratrol binds with BECLIN-1 to induce autophagy. This highlights the role of resveratrol in promoting autophagy and inhibiting tumors. As determined earlier, all miRNAs are not involved in tumor progression. For instance, miR-424-3p is found in low levels in the ovarian cancer cell lines (SKOV3 and OVCAR-3). Resveratrol is found to induce the expression of this miR-424-3p to control the progression of tumors. Thus, activated miR-424-3p by resveratrol is found to induce galectin-3 (GAL-3) degradation. GAL-3 has NF-κB and Akt as its upstream inducers, which promote tumorigenesis and develop resistance against drugs. Thus, resveratrol was also found to enhance sensitivity against cisplatin via degrading GAL-3 to downregulate Bcl-2 and upregulate caspase-3 to induce apoptosis [119]. Despite its promising antitumor property against cancer cells, its clinical efficacy in relation to patient prognosis is disappointing due to its low bioavailability. However, as discussed earlier, analogues of resveratrol that are developed prove to be advantageous in improving bioavailability. For instance, pterostilbene, a demethylated analogue widely used by researchers to treat against ovarian cancer cell lines. Pterostilbene was determined to inhibit STAT3 in ovarian cancer cells. This resulted in downregulation of various antiapoptotic proteins like Bcl-2, Mcl-1, and cyclin D1. Moreover, its effect is dose- and time-dependent. When administered in lower concentrations it induces cell cycle arrest at S phase, but when administered at higher concentrations it induces arrest at G0/G1 phase. It was also found to enhance the efficacy of platinum-based drugs like cisplatin to sensitize tumor cells [120]. Another analogue of resveratrol, 4,4'-trans-dihydroxystilbene (DHS), is found to be more effective than its parent compound as it inhibits ribonucleotide reductase (RNR). RNR is an enzyme that catalyzes the biosynthesis of dNTPS, which is essential for DNA damage repair and replication. It is a solid protein that consists of two subunits RRM1 and RRM2. Moreover, DHS was found to bind directly to RRM2 and cause ubiquitination via cyclin F. It is degraded by proteasome. This resulted in DNA replication inhibition and induced cell cycle arrest at S phase and apoptosis. Additionally, DHS was found to sensitize the resistant tumor cells against gemcitabine in pancreatic cancer and cisplatin in ovarian cancer, as RRM2 increases NF-κB activation [121]. Furthermore, acetyl resveratrol was also found to inhibit NF-κB activation and attenuate secretion of VEGF to inhibit angiogenesis [122]. Researchers recently started developing novel nanocarriers of the drug to the target site to avoid toxicity to the healthy cells and improve the half-life of the drug. Further, with more advanced research techniques, researchers are now developing herbal nanoparticles that show similar therapeutic efficiency as other synthetic chemicals. Nam et al. [123] used *Angelica gigas* Nakai (AGN) and its extracts decursin and decursinol angelate, which showed anticancer properties, as the nanoparticle. They found that when incorporated with resveratrol, AGN exhibited potent anticancer properties and efficient release of resveratrol at physiological pH. More recently, Guo et al. [124] developed a drug dual-carriers delivery system using ferritin (FRT; pH-sensitive cage-like nanostructure), and nanoscale graphene oxide (NGO; drug carrier with vast surface area and efficient photothermal effect). They initially constructed NGO loaded with resveratrol and connected it with the mitochondrion targeted molecule IR780. Later, it is encapsulated with the second carrier FRT. The so formed DDDs INR@FRT is an efficient therapeutic strategy in photothermal-chemotherapy in ovarian cancer cells in vitro and in vivo. It is designed so that the drug is released at the target site without any leakage in the circulatory system to acid and heat stimuli. It targets the mitochondria to release resveratrol and react with the organelle that further induces apoptosis. Thus, this could be an efficient drug delivery system that potentiates inhibition of cancer cells.

4.2 Curcumin

The yellow-colored dietary compound was reported previously by various studies for its antiapoptotic property against several cancers, including ovarian cancer. It was found to induce apoptosis [125] and autophagy. However, autophagy activation is not only to signal pro-death but also to escape from cell death in order to adapt to stress [126–128]. This subsequently resulted in developing resistance against chemotherapy. Liu et al. [129] showed that curcumin induces apoptosis via increasing PARP cleavage and caspase-9 in ovarian cancer cells. Similarly, they also found that curcumin induces autophagy via mTOR/Akt/p70S6K pathway inhibition in a time- and dose-dependent manner. They found that curcumin inhibits phosphorylation of mTOR and its downstream effectors including 4E-BP1 and p70S6K, whose activation promotes angiogenesis and metastasis. Moreover, the inhibition of this cascade further promotes autophagy that weakens chemotherapy via developing resistance in tumor cells. Therefore, they combined curcumin with chloroquine in order to suppress autophagy as well as enhance the antitumor activity of curcumin to inhibit Akt/mTOR/p70S6K. c-Myb is another protooncogene that is found to induce resistance in various cancers against cisplatin. Tian et al. [130] determined that miR-520h is associated with promoting EMT and metastasis through the upregulation of TGF-β and c-Myb. Moreover, their elevated expression was detected in ovarian cancer cells. Additionally, c-Myb was found to upregulate the expression of NF-κB and STAT3, which are determined to induce EMT. They evaluated the usage of EGCG, curcumin, and sulforaphane to downregulate c-Myb expression and to sensitize tumor cells to cisplatin. Among these, EGCG was found more effective in inhibiting ovarian cancer cells. Thus, including these dietary factors, which are part of a traditional Chinese diet, would be a promising option for tumor inhibition. Similarly, Zhao et al. [131] determined dihydroartemisinin (DHA) as an active metabolite of artemisinin. It is extracted from the Chinese herb *Artemisia annua*, which has medicinal value and is used as antimalaria agent [132] with antitumor properties [133–137]. It was found to induce apoptosis and cell cycle arrest to inhibit tumor progression. They combined DHA with curcumin as an alternative medicine to potentiate the clinical therapy of ovarian cancer via inhibiting midkine (MK). MK is characterized as heparin-binding growth factor, whose overexpression in tumors is associated with poor prognosis [138]. They further found that the combination effect showed upregulation of miR-124 that targets MK. miR-124 is the tumor suppressor that induces apoptosis, but it is found downregulated in various cancers [139]. They determined that DHA alone initially induces cell cycle arrest at G2/M phase and curcumin later induces apoptosis through the downregulation of Bcl-2. But it did not show any alteration in the expression of caspase-3.

Moreover, miR-124 overexpression and silencing of MK reduced MDR1 that codes for P-gp expression. Additionally, the use of this combinational therapy showed no side effects and synergistically exhibited efficient antitumor activity. It inhibited tumor progression and induced cell cycle arrest in vitro and in vivo. Furthermore, the Wnt/β-catenin signaling pathway is found to play a crucial role in promoting proliferation and EMT in ovarian cancer cells [140]. This aberrantly acting pathway further upregulates its downstream genes including cyclin D1, survivin, c-Myc, and MMPs [141]. The secreted frizzled related protein (SFRP) is the tumor suppressor glycoprotein that inhibits the Wnt signaling pathway. However, hypermethylation of DNA in the presence of DNA methyltransferase caused SFRP dysregulation that resulted in ovarian cancer in Taiwanese women [142]. Yen et al. [143] therefore combined 5-aza-2′-deoxycytidine (DAC), a DNA methyltransferase inhibitor, with curcumin to inhibit DNA methylation and induce the expression of SFRP5. They found that curcumin inhibited the activity of DNMT and DNMT3a protein expression. Further, this significantly decreased the expression of β-catenin and their downstream genes. Additionally, this adjuvant combination also inhibits migration of ovarian cancer cells. They downregulated vimentin and fibronectin with the upregulation of E-cadherin. Thus, curcumin enhances the chemotherapeutic drugs and is used as a natural inhibitor of DNMT that can be applied for epigenetic therapy in ovarian cancer treatment. Researchers developed a derivative of curcumin to improvise stability and bioavailability as compared to its parent compound. For instance, Koroth et al. [144] synthesized novel curcumin compounds ST03 and ST08 to induce cell apoptosis and inhibit migratory properties. PA1 ovarian cancer cell lines and MDA-MB-231 breast cancer cell lines were used for this study. Both the derivatives of curcumin showed efficient cytotoxicity against PA1 cancer type that are stem cell-like correlated with cancer recurrence [145] and the other metastatic MDA-MB-231 cell lines. Furthermore, they were found to induce intrinsic apoptotic pathway via the upregulation of caspase-3 and caspase-9. Similarly, another derivative, diarylidenyl-piperidone, was studied for its efficiency to be better than its parent compound against ovarian cancer cells that inhibits proliferation and reverses MDR [39, 146–149]. Thus, these compounds were used as novel anticancer agents to potentiate therapy and to inhibit metastasis and recurrence of cancer. Nano-sized particles are also developed for the delivery of curcumin to overcome MDR. For instance, AgNPs that are suggested to have anticancer properties are coated with curcumin and tested against cisplatin-resistant cells in order to sensitize the cells. cAgNPs are further combined with cisplatin and assessed for inhibition. The results revealed increased upregulation of caspase-3/-9 and p53 with the downregulation of the MMP-9 gene. This suggests increased efficiency in inducing apoptosis in cisplatin-resistant cells [150]. Similarly, Zhao et al. [151] grafted polyethylenimine (PEI) and stearic acid as a nanocarrier confirmed via nuclear magnetic resonance. They were further loaded with curcumin and paclitaxel and targeted to CD44 membrane receptors of ovarian cancer cells. This eventually exerted anticancer effects in chemo-sensitive human ovarian cancer cells (SKOV3) and MDR SKOV3-TR30. The molecular mechanism is evidenced with curcumin sensitizing the cells to paclitaxel via reversing drug resistance and reducing P-gp efflux. Thus, this could be a promising strategy with no toxic side effects and efficient antitumor property against resistant ovarian cancer cells. Furthermore, designing novel-liquid driven *co*-flow focusing loaded with curcumin in poly (lactic-*co*-glycolic acid) microparticles is advantageous for intraperitoneal administration. These were suggested for their chemical stability and 90% release of curcumin into the in vitro medium. The pharmacokinetic results also found efficient release of curcumin to the site through intraperitoneal injection. Thus, this study would be supportive for the therapy of peritoneal organ cancers present in the peritoneal cavity like ovarian cancers [152]. Use of nanotechnology along with combination of drugs is a promising therapeutic strategy for ovarian cancer therapy to overcome the hurdle of multidrug resistance.

4.3 Genistein

Phytoestrogen bioactive genistein that is extracted from dietary sources like soya is characterized as an isoflavone. As discussed in the earlier epidemiological studies, these estrogen-like compounds are found to be beneficial for the therapy of various chronic diseases like cardiovascular disease, diabetes, and cancer. Ovarian cancer is a hormone-associated tumor. About 60%–100% of ovarian cancers is believed to be due to the dysregulated estrogen receptors ERα and ERβ that differ due to their ligand-binding property and tumor progression [153]. ERβ is suggested to inhibit cancer cell proliferation and ERα promotes progression of the tumor [154]. Chan et al. [155] investigated the pharmacological effect of genistein, daidzein, and ERB-041(synthetic agonist of ERβ) on ovarian cancer cells for proliferation, progression, apoptosis, invasion, and potentiated ERβ activity. They further determined that genistein induces cell cycle arrest at S and G2/M phases where daidzein was found to induce at G1 phase. Moreover, this was as a result of the inhibition of PI3K/Akt phosphorylation and upregulation of p21 expression by genistein. Additionally, FAK, whose overexpression is associated with ovariancancer and poor prognosis, is also found to be suppressed due to these bioactive compounds and the agonist. Limitation of chemotherapy is mainly due to the resistance developed by the ovarian cancer cells. Importantly, ovarian cancer stem-like cells (OCSLCs) are a subset of cells that develop resistance due to their slow division nature. They could not be targeted by chemo drugs, and targeting these cells remains a crucial step in clinical application [156, 157]. Moreover, IL-8 that is found to activate STAT3 is involved in maintaining an interaction between CSCs and the tumor microenvironment. Thus, these are associated in promoting inflammation, production of ROS, developing MDR, and tumor progression. In order to determine the molecular mechanism of IL-8/STAT3, Ning et al. [158] cocultured SKOV3 cells with OCSLCs and macrophages

(THP-1). They found that stemness was induced by the activation of IL-8/STAT3 activity [159], and blocking IL-8/STAT3 signaling delayed the communication between tumor-associated macrophages. Genistein was known to inhibit tumorigenesis via inhibiting CSCs in various cancer cells from multiple in vitro and in vivo studies [160]. Ning et al. [158] found that when genistein was exposed to THP-1-treated OCSLCs, it interrupted the interaction between them via blocking the STAT3/IL-8 signaling cascade and reversed the polarization. Thus, genistein suppresses the stemness in SKOV3 cells. As discussed earlier, ovarian cancer is one of the estrogen-responsive cancers whose tumor progression and migration are mediated due to EMT activated via 17β-estradiol (E2). Moreover, the endocrine disrupting chemicals (EDCs) including bisphenol A (BPA) and nonylphenol (NP) play a chief role in promoting EMT to cause cancer migration [161]. Kim et al. [162] investigated that EDCs-BPA and NP upregulated vimentin and downregulated E-cadherin to promote metastasis in BG-I ovarian cancer cells. This acted like E2, whose activation was further found to be inhibited when ICI 182,780 antagonist of ER is used, which blocks ER signaling. Later they also identified that genistein acts like an antagonist ICI 182,780 and reverses the metastasis in ovarian cancer cells when treated. It inhibits EMT that is enhanced by BPA, E2, and NP. Additionally, like the antagonist, genistein was also found to activate TGF-β signaling, whose signaling was found to be inhibited in ovarian cancer cells by ER signaling. Poor bioavailability and decreased stability of genistein limits its usage. But structural modification, like inducing CF_3 or HCF_2 in genistein, would promote anticancer efficacy in tumor cells [163]. Ning et al. [164] determined that the newly synthesized genistein derivative 7-difluoromethoxyl-5,4′-di-n-octylygenistein (DFOG) would efficiently inhibit spheroid and colony formation of OVCSLCs in vitro. It suppressed the activity of p-Akt, p-ERK1/2, and NF-κBp65. Moreover, they also found that this derivative potentiates the activation of FOXO3a and FOXM1 expression, which is necessary for the inhibition of NF-κB, PI3K, and Akt pathways. Similarly, DFOG was also found to sensitize the ovarian cancer cells in order to potentiate let-7D, a tumor suppressor. It was found to target c-Myc. Let-7D is negatively correlated with the expression of c-Myc expression, which is associated with poor prognosis. Its downregulation was found to inhibit the PI3K/Akt pathway, thus let-7d activity is associated with the sensitization of ovarian cancer cells by DFOG [165]. In another study Bai et al. [166] demonstrated the therapeutic efficacy of a novel synthetic isoflavone of genistein, HNPG. They determined that the antitumor activity of HNPG in vitro is time- and dose-dependent. It inhibits proliferation, metastasis, and clone formation, and induces apoptosis in ovarian cancer cells. They found that this genistein derivative increases ROS accumulation and reduces mitochondrial membrane potentiality. They further suggested downregulation of Bcl-2 and upregulation of Bax protein with the decrease in their ratio. Moreover, the release of cytochrome c from the mitochondrial membrane promoted caspase cleavage to induce apoptosis. Recently, genistein was formulated by Mittal et al. [167] to synthesize the nanostructure lipid carrier loaded with genistein to sustain efficient release of the drug at the tumor site in ovarian cancer. They developed the nanostructure employing surfactant with TPGS succinate using solvent emulsification and evaporation techniques. The biodistribution and pharmacokinetic studies revealed efficient entrapment of the drug (94.27%) and better zeta potentiality (−20.21). Concentration of the drug was better retained at the tumor site for longer duration, thus determining the superior carrier system of genistein. Future studies are still needed for determining these isoflavones' actions according to the stage and type of cancer.

5. Conclusion

Gynecological cancers are the malignant neoplasms resulting from the dysregulation of various tumor suppressors and oncogenes. The conventional therapies are associated with adverse side effects and high resistance of tumor cells to drugs. Therefore, optimal therapeutic strategies are essential in addition to the usual conventional therapies. Biomarkers play a crucial role in diagnosis and therapy, but lack of precise biomarkers in gynecological cancers is always a hurdle for therapy. Additionally, a proficient understanding of the molecular mechanisms of biomarkers is essential. Genotyping is one such advancement that forms the foundation for personalized medicine that can be formulated according to disease stage. A complete knowledge of gene and molecular pathways involved in dysregulation of genes and the development of adverse side effects should be well understood to eradicate the threat of tumor progression. Phytochemicals are naturally extracted bioactive compounds that can be expected to overcome such a threat. Moreover, multiple hypothetical studies evidenced that the dietary intake of phytochemicals imparts an effect of chemoprevention and tumorigenesis prevention. The infusion of various phytochemicals or synergistic combination of phytochemicals with chemo drugs potentiates the therapy for better clinical outcomes.

Additionally, use of nanoparticles as drug carriers are widely explored in various in vitro and in vivo studies. They are tested alone or in combination with other drugs to improve their efficiency and pharmacokinetic parameters and to enhance the therapeutic nature of chemo drugs used in combination. The current chapter focuses on various dysregulated genes and their role in different signaling pathways involved in tumor progression. Development of combinational therapies including phytochemicals and chemo drugs along with nanoformulations contribute toward better survival and patient quality of, which improves efficacy and minimizes adverse side effects.

References

[1] Ledford LR, Lockwood S. Scope and epidemiology of gynecologic cancers: an overview. In: Seminars in oncology nursing. Elsevier; 2019.
[2] Centers for Disease Control and Prevention. Inside knowledge: get the facts about gynecologic cancer. CDC Publication; 2012.
[3] Siegel RL, Miller KD, Jemal A. Cancer statistics. CA Cancer J Clin 2019;69(1):7–34.

[4] Saslow D, et al. American Cancer Society, American Society for Colposcopy and Cervical Pathology, and American Society for Clinical Pathology screening guidelines for the prevention and early detection of cervical cancer. CA Cancer J Clin 2012;62(3):147–72.

[5] Torre LA, et al. Ovarian cancer statistics. CA Cancer J Clin 2018;68(4):284–96.

[6] Thangapazham RL, Sharma A, Maheshwari RK. Multiple molecular targets in cancer chemoprevention by curcumin. AAPS J 2006;8(3):E443.

[7] Dobbin ZC, Landen CN. The importance of the PI3K/AKT/MTOR pathway in the progression of ovarian cancer. Int J Mol Sci 2013;14(4):8213–27.

[8] Tashiro H, et al. Mutations in PTEN are frequent in endometrial carcinoma but rare in other common gynecological malignancies. Cancer Res 1997;57(18):3935–40.

[9] Enomoto T, et al. Alterations of the p53 tumor suppressor gene and its association with activation of the cK-ras-2 protooncogene in premalignant and malignant lesions of the human uterine endometrium. Cancer Res 1993;53(8):1883–8.

[10] Kim TH, et al. The synergistic effect of conditional Pten loss and oncogenic K-ras mutation on endometrial cancer development occurs via decreased progesterone receptor action. J Oncol 2010;2010:139087.

[11] Samarnthai N, Hall K, Yeh I. Molecular profiling of endometrial malignancies. Obstet Gynecol Int 2010;2010:162363.

[12] Catasus L, Gallardo A, Prat J. Molecular genetics of endometrial carcinoma. Diagn Histopathol 2009;15 (12):554–63.

[13] Matias-Guiu X, Davidson B. Prognostic biomarkers in endometrial and ovarian carcinoma. Virchows Arch 2014;464(3):315–31.

[14] Biesalski HK. Polyphenols and inflammation: basic interactions. Curr Opin Clin Nutr Metab Care 2007;10(6):724–8.

[15] Rahman I, Biswas SK, Kirkham PA. Regulation of inflammation and redox signaling by dietary polyphenols. Biochem Pharmacol 2006;72(11):1439–52.

[16] Das S, Das DK. Anti-inflammatory responses of resveratrol. Inflamm Allergy Drug Targets 2007;6 (3):168–73.

[17] Guo JY, Xia B, White E. Autophagy-mediated tumor promotion. Cell 2013;155(6):1216–9.

[18] Fukuda T, et al. Autophagy inhibition augments resveratrol-induced apoptosis in Ishikawa endometrial cancer cells. Oncol Lett 2016;12(4):2560–6.

[19] Maybin JA, et al. The expression and regulation of adrenomedullin in the human endometrium: a candidate for endometrial repair. Endocrinology 2011;152(7):2845–56.

[20] Cormier-Regard S, Nguyen SV, Claycomb WC. Adrenomedullin gene expression is developmentally regulated and induced by hypoxia in rat ventricular cardiac myocytes. J Biol Chem 1998;273 (28):17787–92.

[21] Nakayama M, et al. Induction of adrenomedullin by hypoxia and cobalt chloride in human colorectal carcinoma cells. Biochem Biophys Res Commun 1998;243(2):514–7.

[22] Garayoa M, et al. Hypoxia-inducible factor-1 (HIF-1) up-regulates adrenomedullin expression in human tumor cell lines during oxygen deprivation: a possible promotion mechanism of carcinogenesis. Mol Endocrinol 2000;14(6):848–62.

[23] Nguyen SV, Claycomb WC. Hypoxia regulates the expression of the adrenomedullin and HIF-1 genes in cultured HL-1 cardiomyocytes. Biochem Biophys Res Commun 1999;265(2):382–6.

[24] Evans J, et al. Adrenomedullin interacts with VEGF in endometrial cancer and has varied modulation in tumours of different grades. Gynecol Oncol 2012;125(1):214–9.

[25] Sexton É, et al. Resveratrol interferes with AKT activity and triggers apoptosis in human uterine cancer cells. Mol Cancer 2006;5(1):45.

[26] Mineda A, et al. Resveratrol suppresses proliferation and induces apoptosis of uterine sarcoma cells by inhibiting the Wnt signaling pathway. Exp Ther Med 2019;17(3):2242–6.

[27] Chen H-Y, et al. Natural antioxidant resveratrol suppresses uterine fibroid cell growth and extracellular matrix formation in vitro and in vivo. Antioxidants 2019;8(4):99.

[28] Cramer SF, Patel A. The frequency of uterine leiomyomas. Am J Clin Pathol 1990;94(4):435–8.

[29] Baird DD, et al. High cumulative incidence of uterine leiomyoma in black and white women: ultrasound evidence. Am J Obstet Gynecol 2003;188(1):100–7.

[30] Arici A, Sozen I. Transforming growth factor-β3 is expressed at high levels in leiomyoma where it stimulates fibronectin expression and cell proliferation. Fertil Steril 2000;73(5):1006–11.

[31] Leppert PC, et al. Comparative ultrastructure of collagen fibrils in uterine leiomyomas and normal myometrium. Fertil Steril 2004;82:1182–7.

[32] Beevers CS, et al. Curcumin disrupts the Mammalian target of rapamycin-raptor complex. Cancer Res 2009;69(3):1000–8.

[33] Yu S, et al. Curcumin inhibits Akt/mammalian target of rapamycin signaling through protein phosphatase-dependent mechanism. Mol Cancer Ther 2008;7(9):2609–20.

[34] Wong TF, et al. Curcumin targets the AKT–mTOR pathway for uterine leiomyosarcoma tumor growth suppression. Int J Clin Oncol 2014;19(2):354–63.

[35] Feng W, et al. Curcumin promotes the apoptosis of human endometrial carcinoma cells by downregulating the expression of androgen receptor through Wnt signal pathway. Eur J Gynaecol Oncol 2014;35(6):718–23.

[36] Sun M, et al. Effects of curcumin on the role of MMP-2 in endometrial cancer cell proliferation and invasion. Eur Rev Med Pharmacol Sci 2018;22(15):5033–41.

[37] Chen Q, et al. Curcumin suppresses migration and invasion of human endometrial carcinoma cells. Oncol Lett 2015;10(3):1297–302.

[38] Sirohi VK, et al. Curcumin exhibits anti-tumor effect and attenuates cellular migration via Slit-2 mediated down-regulation of SDF-1 and CXCR4 in endometrial adenocarcinoma cells. J Nutr Biochem 2017;44:60–70.

[39] Selvendiran K, et al. Safe and targeted anticancer efficacy of a novel class of antioxidant-conjugated difluorodiarylidenyl piperidones: differential cytotoxicity in healthy and cancer cells. Free Radic Biol Med 2010;48(9):1228–35.

[40] Rath KS, et al. HO-3867, a safe STAT3 inhibitor, is selectively cytotoxic to ovarian cancer. Cancer Res 2014;74(8):2316–27.

[41] Tierney BJ, et al. Aberrantly activated pSTAT3-Ser727 in human endometrial cancer is suppressed by HO-3867, a novel STAT3 inhibitor. Gynecol Oncol 2014;135(1):133–41.

[42] Kumar A, et al. Enhanced apoptosis, survivin down-regulation and assisted immunochemotherapy by curcumin loaded amphiphilic mixed micelles for subjugating endometrial cancer. Nanomed Nanotechnol Biol Med 2017;13(6):1953–63.

[43] Xu H, et al. Liposomal curcumin targeting endometrial cancer through the NF-κB pathway. Cell Physiol Biochem 2018;48(2):569–82.

[44] Scambia G, et al. Clinical effects of a standardized soy extract in postmenopausal women: a pilot study. Menopause (New York, NY) 2000;7(2):105–11.

[45] Beck V, et al. Comparison of hormonal activity (estrogen, androgen and progestin) of standardized plant extracts for large scale use in hormone replacement therapy. J Steroid Biochem Mol Biol 2003;84(2–3):259–68.

[46] Hu Y-C, et al. Detection of a negative correlation between prescription of Chinese herbal products containing coumestrol, genistein or daidzein and risk of subsequent endometrial cancer among tamoxifen-treated female breast cancer survivors in Taiwan between 1998 and 2008: a population-based study. J Ethnopharmacol 2015;169:356–62.

[47] Carbonel AF, et al. Soybean isoflavones attenuate the expression of genes related to endometrial cancer risk. Climacteric 2015;18(3):389–98.

[48] Naciff JM, et al. Dose-and time-dependent transcriptional response of Ishikawa cells exposed to genistein. Toxicol Sci 2016;151(1):71–87.

[49] Yeh C-C, et al. Genistein suppresses growth of human uterine sarcoma cell lines via multiple mechanisms. Anticancer Res 2015;35(6):3167–73.

[50] Polkowski K, Mazurek AP. Biological properties of genistein. A review of in vitro and in vivo data. Acta Pol Pharm 2000;57(2):135–55.

[51] Di X, et al. A high concentration of genistein down-regulates activin A, Smad3 and other TGF-β pathway genes in human uterine leiomyoma cells. Exp Mol Med 2012;44(4):281–92.

[52] Bai J, Luo X. 5-Hydroxy-4′-nitro-7-propionyloxy-genistein inhibited invasion and metastasis via inactivating Wnt/β-catenin signal pathway in human endometrial carcinoma Ji endometrial cells. Med Sci Monit 2018;24:3230.

[53] Ripon SH, Bhuiyan NQ. Cervical cancer risk factors: classification and mining associations. APTI-KOM J Comput Sci Inf Technol 2019;4(1):8–18.

[54] American Chemical Society. Cancer facts & figures 2019. vol. 2019. Atlanta: American Chemical Society; 2019.

[55] Bhatla N, Aoki D, Sharma DN, Sankaranarayanan R. Cancer of the cervix uteri. Int J Gynecol Obstet 2018;143:22–36.

[56] Touboul C, et al. Prognostic factors and morbidities after completion surgery in patients undergoing initial chemoradiation therapy for locally advanced cervical cancer. Oncologist 2010;15(4):405.

[57] Brisdelli F, D'Andrea G, Bozzi A. Resveratrol: a natural polyphenol with multiple chemopreventive properties. Curr Drug Metab 2009;10(6):530–46.

[58] Bollmann F, et al. Resveratrol post-transcriptionally regulates pro-inflammatory gene expression via regulation of KSRP RNA binding activity. Nucleic Acids Res 2014;42(20):12555–69.

[59] Khurana S, et al. Polyphenols: benefits to the cardiovascular system in health and in aging. Nutrients 2013;5(10):3779–827.

[60] Chung T, et al. Expression of apoptotic regulators and their significance in cervical cancer. Cancer Lett 2002;180(1):63–8.

[61] Zhang P, et al. Biological significance and therapeutic implication of resveratrol-inhibited Wnt, Notch and STAT3 signaling in cervical cancer cells. Genes Cancer 2014;5(5–6):154.

[62] Zhang P, et al. PIAS3, SHP2 and SOCS3 expression patterns in cervical cancers: relevance with activation and resveratrol-caused inactivation of STAT3 signaling. Gynecol Oncol 2015;139(3):529–35.

[63] Li L, et al. Resveratrol suppresses human cervical carcinoma cell proliferation and elevates apoptosis via the mitochondrial and p53 signaling pathways. Oncol Lett 2018;15(6):9845–51.

[64] Stanley M. Pathology and epidemiology of HPV infection in females. Gynecol Oncol 2010;117(2): S5–S10.

[65] Pääjärvi G, et al. TCDD activates Mdm2 and attenuates the p53 response to DNA damaging agents. Carcinogenesis 2005;26(1):201–8.

[66] Flores-Pérez A, Elizondo G. Apoptosis induction and inhibition of HeLa cell proliferation by alpha-naphthoflavone and resveratrol are aryl hydrocarbon receptor-independent. Chem Biol Interact 2018;281:98–105.

[67] Mukherjee S, et al. Unique synergistic formulation of curcumin, epicatechin gallate and resveratrol, tricurin, suppresses HPV E6, eliminates HPV+ cancer cells, and inhibits tumor progression. Oncotarget 2017;8(37):60904.

[68] Pitti RM, et al. Induction of apoptosis by Apo-2 ligand, a new member of the tumor necrosis factor cytokine family. J Biol Chem 1996;271(22):12687–90.

[69] Zhang L, Fang B. Mechanisms of resistance to TRAIL-induced apoptosis in cancer. Cancer Gene Ther 2005;12(3):228–37.

[70] Wiley SR, et al. Identification and characterization of a new member of the TNF family that induces apoptosis. Immunity 1995;3(6):673–82.

[71] Nakamura H, et al. Therapeutic significance of targeting survivin in cervical cancer and possibility of combination therapy with TRAIL. Oncotarget 2018;9(17):13451.

[72] Bin WH, et al. Pterostilbene (3′, 5′-dimethoxy-resveratrol) exerts potent antitumor effects in HeLa human cervical cancer cells via disruption of mitochondrial membrane potential, apoptosis induction and targeting m-TOR/PI3K/Akt signalling pathway. J BUON 2018;23(5):1384–9.

[73] Lee K-W, et al. Resveratrol analog, N-(4-methoxyphenyl)-3, 5-dimethoxybenzamide induces G2/M phase cell cycle arrest and apoptosis in HeLa human cervical cancer cells. Food Chem Toxicol 2019;124:101–11.

[74] Kim J-Y, et al. Resveratrol analogue (E)-8-acetoxy-2-[2-(3, 4-diacetoxyphenyl) ethenyl]-quinazoline induces G2/M cell cycle arrest through the activation of ATM/ATR in human cervical carcinoma HeLa cells. Oncol Rep 2015;33(5):2639–47.

[75] Tomoaia G, et al. Effects of doxorubicin mediated by gold nanoparticles and resveratrol in two human cervical tumor cell lines. Colloids Surf B Biointerfaces 2015;135:726–34.

[76] Prasad S, et al. Curcumin, a component of golden spice: from bedside to bench and back. Biotechnol Adv 2014;32(6):1053–64.

[77] Theodore M, et al. Multiple nuclear localization signals function in the nuclear import of the transcription factor Nrf2. J Biol Chem 2008;283(14):8984–94.

[78] Stefanson AL, Bakovic M. Dietary regulation of Keap1/Nrf2/ARE pathway: focus on plant-derived compounds and trace minerals. Nutrients 2014;6(9):3777–801.

[79] Asher G, et al. Mdm-2 and ubiquitin-independent p53 proteasomal degradation regulated by NQO1. Proc Natl Acad Sci 2002;99(20):13125–30.

[80] Patiño-Morales CC, et al. Curcumin stabilizes p53 by interaction with NAD (P) H: quinone oxido-reductase 1 in tumor-derived cell lines. Redox Biol 2020;28:101320.

[81] Ghasemi F, et al. Curcumin inhibits NF-kB and Wnt/β-catenin pathways in cervical cancer cells. Pathol Res Pract 2019;215(10):152556.

[82] Shanmugam MK, et al. The multifaceted role of curcumin in cancer prevention and treatment. Molecules 2015;20(2):2728–69.

[83] Aedo-Aguilera V, et al. Curcumin decreases epithelial-mesenchymal transition by a Pirin-dependent mechanism in cervical cancer cells. Oncol Rep 2019;42(5):2139–48.

[84] He G, et al. Effects of notch signaling pathway in cervical cancer by curcumin mediated photodynamic therapy and its possible mechanisms in vitro and in vivo. J Cancer 2019;10(17):4114.

[85] Paramasivam M, et al. High-performance thin layer chromatographic method for quantitative determination of curcuminoids in Curcuma longa germplasm. Food Chem 2009;113(2):640–4.

[86] Liao C-L, et al. Bisdemethoxycurcumin suppresses migration and invasion of human cervical cancer hela cells via inhibition of NF-κB, MMP-2 and-9 pathways. Anticancer Res 2018;38(7): 3989–97.

[87] Lin C-C, et al. demethoxycurcumin suppresses migration and invasion of human cervical cancer HeLa cells via inhibition of NF-κB pathways. Anticancer Res 2018;38(5):2761–9.

[88] Chaudhary M, et al. 4-Bromo-4′-chloro pyrazoline analog of curcumin augmented anticancer activity against human cervical cancer, HeLa cells: in silico-guided analysis, synthesis, and in vitro cytotoxicity. J Biomol Struct Dyn 2019;38(5):1335–53.

[89] Bava SV, et al. Sensitization of taxol-induced apoptosis by curcumin involves down-regulation of nuclear factor-κ B and the serine/threonine kinase Akt and is independent of tubulin polymerization. J Biol Chem 2018;293(31):12283.

[90] Dang Y-P, et al. Curcumin improves the paclitaxel-induced apoptosis of HPV-positive human cervical cancer cells via the NF-κB-p53-caspase-3 pathway. Exp Ther Med 2015;9(4):1470–6.

[91] Bava SV, et al. Akt is upstream and MAPKs are downstream of NF-κB in paclitaxel-induced survival signaling events, which are down-regulated by curcumin contributing to their synergism. Int J Biochem Cell Biol 2011;43(3):331–41.

[92] Sreekanth C, et al. Molecular evidences for the chemosensitizing efficacy of liposomal curcumin in paclitaxel chemotherapy in mouse models of cervical cancer. Oncogene 2011;30(28):3139–52.

[93] Murugesan K, et al. Effects of green synthesised silver nanoparticles (ST06-AgNPs) using curcumin derivative (ST06) on human cervical cancer cells (HeLa) in vitro and EAC tumor bearing mice models. Int J Nanomedicine 2019;14:5257.

[94] Li C, Ge X, Wang L. Construction and comparison of different nanocarriers for co-delivery of cisplatin and curcumin: a synergistic combination nanotherapy for cervical cancer. Biomed Pharmacother 2017;86:628–36.

[95] Upadhyay A, et al. Microbubble-mediated enhanced delivery of curcumin to cervical cancer cells. ACS Omega 2018;3(10):12824–31.

[96] Hussain A, et al. Inhibitory effect of genistein on the invasive potential of human cervical cancer cells via modulation of matrix metalloproteinase-9 and tissue inhibitiors of matrix metalloproteinase-1 expression. Cancer Epidemiol 2012;36(6):e387–93.

[97] Yang Y, et al. Genistein-induced apoptosis is mediated by endoplasmic reticulum stress in cervical cancer cells. Eur Rev Med Pharmacol Sci 2016;20(15):3292–6.

[98] Sundaram MK, et al. Genistein induces alterations of epigenetic modulatory signatures in human cervical cancer cells. Anti Cancer Agents Med Chem 2018;18(3):412–21.

[99] Dhandayuthapani S, et al. Induction of apoptosis in HeLa cells via caspase activation by resveratrol and genistein. J Med Food 2013;16(2):139–46.

[100] Vivanco I, Sawyers CL. The phosphatidylinositol 3-kinase–AKT pathway in human cancer. Nat Rev Cancer 2002;2(7):489–501.

[101] Wong AJ, et al. Increased expression of the epidermal growth factor receptor gene in malignant gliomas is invariably associated with gene amplification. Proc Natl Acad Sci 1987;84(19):6899–903.

[102] Hay N, Sonenberg N. Upstream and downstream of mTOR. Genes Dev 2004;18(16):1926–45.

[103] Sahin K, et al. Sensitization of cervical cancer cells to cisplatin by genistein: the role of NFkB and Akt/mTOR signaling pathways. J Oncol 2012;2012:461562.

[104] Solomon LA, et al. Sensitization of ovarian cancer cells to cisplatin by genistein: the role of NF-kappaB. J Ovarian Res 2008;1(1):9.

[105] Liu H, et al. Effects of genistein on anti-tumor activity of cisplatin in human cervical cancer cell lines. Obstet Gynecol Sci 2019;62(5):322–8.

[106] Xiong P, et al. Design, synthesis, and evaluation of genistein analogues as anti-cancer agents. Anti Cancer Agents Med Chem 2015;15(9):1197–203.

[107] Chen Y, et al. Effects of 7-difluoromethy-5, 4'-dimethoxygenistein on proliferation and apoptosis of human cervical cancer cells and its mechanism. Zhong Nan Da Xue Xue Bao Yi Xue Ban 2016;41(5):463–70.

[108] Cai L, et al. Folate receptor-targeted bioflavonoid genistein-loaded chitosan nanoparticles for enhanced anticancer effect in cervical cancers. Nanoscale Res Lett 2017;12(1):509.

[109] Zhang H, et al. Fabrication of genistein-loaded biodegradable TPGS-b-PCL nanoparticles for improved therapeutic effects in cervical cancer cells. Int J Nanomedicine 2015;10:2461.

[110] Hussein MJ, Salai JS. Clinical and histopathological features of ovarian cancer in Rizgary Hospital/Erbil City from 2014 to 2017. Med J Babylon 2019;16(2):112–8.

[111] Beral V, Hermon C, Peto R, Reeves G, Brinton L, Marchbanks P, Negri E, Ness R, Peeters PHM, Vessey M, Gapstur SM, Patel AV, Dal Maso L, Talamini R, Chetrit A, Hirsh G, Lubin F, Sadetzki S, Allen N, Beral V, Bull D, Callaghan K, Crossley B, Gaitskell K, Goodill A, Green J, Hermon C, Key T, Moser K, Reeves G, Collins R, Doll R, Peto R, Gonzalez CA, Lee N, Marchbanks P, Ory HW, Peterson HB, Wingo PA, Martin N, Pardthaisong T, Silpisornkosol S, Theetranont C, Boosiri B, Jimakorn P, Virutamasen P, Wongsrichanalai C, Tjonneland A, Titus-Ernstoff L, Byers T, Rohan T, Mosgaard BJ, Vessey M, Yeates D, Freudenheim JL, Chang-Claude J, Kaaks R, Anderson KE, Folsom A, Robien K, Rossing MA, Thomas DB, Weiss NS, Riboli E, Clavel-Chapelon F, Cramer D, Hankinson SE, Tworoger SS, Franceschi S, Negri E, Magnusson C, Riman T, Weiderpass E, Wolk A, Schouten LJ, van den Brandt PA, Koetsawang S, Rachawat D, Palli D, Black A, Berrington de Gonzalez A, Brinton LA, Freedman DM, Hartge P, Hsing AW, Lacey Jr JV, Hoover RN, Schairer C, Graff-Iversen S, Selmer R, Bain CJ, Green AC, Purdie DM, Siskind V, Webb PM, McCann SE, Hannaford P, Kay C, Binns CW, Lee AH, Zhang M, Ness RB, Nasca P, Coogan PF, Palmer JR, Rosenberg L, Kelsey J, Paffenbarger R, Whittemore A, Katsouyanni K, Trichopoulou A, Trichopoulos D, Tzonou A, Dabancens A, Martinez L, Molina R, Salas O, Goodman MT, Lurie G, Carney ME, Wilkens LR, Hartman L, Manjer J, Olsson H, Grisso JA, Morgan M, Wheeler JE, Peeters PHM, Casagrande J, Pike MC, Ross RK, Wu AH, Miller AB, Kumle M, Lund E, McGowan L, Shu XO, Zheng W, Farley TMM, Holck S, Meirik O, Risch HA. Ovarian cancer and body size: individual participant meta-analysis including 25,157 women with ovarian cancer from 47 epidemiological studies. PLoS Med 2012;9(4):e1001200.

[112] Collaborative Group on Epidemiological Studies of Ovarian Cancer. Menopausal hormone use and ovarian cancer risk: individual participant meta-analysis of 52 epidemiological studies. Lancet 2015;385(9980):1835–42.

[113] Lauby-Secretan B, et al. Body fatness and cancer—viewpoint of the IARC Working Group. N Engl J Med 2016;375(8):794–8.

[114] Alsop K, et al. BRCA mutation frequency and patterns of treatment response in BRCA mutation–positive women with ovarian cancer: a report from the Australian Ovarian Cancer Study Group. J Clin Oncol 2012;30(21):2654.

[115] Kim TH, Park JH, Woo JS. Resveratrol induces cell death through ROS-dependent downregulation of Notch1/PTEN/Akt signaling in ovarian cancer cells. Mol Med Rep 2019;19(4):3353–60.

[116] Liu Y, et al. Resveratrol inhibits the proliferation and induces the apoptosis in ovarian cancer cells via inhibiting glycolysis and targeting AMPK/mTOR signaling pathway. J Cell Biochem 2018;119(7):6162–72.

[117] Zhang Y, et al. Resveratrol induces immunogenic cell death of human and murine ovarian carcinoma cells. Infect Agents Cancer 2019;14(1):27.

[118] Ferraresi A, et al. Resveratrol inhibits IL-6-induced ovarian cancer cell migration through epigenetic up-regulation of autophagy. Mol Carcinog 2017;56(3):1164–81.

[119] El-kott AF, et al. The apoptotic effect of resveratrol in ovarian cancer cells is associated with down-regulation of galectin-3 and stimulating miR-424-3p transcription. J Food Biochem 2019;43(12): e13072.

[120] Wen W, et al. Pterostilbene suppresses ovarian cancer growth via induction of apoptosis and blockade of cell cycle progression involving inhibition of the STAT3 Pathway. Int J Mol Sci 2018; 19(7):1983.

[121] Chen C-W, et al. DHS (trans-4, 4'-dihydroxystilbene) suppresses DNA replication and tumor growth by inhibiting RRM2 (ribonucleotide reductase regulatory subunit M2). Oncogene 2019;38 (13):2364–79.

[122] Tino AB, et al. Resveratrol and acetyl-resveratrol modulate activity of VEGF and IL-8 in ovarian cancer cell aggregates via attenuation of the NF-κB protein. J Ovarian Res 2016;9(1):84.

[123] Nam S, et al. Development of resveratrol-loaded herbal extract-based nanocomposites and their application to the therapy of ovarian cancer. Nanomaterials 2018;8(6):384.

[124] Guo X, Mei J, Zhang C. Development of drug dual-carriers delivery system with mitochondria-targeted and pH/Heat responsive capacity for synergistic photothermal-chemotherapy of ovarian cancer. Int J Nanomedicine 2020;15:301.

[125] Tork OM, Khaleel EF, Abdelmaqsoud OM. Altered cell to cell communication, autophagy and mitochondrial dysfunction in a model of hepatocellular carcinoma: potential protective effects of curcumin and stem cell therapy. Asian Pac J Cancer Prev 2015;16(18):8271–9.

[126] Aoki H, et al. Evidence that curcumin suppresses the growth of malignant gliomas in vitro and in vivo through induction of autophagy: role of Akt and extracellular signal-regulated kinase signaling pathways. Mol Pharmacol 2007;72(1):29–39.

[127] Kim JY, et al. Curcumin-induced autophagy contributes to the decreased survival of oral cancer cells. Arch Oral Biol 2012;57(8):1018–25.

[128] Li B, et al. Curcumin induces cross-regulation between autophagy and apoptosis in uterine leiomyosarcoma cells. Int J Gynecol Cancer 2013;23(5):803–8.

[129] Liu L-D, et al. Curcumin induces apoptotic cell death and protective autophagy by inhibiting AKT/mTOR/p70S6K pathway in human ovarian cancer cells. Arch Gynecol Obstet 2019;299(6):1627–39.

[130] Tian M, et al. Modulation of Myb-induced NF-kB-STAT3 signaling and resulting cisplatin resistance in ovarian cancer by dietary factors. J Cell Physiol 2019;234(11):21126–34.

[131] Zhao J, et al. Dihydroartemisinin and curcumin synergistically induce apoptosis in SKOV3 cells via upregulation of MiR-124 targeting midkine. Cell Physiol Biochem 2017;43(2):589–601.

[132] Zhang X-G, et al. A review of dihydroartemisinin as another gift from traditional Chinese medicine not only for malaria control but also for schistosomiasis control. Parasitol Res 2014;113(5):1769–73.

[133] Zhang CZ, et al. Dihydroartemisinin exhibits antitumor activity toward hepatocellular carcinoma in vitro and in vivo. Biochem Pharmacol 2012;83(9):1278–89.

[134] Lin R, et al. Dihydroartemisinin (DHA) induces ferroptosis and causes cell cycle arrest in head and neck carcinoma cells. Cancer Lett 2016;381(1):165–75.

[135] Lee J, Zhou H-J, Wu X-H. Dihydroartemisinin downregulates vascular endothelial growth factor expression and induces apoptosis in chronic myeloid leukemia K562 cells. Cancer Chemother Pharmacol 2006;57(2):213–20.

[136] Dong F, et al. Dihydroartemisinin targets VEGFR2 via the NF-κB pathway in endothelial cells to inhibit angiogenesis. Cancer Biol Ther 2014;15(11):1479–88.

[137] Wu B, et al. Dihydroartiminisin inhibits the growth and metastasis of epithelial ovarian cancer. Oncol Rep 2012;27(1):101–8.

[138] Ikematsu S, et al. Serum midkine levels are increased in patients with various types of carcinomas. Br J Cancer 2000;83(6):701–6.

[139] Zhang L, et al. Genomic and epigenetic alterations deregulate microRNA expression in human epithelial ovarian cancer. Proc Natl Acad Sci 2008;105(19):7004–9.

[140] Arend RC, et al. The Wnt/β-catenin pathway in ovarian cancer: a review. Gynecol Oncol 2013;131 (3):772–9.

[141] Barbolina MV, Burkhalter RJ, Stack MS. Diverse mechanisms for activation of Wnt signalling in the ovarian tumour microenvironment. Biochem J 2011;437(1):1–12.

[142] Su HY, et al. An epigenetic marker panel for screening and prognostic prediction of ovarian cancer. Int J Cancer 2009;124(2):387–93.

[143] Yen H-Y, et al. Regulation of carcinogenesis and modulation through Wnt/β-catenin signaling by curcumin in an ovarian cancer cell line. Sci Rep 2019;9(1):1–14.

[144] Koroth J, et al. Investigation of anti-cancer and migrastatic properties of novel curcumin derivatives on breast and ovarian cancer cell lines. BMC Complement Altern Med 2019;19(1):1–16.

[145] Zeuthen J, et al. Characterization of a human ovarian teratocarcinoma-derived cell line. Int J Cancer 1980;25(1):19–32.

[146] Kálai T, et al. Synthesis of N-substituted 3, 5-bis (arylidene)-4-piperidones with high antitumor and antioxidant activity. J Med Chem 2011;54(15):5414–21.

[147] Terlikowska KM, et al. Potential application of curcumin and its analogues in the treatment strategy of patients with primary epithelial ovarian cancer. Int J Mol Sci 2014;15(12):21703–22.

[148] Adams BK, et al. Synthesis and biological evaluation of novel curcumin analogs as anti-cancer and anti-angiogenesis agents. Bioorg Med Chem 2004;12(14):3871–83.

[149] Adams BK, et al. EF24, a novel synthetic curcumin analog, induces apoptosis in cancer cells via a redox-dependent mechanism. Anti-Cancer Drugs 2005;16(3):263–75.

[150] Ramezani T, et al. Sensitization of resistance ovarian cancer cells to cisplatin by biogenic synthesized silver nanoparticles through p53 activation. Iran J Pharm Res 2019;18(1):222.

[151] Zhao M-D, et al. Co-delivery of curcumin and paclitaxel by "Core-Shell" targeting amphiphilic copolymer to reverse resistance in the treatment of ovarian cancer. Int J Nanomedicine 2019;14:9453.

[152] Dwivedi P, et al. Core–shell microencapsulation of curcumin in PLGA microparticles: programmed for application in ovarian cancer therapy. Artif Cells Nanomed Biotechnol 2018;46(Suppl 3): S481–91.

[153] Anderl P, et al. Correlation between steroid hormone receptors, histological and clinical parameters in ovarian carcinoma. Gynecol Obstet Investig 1988;25(2):135–40.

[154] Leung Y-K, et al. Estrogen receptor (ER)-β isoforms: a key to understanding ER-β signaling. Proc Natl Acad Sci 2006;103(35):13162–7.

[155] Chan KK, et al. Estrogen receptor modulators genistein, daidzein and ERB-041 inhibit cell migration, invasion, proliferation and sphere formation via modulation of FAK and PI3K/AKT signaling in ovarian cancer. Cancer Cell Int 2018;18(1):65.

[156] Shibue T, Weinberg RA. EMT, CSCs, and drug resistance: the mechanistic link and clinical implications. Nat Rev Clin Oncol 2017;14(10):611.

[157] Ma J, et al. Combination of a thioxodihydroquinazolinone with cisplatin eliminates ovarian cancer stem cell-like cells (CSC-LCs) and shows preclinical potential. Oncotarget 2018;9(5):6042.

[158] Ning Y, et al. Genistein inhibits stemness of SKOV3 cells induced by macrophages co-cultured with ovarian cancer stem-like cells through IL-8/STAT3 axis. J Exp Clin Cancer Res 2019;38(1):1–15.

[159] Ning Y, et al. Co-culture of ovarian cancer stem-like cells with macrophages induced SKOV3 cells stemness via IL-8/STAT3 signaling. Biomed Pharmacother 2018;103:262–71.

[160] Liu Y, et al. Genistein-induced differentiation of breast cancer stem/progenitor cells through a paracrine mechanism. Int J Oncol 2016;48(3):1063–72.

[161] Hwang K-A, et al. Anticancer effect of genistein on BG-1 ovarian cancer growth induced by 17 β-estradiol or bisphenol A via the suppression of the crosstalk between estrogen receptor alpha and insulin-like growth factor-1 receptor signaling pathways. Toxicol Appl Pharmacol 2013;272 (3):637–46.

[162] Kim Y-S, Choi K-C, Hwang K-A. Genistein suppressed epithelial–mesenchymal transition and migration efficacies of BG-1 ovarian cancer cells activated by estrogenic chemicals via estrogen receptor pathway and downregulation of TGF-β signaling pathway. Phytomedicine 2015;22(11):993–9.

[163] Ning Y, et al. Apoptosis induced by 7-difluoromethoxyl-5, 4′-di-n-octyl genistein via the inactivation of FoxM1 in ovarian cancer cells. Oncol Rep 2012;27(6):1857–64.

[164] Ning Y, et al. Inactivation of AKT, ERK and NF-κB by genistein derivative, 7-difluoromethoxyl-5, 4′-di-n-octylygenistein, reduces ovarian carcinoma oncogenicity. Oncol Rep 2017;38(2):949–58.

[165] Ning Y-X, et al. Let-7d increases ovarian cancer cell sensitivity to a genistein analog by targeting c-Myc. Oncotarget 2017;8(43):74836.

[166] Bai J, Yang BJ, Luo X. Effects of 5-hydroxy-4′-nitro-7-propionyloxy-genistein on inhibiting proliferation and invasion via activating reactive oxygen species in human ovarian cancer A2780/DDP cells. Oncol Lett 2018;15(4):5227–35.

[167] Mittal P, et al. Formulation and characterization of genistein-loaded nanostructured lipid carriers: pharmacokinetic, biodistribution and in vitro cytotoxicity studies. Curr Drug Deliv 2019;16 (3):215–25.

CHAPTER 2

Identification of potential drug candidates for the treatment of triple-negative breast cancer

Leimarembi Devi Naorem, Mathavan Muthaiyan, Ishita Bhattacharyya, Dinakara Rao Ampasala, and Amouda Venkatesan

Centre for Bioinformatics, School of Life Sciences, Pondicherry University, Puducherry, India

Abstract

Triple-negative breast cancer (TNBC) occurs in nearly 20% of patients diagnosed with breast cancer (BC). It is a subtype of BC that lacks expression of estrogen receptors, progesterone receptors, and human epidermal growth factor receptors. TNBC is highly aggressive in nature and is clinically challenging. However, developing new drugs is time consuming with high failure rates. Thus to overcome these difficulties, drug repurposing has been proposed. Drug repurposing is a process of finding new therapeutic indications for existing drugs to improve drug productivity and to utilize their full potential. In our study, we collected and used 1075 differentially expressed genes (DEGs) and significant hub genes of TNBC from our previous research. Drug repurposing tools such as CMap, sscMap, SPIED3, and LINCS-L1000 CDS2 were employed to screen the drug candidates for DEGs. Subsequently, the binding affinity between drug candidates and TNBC-associated hub genes was calculated using molecular docking. In addition, for the shortlisted drugs, we collected the known target genes from STITCH and DrugBank databases and found there to be slight overlap with the hub genes. Further, to study the important functions and the pathways of DEGs, enrichment analysis was performed using the Enrichr tool. On the basis of the scores obtained, 31 drugs were found to be promising to repurpose for the treatment of cancer. Also, it is evident from the docking studies that dabrafenib and MAPK1 exhibited the best binding energy of -9.9 kcal/mol. From the reported literature, it is noted that the aforementioned drugs used for the treatment of different cancers are found to be repurposed for other cancers as well. The analysis of pathways and ontologies showed that DEGs are highly enriched in biological processes such as cell division, proliferation, DNA replication, and cancer-related pathways. Thus, this study may facilitate finding new indications for existing drugs for the treatment of cancer.

Keywords: Triple-negative breast cancer, Drug repurposing, Docking studies.

Abbreviations

ATC	Anatomical Therapeutic Chemical
BC	breast cancer
CMap	Connectivity Map
CRC	colorectal cancer
DEGs	differentially expressed genes
DTI	drug-target interactions
GO	gene ontology

G.P. Nagaraju, R.R. Malla (eds.)
A Theranostic and Precision Medicine Approach for Female-Specific Cancers
ISBN 978-0-12-822009-2
https://doi.org/10.1016/B978-0-12-822009-2.00002-9

STITCH   search tool for interactions of chemicals
TNBC   triple-negative breast cancer

1. Introduction

Triple-negative breast cancer (TNBC) is a subtype of breast cancer (BC) that is very similar to basal-like BC [1]. It is a heterogeneous group of tumors that lack expression of estrogen receptors, progesterone receptors, and human epidermal growth factor receptors. TNBC has very limited treatment options as it does not respond to endocrine therapy and specific targeted therapies. It is aggressive in nature with high recurrence, metastatic rates, and poor prognosis [2]. It is typically treated with a combination of surgery, chemotherapy, and radiation therapy. The most common treatment options are neoadjuvant chemotherapy, poly (ADP-ribose) polymerase inhibitors, immunotherapy, and chemotherapy using cytotoxic agents [3]. Recently, the Food and Drug Administration (FDA) granted accelerated approval to atezolizumab (TECENTRIQ, Genentech Inc.) in combination with protein-bound paclitaxel for adult patients with TNBC whose tumors express PD-L1 (PD-L1 stained tumor-infiltrating immune cells of any intensity covering $\geq 1\%$ of the tumor area) [4]. However, developing new drugs is time consuming with high failure rates. Thus to overcome these difficulties, drug repurposing has been proposed [5].

Drug repurposing or re-profiling is a strategy that helps to identify new uses of existing approved or investigational drugs for the treatment of other diseases that are outside the scope of the original medical indication. Usually, drug repurposing starts with in silico screening of approved or marketed drugs against a particular therapeutic target and further shortlisted drugs are processed in vitro and in vivo. Then the compounds pass through clinical trials for new indications and enter into the market with FDA labeling. Advancement in genomics and computational methods has made drug development and research much easier [6]. Current in silico drug repurposing techniques focus on gene expression or drug-target interactions (DTI). Various developed computational approaches make the process of drug repurposing much easier. For instance, Connectivity Map (CMap) contains more than 7000 expression profiles representing 1309 compounds [7] where users can compare their input signatures with those databases. The connectivity score is the main result upon executing a query, which is measured using Kolmogorov-Smirnov statistic [8]. A high negative score indicates that the corresponding perturbagen reversed the expression of the query signature, while a high positive indicates that the perturbagen induced it. CMap contains small molecules and the ones with Anatomical Therapeutic Chemical (ATC) codes are therapeutic small molecules that can be used to find the existing drug for a new indication. In addition, many other drug-repurposing tools are available, such as sscMap, SPIED3, and LINCS-L1000

CDS2 [9–11]. Also, various levels of DTI are available in public databases such as Drug-Bank, STITCH (search tool for interactions of chemicals) [12, 13], and others.

The aim of this study is to screen the drug candidates for TNBC treatment using in silico analysis. Differentially expressed genes (DEGs) were queried using several drug repurposing tools and both FDA-approved drugs and investigational drugs shortlisted in the screening were analyzed on the basis of their pharmacological actions. The drugs were docked with hub genes of TNBC to check for the target-ligand interaction and stability of the complex. Further, pathway and ontology analysis of DEGs were performed to study the enriched functions and pathways. The identified drug candidates provide novel insights into alternative therapeutic options for TNBC patients.

2. Materials and methods

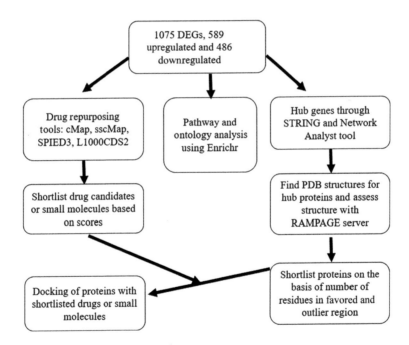

2.1 Data collection

Of 1075 DEGs of TNBC, 589 upregulated and 486 downregulated DEGs as well as 12 upregulated and 4 downregulated hub genes generated from our previous study are considered for the study [14]. The upregulating hub genes are AURKB, CCNB2, CDC20, DDX18, EGFR, ENO1, MYC, NUP88, PLK1, PML, POLR2F, and SKP2. The downregulating hub genes are CCND1, GLI3, SKP1, and TGFB3.

2.2 Hub genes identification using Network Analyst

DEGs were uploaded in Network Analyst to study the protein–drug interaction network. Based on the degree and betweenness measures, seven genes were shortlisted (EGFR, AURKB, PLK1, AURKA, MAPK1, PNMT, GAMT). Of these, three were upregulatory and two were downregulatory, and EGFR, AURKB, and PLK1 were found to be common in the list of hub genes extracted from the previous study [14].

2.3 Target proteins for docking studies

We collected 20 hub proteins from our previous study [14] and the NetworkAnalyst tool for the docking study. Subsequently, structures of the human proteins were retrieved from RCSB-PDB (www.rcsb.org) and analyzed using RAMPAGE (http://mordred. bioc.cam.ac.uk).

2.4 Screening of candidate drugs using repurposing tools

Various bioinformatics tools were used for shortlisting the drug candidates: CMap [15], sscMap [9], SPIED3 (SPIEDw v2.0) [10], and LINCS-L1000CDS2 [11]. The up- and downregulated genes, collectively known as disease signatures, were used as a query for each tool. Input data in a required format was prepared for each tool. The different results of the tools are interpreted separately for screening of drugs. The negatively signed drugs were considered for repurposing. The perturbagens having ATC codes are FDA-approved drugs. The permuted results and detailed results were downloaded from CMap (https://portals.broadinstitute.org/cmap/). Further, the top 100 (rank wise) chemical perturbagens were considered for shortlisting of potential drugs for TNBC. They were filtered on the basis of criteria with negative mean connectivity score <-0.199, enrichment score <-0.700, and specificity >0.50 candidate drugs were obtained.

For sscMap, results were saved as .tab file and potential candidates were selected on the basis of setsize >5 and setscore <-0.085. Only drugs with significance value of 1 were considered. Setsize in sscMap is the number of replicate reference profiles in the database for a compound, and 10 drugs were shortlisted.

From LINCS-L1000CDS2, 50 perturbagens were obtained in reverse mode and 50 in mimic mode. The detailed results for LINCS-L1000 were obtained from Enrichr (https://amp.pharm.mssm.edu/Enrichr/) and drugs with Z-score <-2.000 were considered.

SPIED3, a modified version of SPIEDw (v2.0), has separate datasets comprising drug treatments from CMap, LINCS, and DrugMatrix datasets. The drugs from DRUGScmap and DRUGSlincs were obtained. In the CMap results, 51 drugs have a negative correlation score. In the LINCS results, 50 drugs have a negative correlation score.

By comparing the results, we found 17 drugs from CMap-SPIED3, eight drugs from LINCS-L1000-SPIED3, and nine drugs from sscMap. Finally, 31 unique drugs were shortlisted.

2.5 Extraction of known target genes and indications for the shortlisted drugs

The respective target gene to the resulted 31 drugs was obtained from DrugBank [12] and STITCH [13] databases. Also, the indications and associated conditions were observed from DrugBank and literature.

2.6 Molecular docking studies

The PyRx [16]-Virtual Screening Tool was used to performed the docking studies for screened drugs and the shortlisted hub genes. Each of the 15 proteins was docked with 31 potential drug candidates using AutoDock Vina [17] wizard. The proteins were first viewed using the PyMOL molecular visualization system, and ligands and unnecessary water molecules were removed. To complete the structure by adding missing residues in chains, the modified proteins were uploaded to the WHAT IF web interface to generate a PDB file. Using MarvinSketch, the structure of the drug molecules were drawn and cleaned in 2D and 3D, generating a PDB file. The docking gives different poses for each drug and highlights their interactions with the target protein. Binding energies of $< -5.0\,\mathrm{kcal/mol}$ are considered to be stable for protein-drug complexes. Using PyMOL [18], the complex was visualized.

2.7 Enrichment analysis of DEGs

The 1075 DEGs were uploaded to the Enrichr [19] database for analysis. The pathways and ontologies for humans were interpreted. For the pathways, KEGG (2019) and Wiki-Pathways (2019) databases were studied. For ontologies, GO biological process (2018), GO molecular function (2018), and GO cellular component (2018) results were downloaded and analyzed.

3. Results

3.1 Drug targets

The drug-protein interaction network was studied using the NetworkAnalyst tool (Fig. 1). Based on the degrees and betweenness measures, seven genes—EGFR, AURKB, PLK1, AURKA, MAPK1, PNMT, and GAMT—and 20 hub proteins—AURKB, CCNB2, CDC20, DDX18, EGFR, ENO1, MYC, NUP88, PLK1, PML, POLR2F, SKP2,

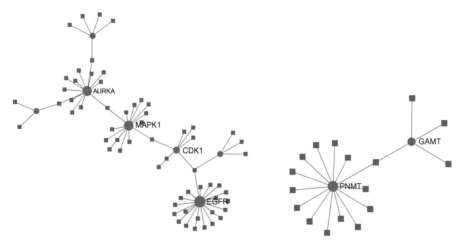

Fig. 1 Protein-drug interaction network using NetworkAnalyst.

Table 1 Shortlisted hub genes for molecular docking studies.

Hub gene	PDB ID	Binding site	RAMPAGE
AURKB	4AF3	83–88	93.8% (0.3%)
PLK1	3FC2	59–64	96.9% (0.0%)
CDC20	4GGA	183–188	97.1% (0.0%)
DDX18	3LY5	223–227	96.6% (0.0%)
EGFR	3GOP	718–723	97.2% (0.0%)
ENO1	3B97	7–11	96.8% (0.4%)
MYC	1NKP	903, 906, 907, 910, 913	97.2% (0.0%)
SKP2	1FS1	15, 16, 17, 19, 20	95.7% (1.3%)
AURKA	1MQ4	211–213	96.1% (0.4%)
MAPK1	1PME	31–35	96.3% (0.0%)
CCND1	2W96	15, 16, 17, 18, 20	91.3% (3.1%)
GLI3	4BLD	501–505	98.1% (0.0%)
TGFB3	1KTZ	34–38	95.1% (0.0%)
PNMT	1HNN	237, 240, 241, 244, 249	95.2% (0.8%)
GAMT	3ORH	18, 19, 20,23, 24	96.7% (0.0%)

CCND1, GLI3, SKP1, AURKA, MAPK1, PNMT, GAMT, and TGFB3—collected from our previous study were shortlisted [14] and considered for further study.

Subsequently, the structure of these human proteins was retrieved from RCSB-PDB (www.rcsb.org) and analyzed using RAMPAGE (http://mordred.bioc.cam.ac.uk). On the basis of percentage of allowed region (>95%) and outlier region (<1.5%), 15 proteins were considered for molecular docking studies (Table 1).

3.2 Potential drug candidates for TNBC using repurposing tools and analysis

The set of collected DEGs was queried in different formats in each tool. For every tool used in this study, negative scores have been considered only for the purpose of drug repurposing. Negative scores indicate that the compound is inhibitory, that is, it inversely connected to the gene signature. Further, to find an existing drug for a new indication, perturbagens having ATC codes were considered. It is found that only therapeutic small molecules have ATC codes.

In CMap 396 up tags and 293 down tags were found, and 6100 instances were processed with 626 positive scoring instances and 1456 negative scoring instances. SPIED3 resulted in 579 upregulated genes and 472 downregulated genes. Based on the different criteria set for each tool, a total of 31 drugs were shortlisted (Table 2) for further analysis.

The pharmacological actions and uses of these drugs were studied in the literature, DrugBank, PubChem [30], and ChEMBL [31, 32]. Trametinib is found to be repurposed for TNBC using the network biology method [33]. Mefloquine, which is an established antimalarial drug, is repurposed for colorectal cancer (CRC) [34]. Dabrafenib, dasatinib, dovitinib, mitoxantrone, and selumetinib are either approved drugs or under Phase III clinical trials for treatment of different types of cancer. Scriptaid [27] and quinostatin are drugs under clinical trials that are said to have anticancer activities, but they do not have any established targets.

Also, using STITCH and DrugBank databases, established target genes (Table 3) for these 31 compounds were obtained. It is seen that there is a very slight overlap of these genes with the hub genes for TNBC. Thus, it is a preliminary indication that there is no significant difference in expression of the drug-associated target genes in TNBC.

3.3 Molecular docking studies for hub genes and drug candidates

The 31 drugs were docked with each of 15 hub genes to study the interaction and stability of the drug-target complex. The best drug-target complex has a binding energy of -9.9 kcal/mol (Fig. 2A), formed by dabrafenib and MAPK1 (Table 4). With the binding energy of -9.8 kcal/mol, the scriptaid-AURKB complex is also very stable (Fig. 2B). Seventeen complexes have binding energy ≤ -9.0 kcal/mol and 429 drug-target complexes have binding energy < -5.0 kcal/mol.

3.4 Pathway and ontology analysis

KEGG and WikiPathways databases were used for pathway analysis of 1075 DEGs. All terms of gene ontology (GO) enrichment, namely, biological process, cellular

Table 2 Shortlisted drugs with ATC codes and uses.

S. No.	Drug	ATC code	Uses (literature and DrugBank)
1	Dabrafenib [20]	L01XE23	Melanoma, anaplastic thyroid cancer
2	Dasatinib [21]	L01XE06	Chronic myelogenous leukemia (CML)
3	Dovitinib	–	Under investigation (Phase III) for multiple myeloma, leukemia, solid tumors
4	Emetine	P01AX02	B-cell lymphoma, acute myeloid leukemia
5	Entinostat (MS-75) [22]	–	Advanced breast cancer
6	Fendiline	C08EA01	Pancreatic cancer
7	Fulvestrant	L02BA03	Hormone receptor-positive metastatic breast cancer
8	Latamoxef [23]	J01DD06	Anaerobic infections in cancer patients
9	LY-294002 [24]	–	Potentiates anticancer effect of oxaliplatin for gastric cancer (experimental)
10	Mefloquine	P01BC02	Malaria, repurposed in colorectal cancer
11	Metergoline	G02CB05	Psychoactive drug that acts as a ligand for serotonin and dopamine receptors
12	Mitoxantrone	L01DB07	Multiple sclerosis, metastatic breast cancer
13	Palbociclib	L01XE33	Advanced and metastatic breast cancer
14	Perhexiline	C08EX02	Coronary vasodilator-angina pectoris
15	Prochlorperazine [25]	N05AB04	Schizophrenia (antipsychotic), nausea, vertigo, repurposed for lung cancer
16	Protriptyline	N06AA11	Antidepressant (nervous system)
17	Puromycin	–	Antibiotic (research ongoing)
18	Pyrimethamine	P01BD01	Antimalarial repurposed for literature-hepatocellular carcinoma)
19	Quinostatin [26]	–	Leukemia
20	Resveratrol	–	In preclinical studies, resveratrol has been found to have potential anticancer properties
21	Rifabutin	J04AB04	Advanced HIV, tuberculosis, late-phase tuberculosis
22	Scriptaid [27]	–	Hepatocellular carcinoma
23	Selumetinib	–	Under investigation (Phase III) for breast cancer, lung cancer
24	Sirolimus [28]	S01XA23, L04AA10	Cancer therapy
25	Tanespimycin	–	Treatment of several types of cancer, solid tumors, or chronic myelogenous leukemia
26	Thioridazine	N05AC02	Antipsychotic
27	Trametinib [29]	L01XE25	Anaplastic thyroid cancer, metastatic melanoma
28	Trichostatin A	–	Anticancer (ongoing research)
29	Valproic acid	N03AG01	Epilepsy, under investigation for the treatment of HIV and various cancers
30	Vorinostat	L01XX38	Cutaneous T-cell lymphoma
31	Wortmannin	–	Melanoma

Table 3 Available targets for the screened drugs.

S. No.	Drug	Target genes
1	Dabrafenib	BRAF, RAF1, SIK1, NEK11, LIMK1, EIF2AK3, MDK
2	Dasatinib	ABL1, SRC, EPHA2, LCK, YES1, KIT, PDGFRB, STAT5B, ABL2, FYN, BTK, NR4A3, BCR, CSK, EPHA5, EPHB4, FGR, FRK, HSPA8, LYN, ZAK, MAPK14, PPAT, HCK
3	Dovitinib	NR1I2, FGFR2, FGFR1, FGFR3, FLT3, PDGFRB, YES1, KIT, LCK, FGR
4	Emetine	MAPK14, FOS, SNRPA, TIMP3, PARP1
5	Entinostat (MS-275)	ROCK1, ERBB2, CDH1, HDAC2, HDAC1, STAT3, HSP90AA1, HSP90AB1, CFLAR, INPP5J
6	Fendiline	HRAS
7	Fulvestrant	ESR1, AR, ESR2, PRL, IGFR1, CYP19A1, TFF1, PGR, CCND1, RB1
8	Latamoxef	ELANE
9	LY-294002	PIM1, PIK3CG, MAPK3, CASP3, MTOR, RPS6KB1, PIK3CA, PIK3CD, AKT1, GSK3B
10	Mefloquine	HBA1, ADORA2A, JUN, CALR, ABCC1, ABCC4, DHFR
11	Metergoline	SCN2A, HTR2C, HTR2A, HTR2B, HTR7, HTR6, HTR1A, HTR1B, HTR1D, HTR1F, HTR1E
12	Mitoxantrone	TOP2A, TOP2B, ABCB1, ABCC1, ABCG2, PIM1, ATM, PTPN2, FAS, CASP3
13	Palbociclib	CDK2, CDK4, CDK6, CDKN2B, TP53, RB1, CCND1, CDKN2A, DRD2, CCNE1
14	Perhexiline	CPT1A, CPT2, KCNH2, ERBB3, CYP2D6, TMPRSS1D
15	Prochlorperazine	DRD2, DRD3, DRD4, ADRA2B, S100A4, ADRA1D, HRH11, HTR2A, ADRA1B, HTR2C
16	Protriptyline	SLC6A2, SLC6A4, ADORA3, BACE1
17	Puromycin	RPL10L, RPL13A, RPL23, RPL15, RPL19, RPL23A, RSL24D1, RPL26L1, RPL8, RPL37, RPL3, RPL11, FOS, RBM3, KLK4
18	Pyrimethamine	DHFR, HEXB, SLC46A1, CHRM1, CYP2C9, PDF, HHEX, STAT3, TYMS, DHFRL1
19	Quinostatin	–
20	Resveratrol	SIRT1, SIRT3, SIRT5, TP53, NOS3, PTGS2, AKT1, ESR1, PTGS1, PPARG, NQO2, CSNK2A1, ALOX15, ALOX5, AHR, P14K2B, ITGA5, ITGB3, APP, SNCA, MTNR1A, MTNR1B, CLEC14A, NR1I2, NR1I3, SLC2A1, CBR1, PPARA, PPARG, KHSRP, YARS
21	Rifabutin	HSP90AA1, HSP90B1, BCL6, CYP3A4, CYP3A5, AADAC, CYP3A7, ABCB11
22	Scriptaid	REN, AGT, CASP3, TP53, XRCC5, HDAC1, HDAC2, HDAC3, HDAC6
23	Selumetinib	DPEP1, EIF2AK3, BCL2L11, FOXO3, STAT3, MAPK1, MAPK3, MAP2K1

Continued

Table 3 Available targets for the screened drugs—cont'd

S. No.	Drug	Target genes
24	Sirolimus (Rapamycin)	MTOR, FKBP1A, FGF2, FKBP3, FKBP5, FKBP4, AKT1, RPTOR, RPS6KB1, IRS1, EIF4EBP1
25	Tanespimycin	HSP90AA1, HSP90AB1, HSPA4, EGFR, ERBB2, TP53, CCND1, VEGFA, AKT1, HSPB1
26	Thioridazine	DRD2, DRD1, ADRA1A, ADRA1B, HTR2A, KCNH2, HTR6, DRD3, HTR1A, CYP2D6, HTR2C, ADRA1A, HRH1
27	Trametinib	MAP2K1, MAP2K2, MAPK1, MAPK3, RPS6KB1, RB1, ATK1, EIF2AK3, GNAQ, GNA11
28	Trichostatin A	HIST4H4, HDAC9, HDAC8, HDAC10, HDAC2, HDAC1, HDAC3, HDAC4, HDAC6, HDAC7
29	Valproic acid	HDAC9, ABAT, ACADSB, OGDH, ALDH5A1, HDAC2, PPARA, PPARD, PPARG, GSK3B, SMN1, ABCB1, CYP2C9, CYP2B6, CYP2A6, TSPO, BDNF, RELN
30	Vorinostat	HDAC1, HDAC2, HDAC3, HDAC6, HDAC8, HDAC7, HSP90AA1, TP53, BCL2L1, H2AFX
31	Wortmannin	PIK3CG, PLK1, PIK3R1, PIK3CA, PIK3CD, PLK3, MYLK, NOS3, AKT1, PIK3C3

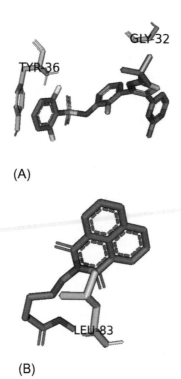

(A)

(B)

Fig. 2 (A) Dabrafenib-MAPK1 complex. (B) Scriptaid-AURKB complex.

Table 4 Top receptor-drug complexes.

Ligand	Target	Binding energy
Dabrafenib	MAPK1	−9.9
Scriptaid	AURKB	−9.8
Metergoline	MAPK1	−9.8
Dovitinib	AURKB	−9.5
Metergoline	AURKB	−9.5
Scriptaid	SKP2	−9.5
Dabrafenib	AURKB	−9.3
Emetine	AURKB	−9.3
Entinostat	AURKB	−9.3
LY-294002	AURKA	−9.3
Dabrafenib	CCND1	−9.3
Latamoxef	AURKB	−9.2
Fendiline	SKP2	−9.1
Palbociclib	MAPK1	−9.1
Metergoline	AURKA	−9
Dabrafenib	PNMT	−9
Quinostatin	PNMT	−9

component, and molecular function, were analyzed. Based on P-value, the following results were found to be significant:

- Small cell lung cancer, acute myeloid leukemia, bladder cancer, and endometrial cancer from KEGG (Fig. 3A).
- Retinoblastoma gene in cancer, gastric cancer network, metastatic brain tumor, imatinib, and chronic myeloid leukemia from WikiPathways (Fig. 3B).
- Mitotic nuclear division, regulation of transcription involved in G1/S transition of mitotic cell cycle, positive regulation of mitotic cell cycle phase transition, and attachment of mitotic spindle microtubules to kinetochore from GO biological process (Fig. 3C).
- Chromosome, centromeric region, mitotic spindle, condensed chromosome, and centromeric region from GO cellular component (Fig. 3D).
- Cytochrome-b5 reductase activity, acting on NAD(P)H, histone serine kinase activity, MAP kinase tyrosine/serine/threonine phosphatase activity, and histone kinase activity from GO molecular function (Fig. 3E).

4. Discussion

In this study, we considered 589 upregulated and 486 downregulated genes out of 1075 DEGs of TNBC collected from our previous study. We also examined 20 hub genes extracted from the previous study as well as via the NetworkAnalyst tool. DEGs are

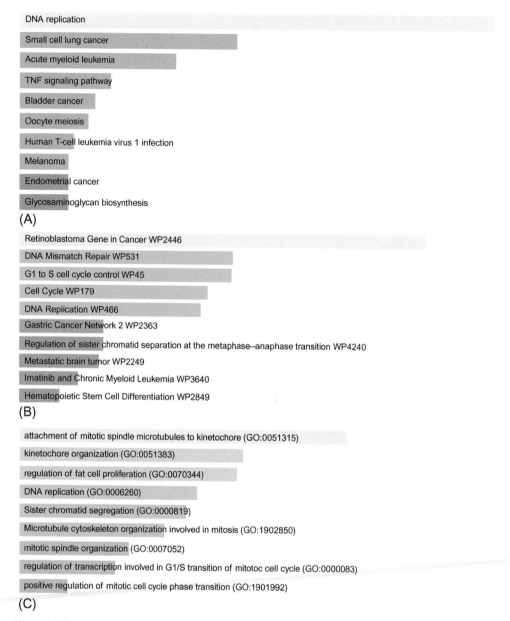

(A)

(B)

(C)

Fig. 3 (A) Bar graph for KEGG pathways analysis. (B) Bar graph for WikiPathways analysis. (C) Bar graph for GO biological process.

(Continued)

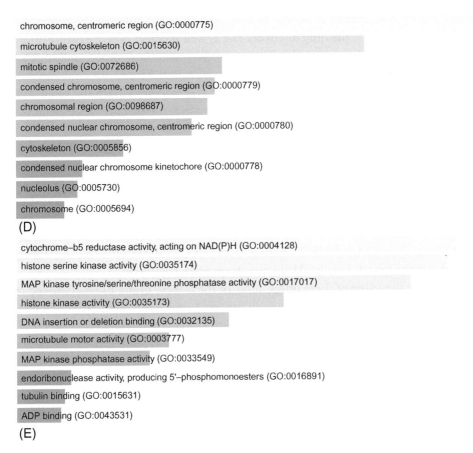

chromosome, centromeric region (GO:0000775)

microtubule cytoskeleton (GO:0015630)

mitotic spindle (GO:0072686)

condensed chromosome, centromeric region (GO:0000779)

chromosomal region (GO:0098687)

condensed nuclear chromosome, centromeric region (GO:0000780)

cytoskeleton (GO:0005856)

condensed nuclear chromosome kinetochore (GO:0000778)

nucleolus (GO:0005730)

chromosome (GO:0005694)

(D)

cytochrome–b5 reductase activity, acting on NAD(P)H (GO:0004128)

histone serine kinase activity (GO:0035174)

MAP kinase tyrosine/serine/threonine phosphatase activity (GO:0017017)

histone kinase activity (GO:0035173)

DNA insertion or deletion binding (GO:0032135)

microtubule motor activity (GO:0003777)

MAP kinase phosphatase activity (GO:0033549)

endoribonuclease activity, producing 5'–phosphomonoesters (GO:0016891)

tubulin binding (GO:0015631)

ADP binding (GO:0043531)

(E)

Fig. 3, cont'd   (D) Bar graph for GO cellular component. (E) Bar graph for GO molecular function.

found to highly enrich in cancer-related pathways and biological processes like cell division, proliferation, and DNA replication. Further, DEGs were mapped to several drug repurposing tools, and 31 drugs were shortlisted. Also, the target genes for the drug candidates were searched through STITCH and DrugBank databases and found to be slightly overlapped with the hub genes. This was a preliminary indication that no significant difference in expression of the drug-associated target genes in TNBC was identified. Furthermore, dabrafenib, scriptaid, and metergoline were considered as significant drug candidates through docking studies.

Therefore, considering the research design, various steps were followed to identify the drug candidates for TNBC. For this, gene signature [35] was used as a query to repurpose existing drugs for the treatment of TNBC using several drug repurposing tools. First, CMap was used to filter the drug candidates by negative mean connectivity score <-0.199, enrichment score <-0.700, and specificity >0. CMap is an online database that could infer the functional connections between small molecules/drugs, genes, and diseases. They can increase the drug discovery rate as well as detect uses of existing drugs. For instance, the research reported by Chen et al. identified new drug candidates for esophageal carcinoma by utilizing CMap [36]. Likewise, in the study, up- and downregulated genes known as disease signatures were used as queries for CMap and filtered on the basis of negative mean connectivity, enrichment scores, and specificity; 50 candidate drugs were obtained.

Further, through sscMap, 10 drugs were shortlisted based on setsize (>5) and setscore (<−0.085) and significance level of 1. From LINCS-L1000CDS2, 50 perturbagens were obtained in reverse mode and 50 in mimic mode with Z-score <−2.000. SPIED3, which is a modified version of SPIEDw (v2.0), has separate datasets comprising drug treatments from CMap, LINCS, and DrugMatrix datasets. CMap results showed 51 drugs with negative correlation score and LINCS results showed 50 drugs. By comparing the results, 17 drugs from CMap-SPIED3, 8 drugs from LINCS-L1000-SPIED3, and 9 drugs from sscMap were found. Finally, 31 unique drugs were shortlisted.

On verifying the usage of these drugs, most of them were found to be approved or experimental treatments for different types of cancer. Some antimalarial drugs like mefloquine and pyrimethamine have been already repurposed to certain types of cancer therapy. Also, it is reported that the drug trametinib was identified as a potential candidate and can be repositioned in TNBC.

Drugs were further analyzed using DrugBank, STITCH, and literature search. It was found that trametinib has already been repurposed for TNBC using different approaches [33]. Also, some drugs are already established for cancer treatment like dabrafenib, mitoxantrone, palbociclib, dasatinib, vorinostat, and fulvestrant. Drugs like dovitinib, selumetinib, and tanespimycin are under clinical trials for treatment of lymphoma, BC, and other types of cancer. Trichostatin A [37] and resveratrol [38] are natural derivatives that have potential anticancer effects and are being studied. Trichostatin A is natural derivative of dienohydroxamic acid isolated from species of the bacterial genus *Streptomyces*. Resveratrol is a phytoalexin derived from grapes and other food products with antioxidant and potential chemopreventive activities. Antimalarial drugs such as pyrimethamine and mefloquine have been successfully repurposed to hepatocellular carcinoma and CRC, respectively.

Moreover, TNBC-associated hub genes identified through the previous study were taken into consideration for further analysis. In addition to bioinformatics database analysis, drug repurposing was also explored further through structural exploration. Thus, docking studies were performed for the hub genes with 31 potential drug candidates using AutoDock Vina. In the study, binding affinity between drug molecules and proteins was identified. It was found that dabrafenib and MAPK1 exhibited the best binding energy of −9.9 kcal/mol followed by scriptaid and meterogline. The greater negative binding energy, the greater the stability of the drug-target complex. Thus, through drug repurposing and docking studies, dabrafenib, scriptaid, and meterogline were predicted to be potential drug candidates that may target the hub genes that influence pathogenesis. Thus, these drugs are potential candidates to be repurposed for TNBC.

5. Conclusion

With limited treatment options for TNBC and increasing need for effective therapy, drug repurposing is one of the best options as it is less time consuming than novel drug discovery. This work establishes possible drug candidates based on their known target genes, pharmacological usage, and gene signatures that can be reused for the treatment of a highly aggressive subtype of BC. We shortlisted 31 drugs and extracted their existing targets. Also, usage of these drugs was seen to be mostly related to anticancer activity. Dabrafenib, scriptaid, and meterogline were predicted to be potential drug candidates that may target TNBC-associated hub genes. The obtained drugs are potential candidates to be repurposed to treat TNBC. Some more new targets can be established for the drugs with the help of docking studies.

Acknowledgments

The authors thank Centre for Bioinformatics, Pondicherry University for providing the computer facility to carry out the work. Leimarembi Devi Naorem acknowledges a Senior Research Fellowship from the Council of Scientific & Industrial Research (CSIR). Mathavan Muthaiyan acknowledges a Senior Research Fellowship from the Rajiv Gandhi National Fellowship (RGNF).

References

[1] Shao F, Sun H, Deng CX. Potential therapeutic targets of triple-negative breast cancer based on its intrinsic subtype. Oncotarget 2017;8(42):73329.

[2] Wahba HA, El-Hadaad HA. Current approaches in treatment of triple-negative breast cancer. Cancer Biol Med 2015;12(2):106.

[3] Jia H, Truica CI, Wang B, Wang Y, Ren X, Harvey HA, Song J, Yang JM. Immunotherapy for triple-negative breast cancer: existing challenges and exciting prospects. Drug Resist Updat 2017;32:1–15.

[4] Cyprian FS, Akhtar S, Gatalica Z, Vranic S. Targeted immunotherapy with a checkpoint inhibitor in combination with chemotherapy: a new clinical paradigm in the treatment of triple-negative breast cancer. Bosn J Basic Med Sci 2019;19(3):227.

[5] Gns HS, Saraswathy GR, Murahari M, Krishnamurthy M. An update on drug repurposing: re-written saga of the drug's fate. Biomed Pharmacother 2019;110:700–16.

[6] Mirza N, Sills GJ, Pirmohamed M, Marson AG. Identifying new antiepileptic drugs through genomics-based drug repurposing. Hum Mol Genet 2017;26(3):527–37.

[7] Luo B, Gu YY, Wang XD, Chen G, Peng ZG. Identification of potential drugs for diffuse large b-cell lymphoma based on bioinformatics and Connectivity Map database. Pathol Res Pract 2018;214 (11):1854–67.

[8] Siavelis JC, Bourdakou MM, Athanasiadis EI, Spyrou GM, Nikita KS. Bioinformatics methods in drug repurposing for Alzheimer's disease. Brief Bioinform 2016;17(2):322–35.

[9] Zhang SD, Gant TW. sscMap: an extensible Java application for connecting small-molecule drugs using gene-expression signatures. BMC Bioinform 2009;10(1):236.

[10] Williams G. SPIEDw: a searchable platform-independent expression database web tool. BMC Genomics 2013;14(1):765.

[11] Duan Q, Reid SP, Clark NR, Wang Z, Fernandez NF, Rouillard AD, Readhead B, Tritsch SR, Hodos R, Hafner M, Niepel M. L1000CDS 2: LINCS L1000 characteristic direction signatures search engine. NPJ Syst Biol Appl 2016;2(1):1–12.

[12] Wishart DS, Knox C, Guo AC, Shrivastava S, Hassanali M, Stothard P, Chang Z, Woolsey J. DrugBank: a comprehensive resource for in silico drug discovery and exploration. Nucleic Acids Res 2006;34(Suppl_1):D668–72.

[13] Kuhn M, von Mering C, Campillos M, Jensen LJ, Bork P. STITCH: interaction networks of chemicals and proteins. Nucleic Acids Res 2007;36(Suppl_1):D684–8.

[14] Naorem LD, Muthaiyan M, Venkatesan A. Integrated network analysis and machine learning approach for the identification of key genes of triple-negative breast cancer. J Cell Biochem 2019;120 (4):6154–67.

[15] Lamb J, Crawford ED, Peck D, Modell JW, Blat IC, Wrobel MJ, Lerner J, Brunet JP, Subramanian A, Ross KN, Reich M. The Connectivity Map: using gene-expression signatures to connect small molecules, genes, and disease. Science 2006;313(5795):1929–35.

[16] Rashidieh B, Madani Z, Azam M, Maklavani SK, Akbari NR, Tavakoli S, Rashidieh G, Madani B, Azam Z, Maklavani MK, Akbari SK, Tavakoli NR, Rigi G. Molecular docking based virtual screening of compounds for inhibiting sortase A in L. monocytogenes. Bioinformation 2015;11(11):501.

[17] Trott O, Olson AJ. AutoDock Vina: improving the speed and accuracy of docking with a new scoring function, efficient optimization, and multithreading. J Comput Chem 2010;31(2):455–61.

[18] Yuan S, Chan HS, Hu Z. Using PyMOL as a platform for computational drug design. Wiley Interdiscip Rev Comput Mol Sci 2017;7(2):e1298.

[19] Kuleshov MV, Jones MR, Rouillard AD, Fernandez NF, Duan Q, Wang Z, Koplev S, Jenkins SL, Jagodnik KM, Lachmann A, McDermott MG. Enrichr: a comprehensive gene set enrichment analysis web server 2016 update. Nucleic Acids Res 2016;44(W1):W90–7.

[20] Long GV, Flaherty KT, Stroyakovskiy D, Gogas H, Levchenko E, De Braud F, Larkin J, Garbe C, Jouary T, Hauschild A, Chiarion-Sileni V. Dabrafenib plus trametinib versus dabrafenib monotherapy in patients with metastatic BRAF V600E/K-mutant melanoma: long-term survival and safety analysis of a phase 3 study. Ann Oncol 2017;28(7):1631–9.

[21] Ongoren S, Eskazan AE, Suzan V, Savci S, Erdogan Ozunal I, Berk S, Yalniz FF, Elverdi T, Salihoglu A, Erbilgin Y, Iseri SA. Third-line treatment with second-generation tyrosine kinase inhibitors (dasatinib or nilotinib) in patients with chronic myeloid leukemia after two prior TKIs: real-life data on a single center experience along with the review of the literature. Hematology 2018;23 (4):212–20.

[22] Connolly RM, Rudek MA, Piekarz R. Entinostat: a promising treatment option for patients with advanced breast cancer. Future Oncol 2017;13(13):1137–48.

[23] Lagast H, Meunier-Carpentier F, Klastersky J. Moxalactam treatment of anaerobic infections in cancer patients. Antimicrob Agents Chemother 1982;22(4):604–10.

[24] Liu J, Fu XQ, Zhou W, Yu HG, Yu JP, Luo HS. LY294002 potentiates the anti-cancer effect of oxaliplatin for gastric cancer via death receptor pathway. World J Gastroenterol 2011;17(2):181.

[25] Ahmedzai S, Carlyle DL, Calder IT, Moran F. Anti-emetic efficacy and toxicity of nabilone, a synthetic cannabinoid, in lung cancer chemotherapy. Br J Cancer 1983;48(5):657–63.

[26] Kong L, Zhang X, Li C, Zhou L. Potential therapeutic targets and small molecular drugs for pediatric B-precursor acute lymphoblastic leukemia treatment based on microarray data. Oncol Lett 2017;14 (2):1543–9.

[27] Liu L, Sun X, Xie Y, Zhuang Y, Yao R, Xu K. Anticancer effect of histone deacetylase inhibitor scriptaid as a single agent for hepatocellular carcinoma. Biosci Rep 2018;38(4):1–9.

[28] Woo HN, Chung HK, Ju EJ, Jung J, Kang HW, Lee SW, Seo MH, Lee JS, Lee JS, Park HJ, Song SY. Preclinical evaluation of injectable sirolimus formulated with polymeric nanoparticle for cancer therapy. Int J Nanomedicine 2012;7:2197.

[29] Knispel S, Zimmer L, Kanaki T, Ugurel S, Schadendorf D, Livingstone E. The safety and efficacy of dabrafenib and trametinib for the treatment of melanoma. Expert Opin Drug Saf 2018;17(1):73–87.

[30] Kim S, Thiessen PA, Bolton EE, Chen J, Fu G, Gindulyte A, Han L, He J, He S, Shoemaker BA, Wang J. BS The PubChem Project. Nucleic Acids Res 2016;44(D1):D1202–13.

[31] Gaulton A, Hersey A, Nowotka M, Bento AP, Chambers J, Mendez D, Mutowo P, Atkinson F, Bellis LJ, Cibrián-Uhalte E, Davies M. The ChEMBL database in 2017. Nucleic Acids Res 2017;45(D1):D945–54.

[32] Mendez D, Gaulton A, Bento AP, Chambers J, De Veij M, Félix E, Magariños MP, Mosquera JF, Mutowo P, Nowotka M, Gordillo-Marañón M. ChEMBL: towards direct deposition of bioassay data. Nucleic Acids Res 2019;47(D1):D930–40.

[33] Vitali F, Cohen LD, Demartini A, Amato A, Eterno V, Zambelli A, Bellazzi R. A network-based data integration approach to support drug repurposing and multi-target therapies in triple negative breast cancer. PLoS One 2016;11(9), e0170363.

[34] Xu X, Wang J, Han K, Li S, Xu F, Yang Y. Antimalarial drug mefloquine inhibits nuclear factor kappa B signaling and induces apoptosis in colorectal cancer cells. Cancer Sci 2018;109(4):1220–9.

[35] Malcomson B, Wilson H, Veglia E, Thillaiyampalam G, Barsden R, Donegan S, El Banna A, Elborn JS, Ennis M, Kelly C, Zhang SD. Connectivity mapping (ssCMap) to predict A20-inducing drugs and their antiinflammatory action in cystic fibrosis. Proc Natl Acad Sci 2016;113(26):E3725–34.

[36] Chen YT, Xie JY, Sun Q, Mo WJ. Novel drug candidates for treating esophageal carcinoma: a study on differentially expressed genes, using connectivity mapping and molecular docking. Int J Oncol 2019;54(1):152–66.

[37] Zhang XF, Yan Q, Shen W, Gurunathan S. Trichostatin A enhances the apoptotic potential of palladium nanoparticles in human cervical cancer cells. Int J Mol Sci 2016;17(8):1354.

[38] Ko JH, Sethi G, Um JY, Shanmugam MK, Arfuso F, Kumar AP, Bishayee A, Ahn KS. The role of resveratrol in cancer therapy. Int J Mol Sci 2017;18(12):2589.

CHAPTER 3

Chemoresistance in female-specific cancers and the associated anti-resistance therapies

S. Sumathi, K. Santhiya, and J. Sivaprabha

Department of Biochemistry, Biotechnology and Bioinformatics, Avinashilingam Institute for Home Science and Higher Education for Women, Coimbatore, Tamil Nadu, India

Abstract

Cancer is a disease in which there is uncontrolled and abnormal growth of cells that invade and spread to other parts of the body. Various types of cancer are known for their substantial behavior and varied therapeutic response. Women are more susceptible to cancers that affect the breasts, ovaries, cervix, endometrium, vulva, and vagina. The most widely adopted treatments include chemotherapy and radiation. Chemoresistance is the phenomenon where tumor cells exhibit resistance to the effects of chemotherapeutic drugs. The following review focuses on distinct mechanisms of chemoresistance shown in female-specific cancers and their treatment modalities.

Keywords: Estrogen, Chemoresistance, Cancer, Progesterone, Metastatic breast cancer, Chemotherapeutics, Anthracycline resistance.

Abbreviations

ABC receptor	ATP-binding cassette
ABC	Advanced Breast Cancer Conference
Akt	protein kinase B
ARID1A	AT–rich interactive domain 1 A
ATF6	activating transcription factor 6
BAX	Bcl–2-like protein 4
Bcl–2	B-cell lymphoma 2
BIM	Bcl–2-like Protein 11
BRAF	proto ongogene B-Raf
BRCA gene	BReast CAncer gene
BUBR1	budding uninhibited by benzimidazole-related 1
CD	cluster of differentiation
C-erbB-2/ErbB2	epidermal growth factor receptor family
COX2	cyclooxygenase 2
CTLA4	cytotoxic T-lymphocyte associated protein 4
CTNNB1	catenin beta-1
DDP	cis diammine dichloroplatinum II
EMA	European Medicines Agency

G.P. Nagaraju, R.R. Malla (eds.)
A Theranostic and Precision Medicine Approach for Female-Specific Cancers
ISBN 978-0-12-822009-2
https://doi.org/10.1016/B978-0-12-822009-2.00003-0

49

EMT	epithelial mesenchymal transition
ER	endoplasmic reticulum
ER	estrogen
ERK	extracellular signal-regulated kinase
FDA	Food and Drug Administration
GLOBOCAN	Global Cancer Incidence, Mortality and Prevalence
GST	glutathione S transferase
HER2/neu	human epidermal growth factor receptor 2
HIF	hypoxia inducible factor
HIPEC	hyperthermic intraperitoneal chemotherapy intra
ICIs	immune checkpoint inhibitors
IL	interleukin
IRE1α	endoribonuclease inositol-requiring enzyme 1α
ITH	intratumor heterogeneity
KRAS	Ki-ras2 Kirsten rat Sarcoma Viral oncogene homolog
MAD2	mitotic arrest deficient 2
MAP	microtubule-associated protein
MDR1	multidrug resistance protein 1
MEK	mitogen-activated protein kinase
MRP1	MDR-associated protein 1
OS	overall survival
P21$^{cip/waf}$	cyclin-dependent kinase inhibitor 1
P34^{cdc2}	protein kinase cell division Cycle 2
PAMAM	polyamidoamine
PARP	poly ADP ribose polymerase
PD-1	programmed cell death protein -1
PFS	progression-free survival
P-gp	P-glycoprotein
PgR	progesterone
PI3K	phosphotidyl inositol-3- kinase
PIK3CA	phosphotidyl inositol-4,5-bisphosphate 3-kinase catalytic subunit alpha
PLD	pegylated liposomal doxorubicin
PPP2R1A	protein phosphatase 2 scaffold subunit alpha
Prosigna	breast cancer prognostic gene signature assay
PTB	polypyrimidine tract binding protein
PTEN	phosphatase and tensin homolog
RAF	rapidly accelerated fibrosarcoma
SAC	spindle assembly checkpoint
SIOG	International Society of Geriatric Oncology
SRp20	serine/arginine-rich protein 20
Topo I	topoisomerase I
Topo II	topoisomerase II
TOR	toremifer
TP53	tumor protein 53
VEGF	vascular endothelial growth factor
XiaP	X-linked inhibitor of apoptosis

1. Breast cancer

1.1 Epidemiology

Breast cancer is a widely prevalent malignancy in women and ranks second after lung cancer in contribution to the global tumor burden. In 2020, 1.7 million people were diagnosed with and about half a million people died due to breast cancer worldwide [1]. As per the GLOBOCAN 2018 reports, the incidence rate of breast cancer is 1.6% and the mortality rate is 6.6% [2]. An accelerated incidence and mortality rate is due to modern lifestyle habits and lack of accessibility to state-of-the-art diagnosis and therapy.

1.2 Molecular classification of breast cancer

Earlier, four clinically relevant breast cancers including the luminal A, luminal B, HER2-enriched, and basal-like molecular subtypes were elucidated. Currently, the techniques of gene copy number and expression analysis have predicted the presence of 10 different molecular subtypes [3]. The four conventional breast tumor classes could be directly assessed using the multigene assays of prosigna (NanoString Technologies) and BluePrint (Agendia) techniques. They are also indirectly indicated by the immunohistochemical analysis of steroid hormone receptors like estrogen (ER), progesterone (PgR), and HER2, and the tumor proliferation marker Ki67, which reveals the tumor biology. The luminal A subtype is ER or PgR positive or both, HER2 negative, and has low proliferation ability. The luminal B subtype is distinguished from the luminal A type with its higher proliferation rate. The HER2 subtype (non-luminal) is HER2 positive and is devoid of ER and PgR. Absence of all three receptors is the characteristic feature of basal-like or triple negative breast cancer (HER2, ER, and PgR negative). The therapeutic decision for the patient depends on the molecular subtype, locoregional tumor load, and the individual's wishes [4, 5].

1.3 Therapeutic approaches to early breast cancer

Early breast cancer is curable as no metastatic effects are observed. The therapeutic interventions require clinical examination of biopsy and breast imaging methods like mammography and ultrasound [6]. Local therapy includes surgery and radiation, while systemic therapy encompasses endocrine therapy and chemotherapy.

1.4 Local therapy on early breast cancer

1.4.1 Surgery

Breast conservation has emerged as an indispensable therapeutic tool established through oncoplastic surgical techniques and primary systemic therapy that improved access to other organ-preserving procedures [7]. Breast conservation is the primary option to over-rule the mastectomy through oncoplastic techniques and neoadjuvant tumor shrinking therapy [8]. Following the neoadjuvant or primary systemic treatment, surgery is performed for breast conservation, which should satisfy the *no ink on tumor rule* with clear margins to prevent recurrence [9]. Sentinel node surgery lessens the prominent side effects of axillary surgery [10]. In some cases, surgical removal of axillary lymph nodes makes no difference in the therapeutic outcome [11].

1.4.2 Radiotherapy

Less invasive radiotherapy approaches include partial breast irradiation or hypofractioned radiotherapy for decreasing the tumor burden [12]. Interoperative or stand-alone radiation therapy tends to decrease the long-term toxic effects. This treatment is commonly effected in the axillar and supraclavicular regions. Modern radiotherapy is known to compromise dosage-related issues with an optimized boost [13]. Nodal irradiation can deprive the positive tumor node even if axillary surgery is not performed, but is still known to produce some toxic effects [14].

1.4.3 Systemic therapy for early breast cancer

HER2 negative luminal tumors are treated with neoadjuvant or adjuvant chemotherapy depending on factors such as tumor proliferation, tumor grade, and the lymph node involved. These are determined by multigene expression assays, which also trace the risk of early and late recurrences [15].

1.4.4 Endocrine therapy

Adjuvant endocrine therapy is administered to early luminal breast cancer patients for a span of 5–10 years with a standardized dose of 20 mg tamoxifen per day in premenopausal women, which has shown to reduce the recurrence in ER-positive disease [16]. Continuous exposure to tamoxifen for 10 years halves the chance of breast cancer–associated mortality in affected patients but has pronounced side effects in later life [17]. Endocrine therapy is very effective as the first-line treatment of choice in HER2-negative luminal metastatic breast cancer since detrimental effects like organ damage alone would recruit the chemotherapy in action [18]. Aromatase inhibitors are the most preferred agent of choice in postmenopausal women. Other drugs involved in these therapeutic conventions are fulvestrant, progestin, and tamoxifen. Endocrine monotherapy or the co-administration of endocrine agents with targeted therapeutics may be suitable for slowly progressing disease [15].

1.4.5 Therapeutic features of metastatic breast cancer

The clinical therapeutic trends do not offer a potential cure for metastatic breast cancer, but rather can extend life and reduce symptoms. This fact is assured by there being less than 5% long-term survivors [19]. According to the proceedings of the biannual Advanced Breast Cancer conference (ABC), systemic therapy is the preferred option for metastatic breast cancer, and locoregional therapy can be adopted in certain conditions like primary metastatic diseases. Histology varies from the primary tumor site to the metastatic site due to tumor heterogeneity, which helps in determining the potential target site for treatment. During the first bone metastases, bone-stabilizing drugs like bisphosphonates or denosumab are given as maintenance therapy [20].

1.4.6 Chemotherapeutics to combat the big trial

Preoperative and postoperative trials of chemotherapy are followed in early breast cancer and generally promote overall disease-free survival [21]. A good pathological complete response and an optimized prognosis are ensured with triple negative and HER2 positive breast cancer [22]. Synergistic effects of anthracycline and other taxanes in terms of four cycles of anthracycline, weekly paclitaxel, and thrice-weekly docetaxel are found to reduce a 10-year mortality rate in breast cancer when administered over a period of 18–24 weeks as a standard chemotherapeutic regimen [23]. Four to six cycles of docetaxel-cyclophosphamide (TC) is an effective substitute for anthracycline, but is not a standard therapy [24]. From the GIM trial, dose-dense administration of anthracycline–taxane is found to increase five-year disease-free survival compared to other standard chemo regimes [25]. In chemotherapy, biological age of the patient is considered as an important factor over chronological age and hence standard drug protocols are preferably followed in fit older people. The International Society of Geriatric Oncology (SIOG) has documented the dosage forms of drugs as per the requirements of elderly people [26]. Pathological complete response is achieved with the addition of platinum to the standard chemo regime in triple negative breast cancer of BRCA 1/2 mutation and wild type BRCA [27]. In metastasis the anti-VEGF antibody bevacizumab can procure a progression-free survival (PFS), however, it could not promise an overall survival with combinatorial agents like paclitaxel or capecitabine. It is approved by the European Medicines Agency (EMA) but not the Food and Drug Administration (FDA) and hence is practiced as a therapeutic measure only in certain countries [28].

1.4.7 Chemoresistance in breast cancer

Cellular chemoresistance is a major contributor to overall survival reduction. It adopts several mechanisms at the molecular levels to accomplish poor prognostic effects.

1.4.8 Role of CD73 in anthracycline resistance

From a set of 6000 patients' gene expression profile analysis, high CD73 expression is correlated with inhibitory actions on the efficacy of anthracycline-based chemotherapy. The prominent effect of CD73 during the neoadjuvant regimen reduced the pathological complete response and led to the disappearance of invasive tumor at surgery. Activation of adenosine receptors led to reduction in antitumor effects as a result of chemoresistance to doxorubicin [29].

1.4.9 Microtubule stabilizing agents

Certain taxanes like paclitaxel and docetaxel act as microtubule stabilizing agents, which disrupt the spindle microtubule dynamics leading to cell death and apoptosis. Chemoresistance to taxanes is studied with respect to the spindle assembly checkpoint (SAC) and dysfunctional regulation of apoptotic signaling. SAC proteins like MAD2, BUBR1, gamma-synuclein, and aurora A are vital markers for taxane resistance. Overexpression of drug efflux pump MDR-1/P-gp altered expression of microtubule-associated proteins (MAPs) such as tau, stathmin, and MAP4 determine the chances of recurrence of the condition and also check the beneficial effects of taxane treatment [30].

1.4.10 β tubulin mutations

These mutations alter microtubule dynamics and stability [31], which in turn resists the binding of anti-mitotic drugs like paclitaxel to β tubulin subunits [32]. Repeated exposure of cancer cells to anti-mitotic drugs can produce a highly resistant phenotype due to the loss of heterozygosity of highly resistant tubulin allele [33]. Seven β tubulin isotypes are present in humans. Microtubules made up of βIII and βIV tubulin isotypes may need to be bound to paclitaxel to induce microtubule stability [34]. They contribute to drug resistance with the altered binding site containing serine/arginine substitution at the 277th position of the βIII tubulin [35].

1.4.11 Microtubule-associated proteins (MAPs)

MAPs involved in microtubule dynamics facilitate the interaction between tubulin polymers and microtubules. They also modulate the cellular response to mitotic inhibitor drugs [36]. Tau is a neuronal MAP, which, on low expression level, renders resistance against paclitaxel binding. The dysfunction of stathmin, which is a microtubule stabilizing agent, leads to taxane resistance [37]. Downregulation of MAP4 heightens the dynamicity of microtubules and affects paclitaxel resistance [36].

1.4.12 Multidrug resistance (MDR)

MDR1 gene encoded P-glycoprotein is a drug efflux pump that is energy dependent. It binds to a variety of hydrophobic drugs like paclitaxel, doxorubicin, vincristine, and vinblastine and subsequently induces resistance [38].

1.4.13 C-erbB-2/HER 2-neu

Two mechanisms are proposed for cellular resistance against paclitaxel due to the corresponding HER2/neu proto-oncogene overexpression. Association or binding of P21$^{waf1/cip-1}$ with the kinase P34^{cdc2} inhibits its paclitaxel-induced activation, and apoptosis at the G2/M phase consequently leads to high expression of HER2/neu [39]. Resistance against paclitaxel-induced apoptosis is brought about by overexpression of

ErbB2 receptor tyrosine kinase, which inhibits the $P34^{cdc2}$ activation in primary breast tumors [40]. High resistance to paclitaxel, doxorubicin, and 5-fluorouracil is advocated by the HER2/PI3K/Akt pathway [41].

1.4.14 Antiapoptotic signaling

Raf/MEK/ERK is a cell survival pathway and involves various pro-apoptotic and anti-apoptotic proteins. Phosphorylation of BIM, a pro-apoptotic protein by ERK, prevents BAX–BAX homodimerization, which is induced by Bcl-2, and substantial anti-apoptotic activity is provoked [42]. Protein tyrosine kinases (PTK) induce drug resistance by regulating the anti-apoptotic signaling pathway phosphotidyl inositol 3-kinase/Akt (PI3K/Akt) [43].

1.5 Overcoming chemoresistance in breast cancer

1.5.1 Biomarkers for bypassing chemoresistance

Hypoxia inducible factors (HIF) and their gene products are generally seen in triple negative breast cancer. Chemotherapy-induced HIF activity effected the proliferation of breast cancer stem cells using IL-6 and IL-8 signals that lead to increased effect of MDR. Combination of digoxin, an HIF inhibitor, and paclitaxel or gemcitabine resulted in tumor control by the abrogation of resistance [44].

1.5.2 Antitumor drug candidates

P-gp drug effluxer is acted upon by celecoxib, which is a specific cyclooxygenase2 (COX2) inhibitor [45]. LY294002 is the specific inhibitor of PI3K, which inactivates P-gp and decreases the concentration of survivin [46]. Accumulation of doxorubicin inside the cell is increased by toremifer (TOR) [47]. MDR is reversed by lobeline.

1.5.3 Combination therapeutics

Tamoxifen along with everolimes greatly decreases the incidence of secondary drug resistance and inhibits cell proliferation or induces apoptosis. Quercetin enhances tumor apoptosis under hyperthermic conditions and could reverse MDR [48].

1.5.4 Novel drug delivery system

A drug delivery system is used for targeted drug release, long hours of blood circulation, and enriched biocompatibility. The system is typically in the nanoscale preparation. Simultaneous delivery of major vault protein-targeted small interfering RNA and doxorubicin is promoted by polyamidoamine (PAMAM), which overcomes multidrug resistance by modulating intracellular drug resistance [49]. Liposome vectors of parthenolide and vinorelbine eliminate breast cancer stem cells [50].

1.5.5 Immunotherapy

Immune checkpoint inhibitors (ICIs) selectively kill cancer cells by utilizing the loopholes of the immune system [51]. ICI nivolumab targets the programmed cell death protein – 1 (PD-1). Inhibition of P-gp by immunomodulators like cyclosporin A can reduce MDR [52]. ICIs, immune potentiators, targeted agents, and epigenetic modifiers like histone deacetylase inhibitors can reverse breast tumor MDR [51].

2. Ovarian cancer

2.1 Ovarian cancer around the world

Ovarian cancer is a widespread gynecological disorder and ranks third after cervical and uterine cancers. The poor prognostics, increased mortality rate rather than the incidence level, subtle growth of tumors, and slow onset of symptoms are the hallmarks of this "silent killer [53, 54]." Ovarian cancer is three times as lethal as breast cancer, and it is estimated that by 2040, the mortality rate will increase [55]. It is a diverse range of malignancies with variation in tumor biology and other molecular characteristics. More than 90% of ovarian tumors are of epithelial origin, with the greatest incidence of serous carcinoma in non–Hispanic whites (5.2 per 100,000) and the lowest incidence in non-Hispanic blacks and Asian–Pacific islanders (3.4 per 100,000). Still, the latter population has high risk rates of endometrioid and clear cell carcinoma. Progressive tumor growth is enhanced by the delayed diagnosis observed during stage III or IV, especially in the case of serous carcinoma [56].

2.2 Distinctive subtypes

Being the seventh most frequently diagnosed cancer type among women, ovarian cancer can be of epithelial or non-epithelial origin. The epithelial class is the most prominent, which is attributed to its five subtypes with distinguishing developmental, histological, molecular, and prognostic patterns and characteristics [57]. These subtypes include high-grade serous carcinoma affecting 70% of the susceptible population, low-grade serous carcinoma affecting <5%, endometrioid cancer affecting 10%, and clear cell and mucinous types affecting 10% and 3% of patients, respectively. These kinds of cancers mimic the tumor morphology of uterine cancer as well as those carcinomas that originate in the fallopian tubes or endometrium, and are often primarily misdiagnosed as the former type of cancers [58]. Based on the tumorigenic profile of epithelial carcinoma, it is categorized into two types. Type I includes low-grade serous, mucinous, endometrioid, and clear cell carcinoma and Brenner tumors, which are marked for high degree of mutations in specialized genes like KRAS, BRAF, PTEN, PIK3CA, CTNNB1, ARID1A, and PPP2R1A. Type II tumors are recognized as high-grade serous and endometrioid that are vigorous with genetic instability [59]. High-grade serous and endometrioid carcinomas are due to TP53 mutations and dysfunction of BRCA1 and/or BRCA2 [60].

2.3 Contributors to ovarian cancer

The greater incidence rate of epithelial ovarian cancer is often observed in postmenopausal women population with a five-year survival rate of 22% compared to 48% in younger women of premenopausal stage. The younger women showed an increased survival advantage at all stages of carcinoma progression, which is determined by independent prognostic factors such as age, activity, and extent of cellular reductive surgery [61]. Another factor of chance for the invasive tumor is preterm labor consequences. Also, giving birth to a male infant can double the risk [62]. Relapse of pelvic inflammatory disease correlates to the risk of ovarian cancer [63]. Familial history of breast tumor is an important genetic predisposal factor [64]. Lifestyle habits like physical inactivity and associated obesity are key components of this life-threatening cancer and decreases the chances of survival [65].

2.4 Therapeutic tacklers

2.4.1 Surgery

Both diagnosis and therapeutic strategy are mediated by debulking or cytoreductive surgery [66] and involves the excision of all naked tumor parts, thereby lowering the risk of relapse [67]. Patients who are platinum-responsive to the tumor can access an advantage, which is confirmed by a 5.6-month increase of PFS following secondary debulking surgery [68]. Optimized effects of hyperthermic intraperitoneal chemotherapy (HIPEC) is also ensured and normally undertaken after cytoreductive surgery [69]. Less affected individuals by surgery can be administered neoadjuvant chemotherapy [70].

2.4.2 Targeted treatments
Anti-angiogenic element

Bevacizumab is used in conjugation with first-line chemotherapy as well as in relapsed patients who are platinum-responsive or platinum-resistant to treatments [71]. This mode of administration using an anti-angiogenic agent like bevacizumab increased PFS and improved the quality of life in patients with stage III or IV epithelial ovarian cancer [72]. After chemotherapy, the sequential regime of bevacizumab for a long period is essential to maintain the PFS stage. Side effects like hypertension and gastrointestinal toxicity were also encountered with the uptake of bevacizumab that may account for just 3% of the cases with these observed side effects. The average PFS was found to be 12.4 months in the platinum-responsive relapsed epithelial cancer patients when bevacizumab was co-administered with the standard carboplatin–gemcitabine compared to 8.4 months PFS of stand-alone chemotherapy. However, the overall survival (OS) remains the same between the two groups [73]. The Italian Association of Medical Oncology [67] suggests a protocol of six cycles of bevacizumab along with carboplatin–paclitaxel in high grade tumors after a debulking surgery.

PARP inhibitors

In BRCA gene mutated cancer cells, chemotherapy-driven DNA damage is overcome with a single strand or alternative DNA repair pathway, which requires the function of a poly ADP ribose polymerase (PARP), and its inhibition can facilitate the process of synthetic lethality [74]. High grade ovarian carcinoma patients with a BRCA mutation can have long PFS by maintenance and monotherapy using olaparib at 400 mg twice daily. Some adverse actions were also seen by the usage of drug in terms of nausea, anemia, and fatigue [75]. Two novel PARP inhibitors, namely niraparib (EMA approved in Nov. 2017) and rucaparib, are yet to be marketed. Nicaparib and rucaparib are known to render a significantly long-term PFS in platinum-responsive relapsed ovarian cancer patients, despite the presence or absence of BRCA mutations and homologous recombination deficiency status [76, 77].

First-line chemotherapy

Platinum is most widely employed in the primary therapy of ovarian cancer and cisplatin is the first platinum-derived drug employed for the cause. However, it has toxic effects related to dosage such as nausea, peripheral neuropathy, and nephrotoxicity. Therefore organo-platinum analogues of cisplatin, namely carboplatin, were undertaken as the primary chemotherapeutic agents [78]. The standard chemotherapeutic combination of carboplatin and pacitaxel is recognized as a first-line strategy, although recurrence rates of 70%–80% are observed within the first 2 years since there are no other agents that could procure the same extraordinary therapeutic effects [67]. Administration of tolerable doses of carboplatin every 3 weeks with a weekly paclitaxel, and a synergistic mixture of bevacizumab with thrice-weekly carboplatin-paclitaxel can also be effective [79, 80].

Second-line chemotherapy

In relapsed patients, the second-line treatment aims to ensure long-term survival and enhance the quality of life with delayed progression of symptoms. Some of the prognostic factors include tumor size, histology, BRCA mutations, and the number of metastases occurred [81]. Therapeutic decision is based on the patient's responsiveness to a platinum-based therapy. Individuals who are considered to be platinum-sensitive, either partially or completely, are subjected to platinum-based combination chemotherapy and they must have a promising platinum-free-interval of >12 or 6–12 months [67].

In platinum-resistant patients, targeted therapies could provide favorable outcomes. Trabectedin–pegylated liposomal doxorubicin (PLD) is known for better results in PFS and OS rates [82].

Resistance to chemotherapeutics

Chemoresistance may be attributed to several reasons such as cancer stem cells, inefficient drug transport, modifications of drug target, alteration of the cellular proteins involved in detoxification, changes associated with DNA repair mechanism, and high tolerance to drug damage [83]. Approximately 50%–70% recurrence in ovarian cancer may be due to chemoresistance and intratumor heterogeneity (ITH) [84].

Problems of drug transport

MDR is a type of cross-resistance gained by the tumor cells in the due course of chemotherapeutics. It commonly involves the overexpression of ATP-binding cassette transporters like P-gp and MDR-associated protein 1 (MRP1) [85]. A high rate of alternative splicing in the MRP1 gene results in numerous splice variants like PTB and SRp20, which show resistance against doxorubicin [86].

Alterations of cellular proteins involved in detoxification

Gluathione and glutathione S-transferases (GST) are involved in detoxifying some alkylating agents of chemotherapy. GST π1 was shown to render resistance to doxorubicin chlorambucil through transfection experiments in yeast [87]. Metallothioneins are a group of low-weight proteins, which protect against cellular DNA damage, oxidative stress, and the associated apoptosis. Overexpression is an important factor and is reported in ovarian cancer [88].

Modifications of the drug target

Topoisomerase I (TopoI) and topoisomerase II (TopoII) are of great importance in the DNA repair metabolism. Hence there is an increased expression of them in malignant tumors and they act as drug targets for many chemotherapeutic agents like camptothecin, etoposide, teniposide, novobiocin, anthracyclines (doxorubicin), and mitoxantrone [89].

Enormous DNA repair activity

One of the ovarian cancer cell lines named A2780 is found to be resistant to *cis*-diamminedichloroplatinum (II) (DDP) due to the removal of platinum-DNA adducts [90]. Similarly, the cisplatin-associated resistance also indulges efficient DNA repair especially at the interstrand crosslink [91]. The platinum-DNA lesions are dislodged by the process of nucleotide excision repair [92].

Tolerance to drug-induced damage

Reduced response to apoptosis is facilitated by increased tolerance to chemotherapy-induced damage and requires the expression of pro-apoptotic factors or tumor suppressor genes as well as the modulation of cell survival elements [92]. X-linked inhibitor of apoptosis (Xiap) expression and Fas ligand downregulation contribute to chemoresistance in ovarian carcinoma [93]. Dysfunction of Fas/Fas ligand assures chemoresistance against cisplatin [94].

2.5 Overcoming chemoresistance

2.5.1 Efflux pump inhibitors

Third-generation inhibitors of drug efflux pumps like P-gp and MDR have minimal inhibitory effect on the cytochrome P450 enzymes in contrast to first- and second-generation inhibitors [95]. These agents, like apatinib and tariquidar, are known to possess increased specificity and potency as well as to reverse resistance to paclitaxel in in vitro and in vivo systems [96, 97]. In mice models of ovarian cancer nano-based agents are chosen that target CD44 to deliver siRNA directed against MDR1, and in combinatorial use with paclitaxel, ensures the specificity to cancer cells [98].

2.5.2 Endothelin receptor and EMT

The epithelial mesenchymal transition (EMT) state that is acquired after chemotherapy leads to chemoresistance in ovarian cancer. The endothelin receptor acts as the precursor of EMT changes and hence blocking it with an antagonist such as zibotentan is believed to reverse the effects of resistance and the subsequent EMT phenotype. This drug is studied for action in clinical trials [99, 100].

2.5.3 Reversal of DNA damage tolerance and enhanced repair

TP53 mutant ovarian cancer cells show resistance to platinum-induced DNA damage and anti-apoptotic effects through improper G1 phase arrest and the subsequent G2 delay, which leads to an escape from genetic damage due to enhanced repair pathways [101]. Wee1 kinase involved in the phosphorylation of CDC2 is a potential drug target for the reversal of platinum resistance in TP53 mutants and acted upon by MK1775 and PD0166285, which are wee1 specific inhibitors [102].

2.5.4 PARP and angiogenic inhibitors

Surpassing the effects of vascular endothelial growth factor (VEGF) block through angiopoietins and neovascularization is a commonly adopted method by cancer cells. Combination of anti-angiogenic agents like bevacizumab and trebananib in platinum-resistant high-grade serous ovarian carcinoma could facilitate an increase in PFS [103, 104]. Resistance to PARP inhibitors is due to the expression of MDR1 efflux pumps where specific drug agents like 6-thioguanine can reverse the antagonistic action of the latter [105].

2.5.5 Checkpoint inhibitors

Increased expression of the PD-1 receptor is observed in high-grade serous ovarian carcinoma, and PD1 inhibitors like pembrolizumab and nivolumab are subjected to FDA trials. Cytotoxic T-lymphocyte–associated protein 4 (CTLA4) is also a checkpoint inhibitor, and associated ipilumab is currently under study [106]. To evade encountering the immune system, some cancer cells will delete the major histocompatibility complex class I and endanger a resistance that can be suitably targeted with further anticipations on checkpoint inhibitors [107].

3. Cervical cancer

Cancer that develops in the cells present in the lower female reproductive system (uterus) leads to cervical cancer. The upper part of the uterus that carries the fetus is connected to the lower part of the vagina via the cervix. The cervix is divided into two regions: the region in the upper part is called the endocervix, which is made up of glandular cells, and the lower region closer to the vagina is called the exocervix, which consists of squamous cells. These two types of cells meet at the transformation zone, which may vary in location depending on the age of women and after giving birth.

Cancer in the cervix usually develops in the cells present in this transformation zone. These normal cells show some pre-cancerous behavior and later may or may not transform into cancer. These pre-cancerous modifications are referred to in various terms, namely, cervical intraepithelial neoplasia, squamous intraepithelial lesion, and dysplasia. Women aged between 35 and 44 years are frequently diagnosed with cervical cancer. It rarely occurs in women younger than 20 years of age; only 15% of cases occur in women older than 65. [108, 109].

3.1 Diagnosis

Women aged 30–49 years should periodically check for the appearance of cancerous tissues. Symptoms related to cervical cancer will also be taken into consideration. A Pap test will be done, and if the results are abnormal the doctor can advise for screening tests like colposcopy, cervical biopsy, colposcopic biopsy, endocervical curettage, and cone biopsy [110].

3.2 Treatment methods

Conventional therapeutic measures include surgery such as cryosurgery, laser surgery, loop electrosurgical excitation procedure, conization, radial trachelectomy, simple/radical hysterectomy, radiation therapy using external beam, chemotherapy, and targeted therapy using monoclonal antibodies.

3.3 Chemotherapeutic drugs and their modes of action

Chemotherapy refers to the usage of chemicals/drugs for treating cancer, cervical cancer in this case. Table 1 gives the list of chemotherapeutic drugs employed for cervical cancer.

3.4 Resistance of cancer cells to chemotherapy

When chemotherapy is used for the treatment of cancer cells, a persistent problem is usually noticed: the progression of resistance of cancer cells to chemotherapeutic drugs used for therapy.

The endoplasmic reticulum (ER) is an indispensable cell organelle involved in the folding of protein. Any disturbances in its normal function results in the accumulation of unfolded or misfolded proteins in the ER lumen. This results in ER stress and eventually causes unfolded protein responses that activate cell death. This strategy is employed

Table 1 Chemotherapeutic drugs and their modes of action on cancer cells [111].

Drugs	Mode of action
Before the 1970s	
Cyclophosphamide	Interferes in the replication of DNA and transcription to RNA
Chlorambucil	Cell cycle arrest and induces apoptosis in cancer cells
Melphalan	Alkylates guanine and inhibits synthesis of DNA and RNA
5-Fluorouracil	Inhibits formation of the nucleotide thymidine and interferes in DNA replication
Methatrexate	Blocks de novo synthesis of the nucleotide thymidine
Vincristine	Inhibits cell division
iBleomycin	Induces DNA strand breaks in cancer cells
Adriamycin	Interferes with the growth and spreading of cancer cells
Mytomycin C	Inhibits cancer cell growth
After the 1970s	
Cisplatin	Interacts with DNA and results in the formation of DNA adducts
Carboplastin	Modifies DNA structure and inhibits DNA synthesis
Ifosfamide	Inhibits DNA synthesis
Palcitaxel	Defective mitotic spindle assembly and inhabits chromosome segregation and cell division
Irinotecan	Inhibits topoisomerase and thus blocks DNA synthesis
Topotecan	Induces DNA damage in cancer cells
Gemcitabine	Inhibits DNA replication in cancer cells
Vinorelbine	Prevents the formation of mitotic spindle, leads to the arrest of tumor cell growth in metaphase
Docetaxel	Disrupts normal formation of microtubules and stops cell division
Doxorubicin	Acts as intercalating agent and inhibits DNA synthesis and RNA transcription
Mitolactol	Interferes in cell division

by several chemotherapeutic drugs to halt the growth of cancer cells [112]. But in certain circumstances, cancer cells develop resistance to chemotherapeutic drugs by employing molecular mechanisms such as altered membrane transport (e.g., overexpression of ABC receptor in cancer cells can promote drug resistance) [113], alteration of drug targets (e.g., MRP-1 becomes resistant to drugs etoposide and doxorubicin by inducing mutation of the topo II gene resulting in decreased accumulation of drugs in cancer cells) [114, 115], activation of antioxidants and detoxification system (e.g., increase in glutathione S-transferase causes apoptosis resistance in cancer cells) [116], and ER stress response (e.g., ATF6, IRE1, and PERK are usually deregulated in cancer cells, which leads to the activation of oncogenes and suppression of tumor suppressor genes) [113, 117, 118].

The chemotherapeutic drug cisplatin (alone or in combination with palcitaxel) is widely employed in the treatment of cervical cancer. Sometimes cancer cells develop resistance to cisplatin and thus reduce the efficacy of the drug to kill cancer cells. This resistance to cisplatin can develop by mechanisms like reduction of platinum deposition in the cancer cells, increased DNA damage repair, apoptosis block, activation of EMT, changes in DNA methylation, cancer stem cell characteristics, and stress response chaperones [119].

3.5 Overcoming chemotherapy resistance

In order to overcome MDR several strategies has been employed. Some of them include the MDR of cancer cells that usually involves overexpression of ABC transporters and chemosensitizers, which has been developed to block the function of ABC transporter to reverse drug resistance (e.g., catechins, carotenes) [120]. Reduction of the intervals between each chemotherapy treatment suppresses MDR in cancer cells [121].

4. Conclusion

Chemotherapy is one of the most widely used treatment modalities for female-specific metastatic cancers. However, cancer cells exhibit drug resistance through various molecular mechanisms as discussed in this chapter. Hence this aspect opens a new research domain that focuses more on the effective chemotherapeutic drugs that overcome this resistance or an altered method of treatment that suits the demand and welfare of the patient undergoing treatment.

References

[1] Ferlay J, Soerjomataram I, Dikshi R, et al. Cancer incidence and mortality worldwide: sources, methods and major patterns in GLOBOCAN 2012. Int J Cancer 2015;136:E359–86.

[2] Bray F, Ferlay J, Soerjomataram I, Siegel RL, Torre LA, Jemal A. Global Cancer Statistics 2018: GLOBOCAN estimates of incidence and mortality worldwide for 36 cancers in 185 countries. CA Cancer J Clin 2018;68:394–424.

[3] Curtis C, Shah SP, Cgin SF, et al. The genomic and transcriptomic architecture of 2000 breast tumours reveals novel subgroups. Nature 2012;486:346–52.

[4] Goldhirsch A, Winer EP, Coats AS, et al. Personalizing the treatment of women with early breast cancer: highlights of St. Gallen International Expert Consensus on the primary therapy of early breast cancer 2013. Ann Oncol 2013;24:206–23.

[5] Coats AS, Winer EP, Goldhirsch A, et al. Tailoring therapies-improving the management of early breast cancer, St. Gallen International Expert Consensus on the primary therapy of early breast cancer 2015. Ann Oncol 2015;26:1533–46.

[6] AGO Breast Commission Recommendations. Diagnosis and therapy of primary and metastatic breast cancer, 2016. Available from: http://www.ago-online.de/en/guidelines-mamma/. (Accessed 15 September 2019).

[7] McLaughlin SA. Surgical management of the breast: breast conservation therapy and mastectomy. Surg Clin North Am 2013;93:411–28.

[8] Haloua MH, Krekel NM, Winters HA, et al. A systematic review of oncoplastic breast-conserving surgery: current weaknesses and future prospects. Ann Surg 2013;257:609–20.

[9] Kümmel S, Holtschmidt J, Loibl S. Surgical treatment of primary breast cancer in the neoadjuvant setting. Br J Surg 2014;101:912–24.

[10] Krag DN, Julian TB, Harlow SP, et al. NSABP-32: phase III, randomized trial comparing axillary resection with sentinel lymph node dissection: a description of the trial. Ann Surg Oncol 2004;11:208S–10.

[11] Giuliano AE, Hunt KK, Ballman KV, et al. Axillary dissection vs no axillary dissection in women with invasive breast cancer and sentinel node metastasis: a randomized clinical trial. JAMA 2011;305:589.

[12] Haviland JS, Owen JR, Dewar JA, et al. The UK Standardisation of Breast Radiotherapy (START) trials of radiotherapy hypofractionation for treatment of early breast cancer: 10-year follow-up results of two randomised controlled trials. Lancet Oncol 2013;14:1086–94.

[13] Franco P, Cante D, Sciacero P, Girelli G, La Porta MR, Ricardi U. Tumor bed boost integration during whole breast radiotherapy: a review of the current evidence. Breast Care (Basel) 2015;10:44–9.

[14] Brown LC, Mutter RW, Halyard MY. Benefits, risks, and safety of external beam radiation therapy for breast cancer. Int J Womens Health 2015;7:449–58.

[15] Harbeck N, Gnant M. Breast cancer. Lancet 2017;389:1134–50.

[16] Early Breast Cancer Trialists' Collaborative Group (EBCTCG). Relevance of breast cancer hormone receptors and other factors to the efficacy of adjuvant tamoxifen: patient-level meta-analysis of randomised trials. Lancet 2011;378:771–84.

[17] Fisher B, Costantino JP, Wickerham DL, et al. Tamoxifen for prevention of breast cancer: report of the National Surgical Adjuvant Breast and Bowel Project P-1 study. J Natl Cancer Inst 1998;90:1371–88.

[18] Cardoso F, Costa A, Norton L, et al. ESO-ESMO 2nd international consensus guidelines for advanced breast cancer (ABC2). Breast 2014;23:489–502.

[19] Greenberg PA, Hortobagyi GN, Smith TL, Ziegler LD, Frye DK, Buzdar AU. Long-term follow-up of patients with complete remission following combination chemotherapy for metastatic breast cancer. J Clin Oncol 1996;14:2197–205.

[20] Wang X, Yang KH, Wanyan P, Tian JH. Comparison of the efficacy and safety of denosumab versus bisphosphonates in breast cancer and bone metastases treatment: a meta-analysis of randomized controlled trials. Oncol Lett 2014;7:1997–2002.

[21] Rastogi P, Anderson SJ, Bear HD, et al. Preoperative chemotherapy: updates of National Surgical Adjuvant Breast and Bowel Project Protocols B-18 and B-27. J Clin Oncol 2008;26:778–85.

[22] Cortazar P, Zhang L, Untch M, et al. Pathological complete response and long-term clinical benefit in breast cancer: the CTNeoBC pooled analysis. Lancet 2014;384:164–72.

[23] Sparano JA, Zhao F, Martino S, et al. Long-term follow-up of the E1199 phase iii trial evaluating the role of taxane and schedule in operable breast cancer. J Clin Oncol 2015;33:2353–60.

[24] Joensuu H, Kellokumpu-Lehtinen PL, Huovinen R, et al. Adjuvant capecitabine, docetaxel, cyclophosphamide, and epirubicin for early breast cancer: final analysis of the randomized FinXX trial. J Clin Oncol 2012;30:11–8.

[25] Del Mastro L, De Placido S, Bruzzi P, et al. Fluorouracil and dose-dense chemotherapy in adjuvant treatment of patients with early-stage breast cancer: an open-label, 2 × 2 factorial, randomised phase 3 trial. Lancet 2015;385:1863–72.

[26] Biganzoli L, Aapro M, Loibl S, Wildiers H, Brain E. Taxanes in the treatment of breast cancer: have we better defined their role in older patients? A position paper from a SIOG Task Force. Cancer Treat Rev 2016;43:19–26.

[27] Alba E, Chacon JI, Lluch A, et al. A randomized phase II trial of platinum salts in basal-like breast cancer patients in the neoadjuvant setting. Results from the GEICAM/2006–03, multicenter study. Breast Cancer Res Treat 2012;136:487–93.

[28] Miles DW, Diéras V, Cortés J, Duenne AA, Yi J, O'Shaughnessy J. First-line breast cancer: pooled and subgroup analyses of data from 2447 patients. Ann Oncol 2013;24:2773–80.

[29] Loi S, Pommey S, Haibe-Kains B, Beavis PA, Darcy PK, Smyth MJ, Stagg J. CD73 promotes anthracycline resistance and poor prognosis in triple negative breast cancer. PNAS 2013;110:11091–6.

[30] McGrogan BT, Gilmartin B, Carney DN, McCann A. Taxanes, microtubules and chemoresistant breast cancer. Biochimica et Biophysica Acta 2008;1785:96–132.

[31] Gonzalez-Garay ML, Chang L, Blade K, Menick DR, Cabral F. A beta-tubulin leucine cluster involved in microtubule assembly and paclitaxel resistance. J Biol Chem 1999;274:23875–82.

[32] Berrieman HK, Lind MJ, Cawkwell L. Do beta-tubulin mutations have a role in resistance to chemotherapy? Lancet Oncol 2004;5:158–64.

[33] Wang Y, O'Brate A, Zhou W, Giannakakou P. Resistance to microtubule-stabilizing drugs involves two events, beta-tubulin mutation in one allele followed by loss of the second allele. Cell Cycle 2005;4:1847–53.

[34] Derry W, Wilson L, Khan IA, Ludena RF, Jordan MA. Taxol differentially modulates the dynamics of microtubules assembled from unfractionated and purified beta-tubulin isotypes. Biochemistry 1997;36:3554–62.

[35] Ferlini C, Raspaglio G, Mozzetti S, Cicchillitti L, Filippetti F, Gallo D, Fattorusso C, Campiani G, Scambia G. The seco-taxane IDN5390 is able to target class III beta-tubulin and to overcome paclitaxel resistance. Cancer Res 2005;65:2397–405.

[36] Orr G, Verdier-Pinard P. Mechanisms of taxol resistance related to microtubules. Oncogene 2003;22:7280–95.

[37] Honore S, Pasquier E, Braguer D. Understanding microtubule dynamics for improved cancer therapy. Cell Mol Life Sci 2005;62:3039–56.

[38] Schinkel AH, Mayer U, Wagenaar E, Mol CA, Van Deemter L, Smit JJ, Van Der Valk MA, Voordouw AC, Spits H, Van Tellingen O, Zijlmans JM, Fibbe WE, Borst P. Normal viability and altered pharmacokinetics in mice lacking Mdr1-Type (drug-transporting) P-glycoproteins. Proc Natl Acad Sci U S A 1997;94:4028–33.

[39] Yu D, Jing T, Liu B, Yao J, Tan M, Mc Donnell TJ, Hung MC. Overexpression of ErbB2 blocks taxol-induced apoptosis by upregulation of p21cip1, which inhibits p34Cdc2 kinase. Mol Cell 1998;2:581–91.

[40] Tan M, Jing T, Lan K-H, Neal CL, Li P, Lee S, Fang D, Nagata Y, Liu J, Arlinghaus R, Hung M-C, Yu D. Phosphorylation on tyrosine-15 of p34Cdc2 activation and is involved in resistance to taxol-induced Apoptosis. Mol Cell 2002;9:993–1004.

[41] Knuefermann C, Lu Y, Liu B, Jin W, Liang K, Wu L, Schmidt M, Mills GB, Mendelsohn J, Fan Z. HER2/PI-3K/Akt activation leads to a multidrug resistance in human breast adenocarcinoma cells. Oncogene 2003;22:3205–12.

[42] Harada H, Quearry B, Ruiz-Vela A, Korsmeyer SJ. Survival factor induced extracellular signal-regulated kinase phosphorylates BIM, inhibiting its association with BAX and proapoptotic activity. Proc Natl Acad Sci U S A 2004;101:15313–7.

[43] Fresno Vara JA, Casado E, De Castro J, Cejas P, Belda-Iniesta C, Gonzalez-Baron M. PI3K/Akt signaling pathway and cancer. Cancer Treat Rev 2004;30:193–204.

[44] Samanta D, Gilkes DM, Chaturvedi P, Xiang L, Semenza GL. Hypoxia-inducible factors are required for chemotherapy resistance of breast cancer stem cells. Proc Natl Acad Sci U S A 2014;111(50): E5429–38. https://doi.org/10.1073/pnas.1421438111.

[45] Kang HK. Cyclooxygenase-independent down-regulation of multidrug resistance associated protein-1 expression by celecoxib in human lung cancer cells. Mol Cancer Ther 2005;4:1358–63.

[46] Zhang W, Ding W, Chen Y, Feng M, Ouyang Y, Yu Y, He Z. Up-regulation of breast cancer resistance protein plays a role in HER2-mediated chemoresistance through PI3K/Akt and nuclear factor-kappa B signaling pathways in MCF7 breast cancer cells. Acta Biochim Biophys Sin (Shangai) 2011;43:647–53.

[47] Mubashar M, Harrington KJ, Chaudhary KS, El Lalani N, Stamp GW, Peters AM. P-glycoprotein-expressing breast and head and neck cancer cell lines. Acta Oncol 2004;43:443–52.

[48] Bachelot T, Bourgier C, Cropet C, Ray-Coquard I, Ferrero J, Freyer GM, Abadie-Lacourtoisie S, Eymard JC, Debled M, Spaëth D, Legouffe E, Allouache D, El Kouri C, Pujade-Lauraine E. Randomized phase II trial of everolimus in combination with tamoxifen in patients with hormone receptor-positive, human epidermal growth factor receptor 2-negative metastatic breast cancer with prior exposure to aromatase inhibitors: a GINECO study. J Clin Oncol 2012;30:2718–24.

[49] Han M, Lv Q, Tang X, Hu YL, Xu DH, Li FZ, Liang WQ, Gao JQ. Overcoming drug resistance of MCF-7/ADR cells by altering intracellular distribution of doxorubicin via MVP knockdown with a novel siRNA polyamidoamine-hyaluronic acid complex. J Control Release 2012;163:136–44.

[50] Liu Y, Lu WL, Guo J, Du J, Li T, Wu JW, Wang GL, Wang JC, Zhang X, Zhang Q. A potential target associated with both cancer and cancer stem cells: a combination therapy for eradication of breast cancer using vinorelbine stealthy liposomes plus parthenolide stealthy liposomes. J Control Release 2008;129:18–25.

[51] Sharma P, Hu-Lieskovan S, Wargo JA, Ribas A. Primary, adaptive, and acquired resistance to cancer immunotherapy. Cell 2017;168:707–23.

[52] Muraro E, Furlan C, Avanzo M, Martorelli D, Comaro E, Rizzo A, Fae' DA, Berretta M, Militello L, Del Conte A, Spazzapan S, Dolcetti R, Trovo' M. Local high-dose radiotherapy induces systemic immunomodulating effects of potential therapeutic relevance in oligometastatic breast cancer. Front Immunol 2017;8:1476. https://doi.org/10.3389/fimmu.2017.01476.

[53] Coburn S, Bray F, Sherman M, Trabert B. International patterns and trends in ovarian cancer incidence, overall and by histologic subtype. Int J Cancer 2017;140(11):2451–60.

[54] Jacobs IJ, Menon U. Progress and challenges in screening for early detection of ovarian cancer. Mol Cell Proteomics 2004;3(4):355–66.

[55] Momenimovahed Z, Tiznobaik A, Taheri S, Salehiniya H. Ovarian cancer in the world: epidemiology and risk factors. Int J Womens Health 2019;11:287–99.

[56] Lindsey AT, Trabert B, DeSantis CE, Miller KD, Samimi G, Runowicz CD, Gaudet MM, Ahmedin Jemal DVM, Siegel RL. Ovarian cancer statistics, 2018. CA Cancer J Clin 2018;68:284–96.

[57] Reid BM, Permuth JB, Sellers TA. Epidemiology of ovarian cancer: a review. Cancer Biol Med 2017;14(1). https://doi.org/10.20892/j.issn.2095-3941.2016.0084.

[58] Prat J. Ovarian carcinomas: five distinct diseases with different origins, genetic alterations, and clinicopathological features. Virchows Arch 2012;460(3):237–49.

[59] Gadducci A, Guarneri V, Alessandro Peccatori F, Ronzino G, Scandurra G, Zamagni C, Zola P, Salutari V. Current strategies for the targeted treatment of high-grade serous epithelial ovarian cancer and relevance of BRCA mutational status. J Ovarian Res 2019;12:9. https://doi.org/10.1186/s13048-019-0484-6.

[60] Bell DA. Origins and molecular pathology of ovarian cancer. Mod Pathol 2005;18(S2):S19. https://doi.org/10.1038/modpathol.3800306.

[61] Chan JK, Loizzi V, Lin YG, et al. IV invasive epithelial ovarian carcinoma in younger versus older women: what prognostic factors are important? Obstet Gynecol 2003;102(1):156–61.

[62] Jordan SJ, Green AC, Nagle CM, et al. Beyond parity: association of ovarian cancer with length of gestation and offspring characteristics. Am J Epidemiol 2009;170(5):607–14.

[63] Risch HA, Howe GR. Pelvic inflammatory disease and the risk of epithelial ovarian cancer. Cancer Epidemiol Biomarkers Prev 1995;4(5):447–51.

[64] Kazerouni N, Greene MH, Lacey Jr. JV, Mink PJ, Schairer C. Family history of breast cancer as a risk factor for ovarian cancer in a prospective study. Cancer 2006;107(5):1075–83.

[65] Bandera EV, Lee VS, Qin B, Rodriguez-Rodriguez L, Powell CB, Kushi LH. Impact of body mass index on ovarian cancer survival varies by stage. Br J Cancer 2017;117(2):282–9.

[66] Jelovac D, Armstrong DK. Recent progress in the diagnosis and treatment of ovarian cancer. CA Cancer J Clin 2011;61(3):183–203.

[67] Associazione Italiana di Oncologia Medica (AIOM): Linee guida AIOM: tumoridell'ovaio. Edizione (2017) Available from: http://www.aiom.it/ [Accessed 22 September 2019].

[68] Du Bois A, Vergote I, Ferron G, Reuss A, Meier W, Greggi S, et al. Randomized controlled phase III study evaluating the impact of secondary cytoreductive surgery in recurrent ovarian cancer: AGO DESKTOP III/ENGOT ov20. J Clin Oncol 2017;35(suppl) Abstract 5501.

[69] van Driel WJ, Koole SN, Sikorska K, Schagen van Leeuwen JH, Schreuder HWR, Hermans RHM, et al. Hyperthermic intraperitoneal chemotherapy in ovarian cancer. N Engl J Med 2018;378(3): 230–40.

[70] May T, Comeau R, Sun P, Kotsopoulos J, Narod SA, Rosen B, et al. A comparison of survival outcomes in advanced serous ovarian cancer patients treated with primary debulking surgery versus neoadjuvant chemotherapy. Int J Gynecol Cancer 2017;27(4):668–74.

[71] Perren TJ, Swart AM, Pfisterer J, Ledermann JA, Pujade-Lauraine E, Kristensen G, et al. A phase 3 trial of bevacizumab in ovarian cancer. N Engl J Med 2011;365(26):2484–96.

[72] Burger RA, Brady MF, Bookman MA, Fleming GF, Monk BJ, Huang H, et al. Incorporation of bevacizumab in the primary treatment of ovarian cancer. N Engl J Med 2011;365(26):2473–83.

[73] Aghajanian C, Blank SV, Goff BA, Judson PL, Teneriello MG, Husain A, et al. OCEANS: a randomized, double-blind, placebo-controlled phase III trial of chemotherapy with or without bevacizumab in patients with platinumsensitive recurrent epithelial ovarian, primary peritoneal, or fallopian tube cancer. J Clin Oncol 2012;30(17):2039–45.

[74] Kaelin Jr. WG. The concept of synthetic lethality in the context of anticancer therapy. Nat Rev Cancer 2005;5(9):689–98.

[75] Ledermann J, Harter P, Gourley C, Friedlander M, Vergote I, Rustin G, et al. Olaparib maintenance therapy in patients with platinum-sensitive relapsed serous ovarian cancer: a preplanned retrospective analysis of outcomes by BRCA status in a randomised phase 2 trial. Lancet Oncol 2014; 15(8):852–61.

[76] Mirza MR, Monk BJ, Herrstedt J, Oza AM, Mahner S, Redondo A, et al. Niraparib maintenance therapy in platinum-sensitive, recurrent ovarian cancer. N Engl J Med 2016;375(22):2154–64.

[77] Coleman RL, Oza AM, Lorusso D, Aghajanian C, Oaknin A, Dean A, et al. Rucaparib maintenance treatment for recurrent ovarian carcinoma after response to platinum therapy (ARIEL3): a randomised, double-blind, placebo-controlled, phase 3 trial. Lancet 2017;390(10106):1949–61.

[78] Kim S, Han Y, Kim SI, Kim HS, Kim SJ, Song YS. Tumor evolution and chemoresistance in ovarian cancer. NPJ Precis Oncol 2018;2:20. https://doi.org/10.1038/s41698-018-0063-0.

[79] Chan JK, Brady MF, Penson RT, Huang H, Birrer MJ, Walker JL, et al. Weekly vs. every-3-week paclitaxel and carboplatin for ovarian cancer. N Engl J Med 2016;374(8):738–48.

[80] Katsumata N, Yasuda M, Takahashi F, Isonishi S, Jobo T, Aoki D, et al. Dosedense paclitaxel once a week in combination with carboplatin every 3 weeks for advanced ovarian cancer: a phase 3, open-label, randomised controlled trial. Lancet 2009;374(9698):1331–8.

[81] Rustin GJ, van der Burg ME, Griffin CL, Guthrie D, Lamont A, Jayson GC, et al. Early versus delayed treatment of relapsed ovarian cancer (MRC OV05/EORTC 55955): a randomised trial. Lancet 2010;376(9747):1155–63.

[82] Poveda A, Vergote I, Tjulandin S, Kong B, Roy M, Chan S, et al. Trabectedin plus pegylated liposomal doxorubicin in relapsed ovarian cancer: outcomes in the partially platinum-sensitive (platinum-free interval 6-12 months) subpopulation of OVA-301 phase III randomized trial. Ann Oncol 2011; 22(1):39–48.

[83] el-Deiry WS. Role of oncogenes in resistance and killing by cancer therapeutic agents. Curr Opin Oncol 1997;9:79–87.

[84] Swanton C. Intratumor heterogeneity: evolution through space and time. Cancer Res 2012;72: 4875–82.

[85] Gottesman MM, Fojo T, Bates SE. Multidrug resistance in cancer: role of ATP-dependent transporters. Nat Rev Cancer 2002;2:48–58.

[86] He X, Ee PL, Coon JS, Beck WT. Alternative splicing of the multidrug resistance protein 1/ATP binding cassette transporter subfamily gene in ovarian cancer creates functional splice variants and is associated with increased expression of the splicing factors PTB and SRp20. Clin Cancer Res 2004;10:4652–60.

[87] Black SM, Beggs JD, Hayes JD, Murarratsu M, Sakai M, Wolfe CR. Expression of human glutathione S-transferase in *S. cerevisiae* confers resistance to the anticancer drugs adriamycin and chlorambucil. Biochem J 1990;268:309–15.

[88] Cherian MG, Jayasurya A, Bay BH. Metallothioneins in human tumors and potential roles in carcinogenesis. Mutat Res 2003;533:201–9.

[89] Kellner U, Sehested M, Jensen P, et al. Culprit and victim- DNA topoisomerase II. Lancet Oncol 2002;3:235–43.

[90] Masuda H, Tanaka T, Matsuda H, Kusaba I. Increased removal of DNA-bound platinum in a human ovarian cancer cell line resistant to cis-diamminedichloroplatinum (II). Cancer Res 1990; 50:1863–6.

[91] Zhen W, Link CJ, O'Connor Jr. PM, et al. Increased gene specific repair of cisplatininterstrand crosslinks in cisplatinresistanthuman ovarian cancer cell lines. Mol Biol Cell 1992;12:3689–98.

[92] Ferry KV, Hamilton TC, Johnson SW. Increased nucleotide excision repair in cisplatin resistant ovarian cancer cells. Biochem Pharmacol 2000;60:1305–13.

[93] Mansouri A, Zhang Q, Ridgway LD, Tian L, Claret FX. Cisplatin resistance in an ovarian carcinoma is associated with a defect in programmed cell death control through XIAP regulation. Oncol Res 2003;13:399–404.

[94] Fraser M, Leung B, Jahani-Asl A, Yan X, Thompson WE, Tsang BK. Chemoresistance in human ovarian cancer: the role of apoptotic regulators. Reprod Biol Endocrinol 2003;1:66.

[95] Binkhathlan Z, Lavasanifar A. P-glycoprotein inhibition as a therapeutic approach for overcoming multidrug resistance in cancer: current status and future perspectives. Curr Cancer Drug Targets 2013;13:326–46.

[96] Mi YJ, Liang YJ, Huang HB, Zhao HY, Wu CP, Wang F, Tao LY, Zhang CZ, Dai CL, Tiwari AK, et al. Apatinib (YN968D1) reverses multidrug resistance by inhibiting the efflux function of multiple ATP-binding cassette transporters. Cancer Res 2010;70:7981–91.

[97] Chung FS, Santiago JS, De Jesus MFM, Trinidad CV, See MFE. Disrupting P-glycoprotein function in clinical settings: what can we learn from the fundamental aspects of this transporter? Am J Cancer Res 2016;6:1583–98.

[98] Yang X, Iyer AK, Singh A, Milane L, Choy E, Hornicek FJ, Amiji MM, Duan Z. Cluster of differentiation 44 targeted hyaluronic acid based nanoparticles for MDR1 siRNA delivery to overcome drug resistance in ovarian cancer. Pharm Res 2015;32:2097–109.

[99] Davidson B, Tropé CG, Reich R. Epithelial-mesenchymal transition in ovarian carcinoma. Front Oncol 2012;2:33.

[100] Tomkinson H, Kemp J, Oliver S, Swaisland H, Taboada M, Morris T. Pharmacokinetics and tolerability of zibotentan (ZD4054) in subjects with hepatic or renal impairment: two open-label comparative studies. BMC Clin Pharmacol 2011;11:3.

[101] Selvakumaran M, Pisarcik DA, Bao R, Yeung AT, Hamilton TC. Enhanced cisplatin cytotoxicity by disturbing the nucleotide excision repair pathway in ovarian cancer cell lines. Cancer Res 2003;63:1311–6.

[102] Leijen S, Beijnen JH, Schellens JHM. Abrogation of the G2 checkpoint by inhibition of Wee-1 kinase results in sensitization of p53-deficient tumor cells to DNA-damaging agents. Curr Clin Pharmacol 2010;5:186–91.

[103] Pujade-Lauraine E, Hilpert F, Weber B, Reuss A, Poveda A, Kristensen G, Sorio R, Vergote I, Witteveen P, Bamias A, et al. Bevacizumab combined with chemotherapy for platinum-resistant recurrent ovarian cancer: the AURELIA open-label randomized phase III trial. J Clin Oncol 2014;32:1302–8.

[104] Monk BJ, Poveda A, Vergote I, Raspagliesi F, Fujiwara K, Bae DS, Oaknin A, Ray-Coquard I, Provencher DM, Karlan BY, et al. Anti-angiopoietin therapy with trebananib for recurrent ovarian cancer (TRINOVA-1): a randomised, multicentre, double-blind, placebo-controlled phase 3 trial. Lancet Oncol 2015;15:799–808.

[105] Issaeva N, Thomas HD, Djurenovic T, Jaspers JE, Stoimenov I, Kyle S, Pedley N, Gottipati P, Zur R, Sleeth K, et al. 6-thioguanine selectively kills BRCA2-defective tumors and overcomes PARP inhibitor resistance. Cancer Res 2010;70:6268–76.

[106] Buchbinder EI, Desai A. CTLA-4 and PD-1 pathways: similarities, differences, and implications of their inhibition. Am J Clin Oncol 2016;39:98–106.

[107] Aust S, Felix S, Auer K, Bachmayr-Heyda A, Kenner L, Dekan S, Meier SM, Gerner C, Grimm C, Pils D. Absence of PD-L1 on tumor cells is associated with reduced MHC I expression and PD-L1 expression increases in recurrent serous ovarian cancer. Sci Rep 2017;7:42929.

[108] Cancer. n.d. Available from: www.cancer.org/cancer/cancer-basics/what-is-cancer.html

[109] Cancer. n.d. Available from: www.cancer.org/cancer/cervical-cancer/causes-risks-prevention/prevention.html

[110] Cancer. n.d. Available from: www.cancer.org/cancer/cervical-cancer/prevention-and-early-detection.html

[111] Kamura T, Ushijima K. Chemotherapy for advanced or recurrent cervical cancer. Taiwan J Obstet Gynecol 2013;52:161–4.

[112] Bahar E, Kim J, Yoon H. Chemotherapy resistance explained through endoplasmic reticulum stress-dependent signaling. Cancer 2019;11: https://doi.org/10.3390/cancers11030338.

[113] Norouzi-Barough L, Sarookhani MR, Sharifi M, Moghbelinejad S, Jangjoo S, Salehi R. Molecular mechanisms of drug resistance in ovarian cancer. J Cell Physiol 2018;233:4546–62.

[114] Hait WN, Choudhury S, Srimatkandada S, Murren JR. Sensitivity of K562 human chronic myelogenous leukemia blast cells transfected with a human multidrug resistance cDNA to cytotoxic drugs and differentiating agents. J Clin Investig 1993;91:2207–15.

[115] Hoffmeyer S, Burk O, von Richter O, Arnold HP, Brockmoller J, Johne A, Cascorbi I, Gerloff T, Roots I, Eichelbaum M. Functional polymorphisms of the human multidrug-resistance gene: multiple sequence variations and correlation of one allele with P-glycoprotein expression and activity *in vivo*. Proc Natl Acad Sci U S A 2000;97:3473–8.

[116] Cumming RC, Lightfoot J, Beard K, Youssoufian H, O'Brien PJ, Buchwald M. Fanconianemia group C protein prevents apoptosis in hematopoietic cells through redox regulation of GSTP1. Nat Med 2001;7:814–20.

[117] Dufey E, Sepulveda D, Rojas-Rivera D, Hetz C. Cellular mechanisms of endoplasmic reticulum stress signaling in health and disease. 1. An overview. Am J Physiol Cell Physiol 2014;307:C582–94.

[118] Wang M, Kaufman RJ. The impact of the endoplasmic reticulum protein-folding environment on cancer development. Nat Rev Cancer 2014;14:581–97.

[119] Zhu H, Luo H, Zhang W, Shen Z, Hu X, Zhu X. Molecular mechanisms of cisplatin resistance in cervical cancer. Drug Des Devel Ther 2016;10:1885.

[120] Hamed AR, Abdel-Azim SN, Shams AK, Hammouda MF. Targeting multidrug resistance in cancer by natural chemosensitizers. Bull Natl Res Cent 2019;43: https://doi.org/10.1186/s42269-019-0043-8.

[121] De Sauza R, Zahedi P, Badame MR, Allen C, Piquette-Miller M. Chemotherapy dosing schedule influences drug resistance development in ovarian cancer. Mol Cancer Ther 2011;10:1289–99.

CHAPTER 4

Epidemiology of cancers in women

Saritha Vara, Manoj Kumar Karnena, and Bhavya Kavitha Dwarapureddi
Department of Environmental Science, GITAM Institute of Science, GITAM (Deemed to be University), Visakhapatnam, Andhra Pradesh, India

Abstract

Epidemiology of cancer in women has been presented in only a few published reports in the literature. This chapter attempts to summarize contemporary burden, trends, and risk factors of cancers in four sites (uterine corpus, ovary, cervix, and breast) that account for more than 60% of cancer burden among women. Deaths due to cancer are generally higher in women from low- and middle-income countries despite lower incidence rates. This burden can be substantially reduced through perception of the risk factors discussed in this chapter.

Keywords: Epidemiology, Breast cancer, Cervical cancer, Ovarian cancer, Uterine cancer.

Abbreviations

BMI body mass index
DCIS ductal carcinoma in situ
ER estrogen receptor
ERT estrogen replacement therapy
HIV human immunodeficiency virus
HPV human papillomavirus
PCOS polycystic ovary syndrome

1. Introduction

The study of how often and why ailments arise in different groups of the population is called epidemiology. The importance of epidemiological information lies in its use for scheduling and assessing approaches to avert ailments and as a guide in handling patients with developed disease. Similar to pathology and clinical findings, disease epidemiology is an integral part of the disease's basic description. Epidemiology emerged as scientific discipline during the 17th century and it has distinct techniques in collecting, analyzing, and interpreting population data. During the 18th century, population comparison studies were performed. Since then, organized concepts and methods for performing population studies have been established, polished, and hypothesized. Fundamentally, epidemiology is multidisciplinary incorporating knowledge inputs from sociology, biology, statistics, and other fields [1]. This comprises the genuine core of the scientific discipline known currently as epidemiology.

G.P. Nagaraju, R.R. Malla (eds.)
A Theranostic and Precision Medicine Approach for Female-Specific Cancers
ISBN 978-0-12-822009-2
https://doi.org/10.1016/B978-0-12-822009-2.00004-2

Epidemiology of cancer studies distribution, determinants, and frequency of malignant disease in precise populations [2]. The objective of these studies is to describe causative factors in formulating preventive strategies toward control of the disease. The clinician is provided with quantification of cancer risk from epidemiological assessment, which will enables him to outline a basis for screening modalities in high-risk populations and to determine efficiency of preventive intervention. Cancer epidemiological research comprises three types: descriptive, analytical, and clinical. Descriptive epidemiology emphasizes trends and rates of disease in a specific population and analytical epidemiology deals in recognizing causes and prejudicing associated risks in disease development. Outlining screening programs, evaluating impacts of prevention strategies on total outcome is dealt with in clinical epidemiology [3]. This chapter examines the epidemiological factors that cause the most common cancers in women, including breast cancer, cervical cancer, ovarian cancer, and uterine cancer.

2. Breast cancer

Breast cancer is the most lethal malignancy in women around the world. The American Cancer Society estimates that 268,600 novel cases of invasive breast cancer are expected to be detected among women. In addition, 48,100 projected cases of ductal carcinoma in situ (DCIS) will also be diagnosed. Furthermore, nearly 41,760 women are projected to die from breast cancer in 2019. Moreover, the seriousness of the risk can be understood from the projections that 13% of women (i.e., nearly one in eight women) will be diagnosed with invasive breast cancer in their lifetime and around 3% (i.e., one in 39 women) will die from breast cancer [4]. This burden of breast cancer calls for continuous attention toward precise understanding of the disease, including its epidemiology.

2.1 Incidence

DCIS and invasive breast cancer incidence increased rapidly between 1980 and 1990, especially in women 50 years of age and older. This is attributed to enhanced prevalence of mammography screening, which increased from 29% during 1987 to 70% during 2000 [5]. Rates of DCIS in women aged 50 and older have increased more than 11-fold since 1980–2008, whereas a sharp decline (13%) in invasive breast cancer was noted between 1999 and 2004. This is attributed largely to the reduced intake of menopausal hormones and small declines in mammography screening since 2000 [6–8]. During the last decade (between 2012 and 2016), rates of DCIS reduced by 2.1% yearly [9], while rates of invasive breast cancer enhanced by 0.3% yearly. Further, it is observed that there has been incidence in the rates of invasive breast cancer [10]. Studies have also related enhanced body mass index (BMI) and reduced average number of births to recent increases in incidence [11].

2.2 Age

Incidence and death rates of breast cancer increase with age up to 70 years. This is attributed to estrogen receptors (ER), which increase with age; hence breast cancers are more likely to occur during postmenopausal ages [12]. However, there is decreased incidence in women 80 years of age and older due to lesser rates of screening, detection by mammography, or incomplete detection [12]. The median age of women with a breast cancer diagnosis during the period 2012–2016 was 62 years [10]. Nevertheless, breast tumors identified in younger women were observed to be larger in size with positive lymph nodes, advanced stages, and reduced survival rates [13].

2.3 Race

Many factors might account for racial differences in mortality from breast cancer, including access to mammography, socioeconomic factors, timely treatment, and biological factors. The mortality rate due to breast cancer is reported to be higher in black women in comparison to white women, which is frequently attributed to ER-negative tumors [14]. Both incidence and mortality rates are usually low among women of other ethnic/racial groups [4]. Triple negative tumors (ER-negative, HER2-negative, and progesterone-negative) are observed to be more common in black women than white women during the premenopausal stage. HR+/HER2− are the common subtypes of breast cancer reported in women of all ethnicities. Moreover, recent studies have stated that race/ethnicity has been replaced by socioeconomic status as a forecaster of worse outcome [15].

2.4 Diagnosis

Invasive breast cancer prognosis is sturdily governed by stage of disease at the time of diagnosis. Although most breast cancers are localized, the trend varies with race. Approximately 64% of patients present with local stage, 27% with regional stage, and 6% with distant stage breast cancer at initial diagnosis. During 2012–2016, there was a 1.1% annual increase in incidence rate of local stage disease, while there was a 0.8% annual decline in early-stage diagnosis [16].

2.5 Breastfeeding

Many previous studies have positively related breastfeeding with overall reduced risk of breast cancer, suggesting that extended durations of breastfeeding will result in greater reduced risk. By suppressing ovulation breastfeeding reduces cancer risk by reducing lifetime exposure to estrogen in women [17]. Further, lower levels of estrogen in breast fluids are reported during breastfeeding, which are autonomous of estrogen levels in serum [18]. Breastfeeding might result in terminal disparity of breast epithelial cells at the cell level making these cells less prone to mutations or carcinogenic effects during cell division [19]. A 4% reduction in breast cancer is attributed to 12 months of breast-feeding in 30 countries as per the data from 47 studies [20]. This protective effect might be sturdier for or even limited to cancers of the triple negative type [21–23].

2.6 Hormone therapy

Menopausal hormones containing estrogen and progestin enhance the risk of HR+ breast cancer; this risk is directly proportional to the duration of use of these hormones [20, 24–27]. Though menopausal hormones are discontinued, elimination of breast cancer is not possible [28]. A study by the Women's Health Initiative [29] found that use of estrogen-only therapy by women for an average duration of 6 years confers a 25% decreased risk of developing breast cancer, which was contradictory to observational studies reporting a slight increase in estrogen therapy users, specifically in lean women and in women who began the therapy soon after menopause [30, 31]. A 5%–10% reduction in incidence of breast cancer was observed after 2002 in countries in the Unites States, Europe, and Canada after reduced use of hormone therapy [32].

2.7 Use of contraceptive hormones

Use of contraceptives with exogenous hormones like estrogen and progestin are related to a small increased risk of breast cancer. An enhanced risk up to 24% is associated with current use of oral contraceptive pills in comparison to women that have never used contraceptive pills [33, 34]. After stopping the use of contraceptives, the risk of developing cancer diminishes and will be similar to the risk in those woman who have never taken contraceptives after 10 years [35].

2.8 Dietary factors

Several studies have inspected the relation between diet and breast cancer and presented diverse results. Although no relation has been concluded between breast cancer and dietary fat from recent meta-analysis, consumption of soy has been associated to reduced breast cancer risk, particularly in Asian women with high intake of soy. This is not true of women in Western populations [36, 37]. Though limited, a growing evidence supports that greater intake of fruits and vegetables contributes to reduce HR-breast cancer [38, 39]. A similar trend is observed with the consumption of foods rich in calcium [40]. Meta-analysis of prospective studies by Dong and Qin presented an inverse relationship between consumption of soy isoflavones and breast cancer.

2.9 Weight of the individual

Postmenopausal weight gain has been linked with two times greater risk of HR+ breast cancer [41]. Higher levels of body fat, even within normal BMI range, are related to enhanced breast cancer postmenopause [42]. This increased risk is attributed to fat tissue being a chief source of estrogen in women after menopause and other trends like greater levels of insulin in women with excess body weight [43]. Contradictory reports have been presented stating that weight loss during onset of adulthood and postmenopause reduces risk of breast cancer in some studies, while other studies have observed that excess weight actually protects from premenopausal breast cancer [44]. A large meta-analysis found a 14% reduction in risk of breast cancer in overweight women and a 26% reduction in obese women in comparison to normal weight women in the aged 40–49 years [45].

2.10 Consumption of alcohol

There exists several confirmed studies relating consumption of alcohol to risk of breast cancer in women. Women consuming one drink per day have a risk of 7%–10%, while those consuming two to three drinks per day have a 20% higher risk [46, 47]. These studies show a linear dose-response relationship. A possible hypothesis that is not well understood is that alcohol consumption increases risk indirectly through enhanced estrogen and other hormone levels.

2.11 Physical activity

Higher level of physical activity is inversely proportional to the risk of breast cancer [46–48]. Probable biological mechanisms include impact of physical activity on body composition, insulin resistance, hormone levels, systemic inflammation, and energy balance [51].

Further, other epidemiological factors including genetics [52, 53], exposure to radiation [54, 55], environmental and occupational carcinogens [56–58], and shift work [59–61] have shown a positive correlation to risk of breast cancer.

3. Cervical cancer

Ranking as the fourth most common cancer in women worldwide, cervical cancer predominates in developing countries, accounting for 12% of all cancers in women; it is largely a cancer found in middle-aged women [62, 63]. If diagnosed at stage I, there is a nearly 96% 5-year survival rate. A model of viral carcinogenesis, cervical cancer is also attributed to persistent infection with an oncogenic human papillomavirus (HPV) type along with a number of cofactors including large number of sexual partners, early sexual debut, extended use of oral contraceptives, smoking, high parity, dietary factors, and coinfection with other sexually transmitted agents like *Chlamydia trachomatis* and human immunodeficiency virus (HIV) [64–67]. Important symptoms of cervical cancer include abnormal vaginal discharge, weight loss, severe pelvic pain, and anorexia [68].

3.1 Human papillomavirus

By now it is quite clear that the chief reason for precancerous and cancerous lesions is infection with oncogenic subtypes of HPV (HPV 16 and HPV 18), which are essential agents in pathogenesis of cervical cancer [69, 70]. Sexual contact, being the mode of transmission of this infection, causes squamous intraepithelial lesions disappearing 6–12 months after immunological intervention; a small percentage of the remaining lesions can cause cancer [71]. The greatest prevalence of HPV is observed in women 25 years of age and is attributed to variations in sexual behavior, according to the results of a meta-analysis [72, 73]. The principal mechanisms involved in contributing carcinogenesis by HPV is activity of E6 and E7 viral oncoproteins interfering with P53 and retinoblastoma tumor-suppressor genes. Further, it is also related to changes in DNA of the host and virus leading to DNA methylation bringing variations in chief cellular pathways that control genetic integrity, apoptosis, cell adhesion, cellular control, and immune response [74]. Epidemiological studies at the molecular level indicate that HPV infection along with lower doses of nitrosamines in cervical mucosa of smokers might accentuate carcinogenesis [75].

3.2 Human immunodeficiency virus

It is reported that HIV-positive women possess greater risk of developing infection from HPV types [76, 77]. The persistent nature of infection by HPV 16 or HPV 18 has been commonly observed with HIV-positive women [78, 79]. Moreover, synergistic interaction among oncogenic HPV 16 and HIV infection results in compromising the immune system and predisposing sexually active women to persistent nature of coinfection by HPV 16 [80].

3.3 *Chlamydia trachomatis*

Previous infection or coinfection of HPV with *Chlamydia trachomatis* has been related to increased risk of cervical squamous cell carcinoma [81]. Case control study from seven countries showed that *C. trachomatis* serum antibodies are associated to a 1.8-fold increase in squamous cell carcinoma by enhancing host susceptibility to HPV. Inflammation from chronic *C. trachomatis* infection results in release of reactive oxygen species leading to impairment of DNA and increased risk of HPV-associated carcinogenesis. Moreover, women infected with *C. trachomatis* possess lesser ability in clearing HPV infection [82]. Cellular deformity might result through inhibition of apoptosis leading to excessive expression of oncogenes E6/E7 [83].

3.4 Reproductive trends

Epidemiological evidences in support of a relation between invasive cervical cancer and multiparity have been presented [84, 85], while an inverse relationship between mother's age at the first pregnancy and cervical cancer has also been shown [86]. Delivery was observed as a predictor of CIN 3, especially in women with greater risk of HPV infection [87]. Hypotheses that are presented to explain probable biological mechanisms governing the risk of cervical cancer include changes in immunity, hormones, and nutrition during pregnancy or cervical trauma during parturition.

3.5 Sexual behavior

Sexual behavior associated with multiple partners has been related to enhanced threat of cervical cancer [88, 89], attributed to increased risk of HPV infection. Further, first intercourse at an early age is also attributed to enhanced cervical cancer risk. A protective effect has been demonstrated by the use of condoms to incidence of both genital HPV infection and cervical intraepithelial lesions [90, 91].

3.6 Oral contraceptives

Recent and contemporary use of oral contraceptives has been related to an upsurge in threat of cervical cancer [84, 92]. An increased risk is also attributed to prolonged use of oral contraceptives [93].

3.7 Obesity

Occurrence of cervical adenocarcinoma linked with hormonal risk factors is related to excess body weight, leading to cervical carcinoma [94–96]. A twofold increase in cervical cancer was reported in obese women and overweight women. Obesity, particularly after menopause, acts as a marker of enhanced sex hormone levels because of conversion of androgen to estrogen in peripheral adipose tissue [94].

3.8 Diet

It is understood that some foods and nutrients might confer protection against the development of cervical cancer. Studies have presented evidence that consumption of retinol, carotenoids, folate, vitamins C and E, fruits, and vegetables might reduce the threat of cervical cancer [97]. A study by Tomita et al. [98] presented that greater concentrations of α- and γ-tocopherol along with enhanced nutritious consumption of green vegetables and yellow fruits were associated with a 50% reduction in cervical cancer. Significant correlation between enhanced consumption of vitamins E, C, A, folate, and lycopene and cervical cancer was reported by Ghosh et al. [99]. Potential biological mechanisms include the essential role played by vitamins C and E in enhancing mucosal reaction to infection thus protecting against free radicals and oxidants. These vitamins also inhibit development of DNA adducts caused by cigarette smoking.

3.9 Smoking

The relation between smoking history and cervical cancer has been well established [84, 100]. A 50% reduction in risk was observed in women who have stopped smoking for 10 years. One of the mechanisms that increases the risk of cervical cancer in women who smoke is the resident initiation of immune suppression by tobacco metabolites, which can cause DNA damage in squamous cells by nicotine and its metabolites [101]. Increased risk of cervical cancer has been reported in women exposed to passive smoking [102]. Several studies have presented a synergistic increase in threat of cervical cancer in smoking women with HPV 16 infection [75].

4. Ovarian cancer

Ovarian cancer ranks third after cervical and uterine cancers [103]. This cancer accounts for nearly 3.4% of all female cancers and 4.4% of cancer deaths among women [103, 104]. Though it has lower prevalence than breast cancer, its lethality is three times greater [105] and it is projected that by 2040 there will be a significant increase in mortality rate. Mortality rates are attributed to asymptomatic and secret growth of the tumor, lack of proper screening, and delayed onset of symptoms, hence ovarian cancer is referred to as the "Silent Killer." Due to the ovarian epithelium containing the same cell membranes as that of the ductal epithelium of mammary glands, risk factors for ovarian cancer are similar to those for breast cancer.

4.1 Age

Epithelial ovarian cancer is a postmenopausal disease related to age [106], and is more pronounced among women older than 60 years [107]. Being associated with older age, the disease is more intense with lower survival rates [108, 109]. The link between age and ovarian cancer is uncertain.

4.2 Genetic and family history

Risk of ovarian cancer increases three- to fivefold with a strong family history (having ovarian, breast, endometrial, and colorectal cancers) resulting in early onset and excess bilateral disease. Approximately one-fifth of ovarian cancers are attributed to mutations in tumor suppressor genes [110], while 65%–85% result due to germline mutations in BRCA genes [111]. Another genetic condition that predisposes to ovarian cancer is hereditary nonpolyposis colorectal cancer, also known as Lynch syndrome.

Ovarian cancers associated with Lynch syndrome are usually nonmucinous with nearly 84% in stage I or II [112]. The cause of Lynch syndrome is hereditary mutation among four mismatch repair genes: MHL1, MSH2, 6, and PMS2, of which MSH2 and MLH1 are most general mutations [113].

4.3 Fertility agents

Relation between fertility drugs and development of ovarian cancer has been presented in several case-control studies. Fertility drugs like clomiphene citrate stimulate secretion of LH and FSH by the pituitary gland, helping in ovulation stimulation. However, controversy exists regarding the precise nature of the enhanced risk as it is hard to reconcile whether the risk is due to use of the fertility agent or other factors like underlying pathologies of infertility [114].

4.4 Oral contraceptives

Several studies indicate use of oral contraceptives to be related with decreased threat of all histological kinds of ovarian cancers [115–117]. Nevertheless, the protective nature of oral contraceptives is to be understood in detail with reference to their time of use, age of use, and duration of consumption [115, 118], as the protective nature might reduce after discontinuation of pills [119].

4.5 Hormone replacement therapy

A 20% enhanced risk was reported upon use of exogenous hormones like estrogen replacement therapy (ERT) [120]. Studies suggested that use of estrogenic methods for 10 or more years enhances risk [121]. Biologically, cell membrane receptors on ovarian surface epithelium are responsive to estrogen. Stimulation of cell proliferation by ERT along with other cellular events might lead to carcinogenesis through modulation of ER [122]. Mørch et al. [123] trusted that irrespective of duration of use, dose, preparation, and method of use of estrogen therapy is linked with amplified risk of ovarian cancer.

4.6 Diet

A positive association between diurnal fish consumption and threat of ovarian cancer has been reported, while a negative correlation has been reported with daily consumption of milk [124, 125]. The protective role of phytoestrogens has been related to ovarian cancer, pointing out the importance of plant-based diets in reducing hormone-related cancers [126, 127]. Studies have revealed enhanced concentrations of vitamin D in plasma to less ovarian cancer, which was similar with calcium and lactose intake [124].

4.7 Obesity

Obesity is directly related to the threat of ovarian cancer as it is associated with central adiposity indicating transformation of androgen in bordering tissues [128, 129]. Waist-hip ratio association with enhanced risk of breast cancer was reported by Anderson et al. [130]. Menopause has been related to obesity and threat of ovarian cancer by Beehler et al. [131]. Epidemiological findings suggest that long-term adiposity increases aromatase-catalyzed estrogen biosynthesis, which in turn enhances genesis of ovarian cancer.

5. Cancer of the corpus uteri

Corpus uteri cancer cases account for nearly 4.4% of all female-specific cancers, with 2.4% of these leading to death [132]. Almost 90% of uterine cancers are endometrial, which originate from the epithelium, while the rest are mesenchymal, originating from myometrial muscle (i.e., endometrial stroma). Owing to the shared risk factors enhancing exposure to both exogenous and endogenous estrogens, global pattern of corpus uteri cancer exhibits similarities with breast and ovarian cancers. The global burden of corpus uteri cancer increased sharply during the 21st century, with reporting showing nearly doubled numbers of new cases and deaths [103, 133, 134]. Multiple factors are related to this increase, including enhanced exposure to risk factors like lack of breastfeeding, giving birth at a late age, consumption of exogenous hormones, diets with high calories and fat, obesity, lack of exercise, and diabetes mellitus [135].

5.1 Age

As per the reports, the majority of women diagnosed are in the age range of 50–65 years (i.e., postmenopausal age) [136], showing that the risk is directly proportional to increasing age.

5.2 Reproductive trends

Factors including menarche at early age, nulliparity, and late menopause are significantly related with increased risk of endometrial cancer, while the opposite of these factors is observed to be protective [136]. These factors are confirmed owing to their association with lower cumulative exposure to estrogen, supporting the significant role of hormonal mechanisms in endometrial carcinogenesis [137].

5.3 Polycystic ovary syndrome

About 30% of premenopausal endometrial patients are reported to have polycystic ovary syndrome (PCOS) [138]. A fourfold increase in the threat of endometrial cancer is stated among women with PCOS [139]. Increased exposure to estrogen by endometrium unopposed by progesterone during anovulatory menstrual cycles is one of the mechanisms underlying the association [140].

5.4 Granulosa cell ovarian tumors

Granulosa cell tumors of the ovary rarely occur in relation to heterogeneous symptoms of the genital tract comprising irregularities in menstrual cycle, infertility, pelvic pain, and uterine pathologies. Though these tumors are scarce, accounting for only 2%–5% of ovarian malignancies, these manifest marked enhancement in circulating estrogens, in turn stimulating ERs in other tissues like the endometrium [141].

5.5 Obesity

Relation of obesity and endometrial cancer has been strongly established [142]. The biological mechanism underlying this relation is that obese women possess higher endogenous estrogen levels due to aromatization of androstenedione in adipose tissue [143].

5.6 Diabetes mellitus

Women with diabetes have an increased risk of endometrial cancer, with a twofold increase for type 2 diabetes and a threefold increased for type 1 diabetes [144]. This is attributed to hyperinsulinemia due to insulin resistance, that is, chronically enhanced levels of circulating insulin might stimulate proliferation of endometrial epithelium through binding to insulin receptors in endometrial tissues [145]. Obesity with type 2 diabetes is found to be synergistic in increasing risk of endometrial cancer [146].

5.7 Hormone replacement

Epidemiological studies have revealed significant increased risk of endometrial cancer among women receiving ERT. The trend observed is direct relation of risk with prolonged usage. A 40% increased risk has been reported over 1 year of estrogen use. The risk is also associated with estrogen dosage [147].

5.8 Physical activity

While sedentary behavior contributes to development of endometrial cancer, physical activity might be protective. A 25% reduction in risk was reported among women with high physical activity [148]. Physical activity results in reduction of central adiposity, leading to favorable changes in balance of endogenous estrogens and androgens.

Further factors like Lynch syndrome and early menarche are also related with increased risk of endometrial cancer [149, 150].

6. Conclusion

The objective of the chapter was to bring into the spotlight the epidemiology of cancers in women. It can be concluded that the four cancer sites detailed possess some similarities. Exogenous and endogenous hormones seem to play a significant role in cancer risk of women. Use of hormone replacement therapy seems to increase risk of endometrial and breast cancer. From the current knowledge of epidemiology, some possible preventive strategies include quitting smoking and drinking alcohol, increased consumption of fruits and vegetables, and vaccination against HPV, all of which could significantly reduce incidence of cancers. This chapter thus presents similarities and dissimilarities of cancer in women, which will enhance one's perception leading to better understanding and taking necessary steps for prevention.

Common risk factors of cancers in four sites.

Site of cancer	Age	Race	Dietary factors	Weight/ obesity	Reproductive trends	Alcohol consumption / smoking	Physical activity	Factors beyond common
								Associated risk factors
Breast	✓	✓	✓	✓	–	✓	✓	–
Cervix	–	–	✓	✓	✓	✓	–	Human papillomavirus, sexual behavior, *Chlamydia trachomatis*, HIV
Ovary	✓	–	✓	✓	–	–	–	Genetic and family history
Uterus	✓	–	–	✓	✓	–	–	Diabetes mellitus, PCOS

References

[1] National Research Council. Epidemiologic studies. In: Analysis of cancer risks in populations near nuclear facilities: phase I. USA: National Academies Press; 2012.

[2] Hennekens CH, Buring JE. Epidemiology in medicine. Boston: Little Brown; 198773–98.

[3] Nazario LA, Macheledt JE, Vogel VG. Epidemiology of cancer and prevention strategies. Lung 1995;169:157–400.

[4] American Cancer Society. Breast cancer facts & figures 2019–2020. Atlanta: American Cancer Society, Inc; 2019.

[5] Breen N, Gentleman JF, Schiller JS. Update on mammography trends: comparisons of rates in 2000, 2005, and 2008. Cancer 2011;117(10):2209–18.

[6] Coombs NJ, Cronin KA, Taylor RJ, Freedman AN, Boyages J. The impact of changes in hormone therapy on breast cancer incidence in the US population. Cancer Causes Control 2010;21(1):83–90.

[7] DeSantis C, Howlader N, Cronin KA, Jemal A. Breast cancer incidence rates in US women are no longer declining. Cancer 2011;20(5):733–9.

[8] Ravdin PM, Cronin KA, Howlader N, Berg CD, Chlebowski RT, Feuer EJ, … Berry DA. The decrease in breast-cancer incidence in 2003 in the United States. N Engl J Med 2007;356(16):1670–4.

[9] SEER*Stat Databases: NAACCR Incidence Data-CiNA Analytic File, 1995–2016, for NHIAv2 Origin and for Expanded Races, Custom File With County, ACS Facts and Figures projection Project (which includes data from CDC's National Program of Cancer Registries (NPCR), CCCR's Provincial and Territorial Registries, and the NCI's Surveillance, Epidemiology and End Results (SEER) Registries), certified by the North American Association of Central Cancer Registries (NAACCR) as meeting high-quality incidence data.

[10] Howlader N, Noone AM, Krapcho M, Miller D, Brest A, Yu M, … Chen HS. SEER cancer statistics review, 1975–2016. Bethesda, MD: National Cancer Institute; 2019.

[11] Pfeiffer RM, Webb-Vargas Y, Wheeler W, Gail MH. Proportion of US trends in breast cancer incidence attributable to long-term changes in risk factor distributions. Cancer 2018;27(10):1214–22.

[12] Yasui Y, Potter JD. The shape of age–incidence curves of female breast cancer by hormone-receptor status. Cancer Causes Control 1999;10(5):431–7.

[13] Metcalfe KA, Finch A, Poll A, Horsman D, Kim-Sing C, Scott J, … Narod SA. Breast cancer risks in women with a family history of breast or ovarian cancer who have tested negative for a BRCA1 or BRCA2 mutation. Br J Cancer 2009;100(2):421–5.

[14] Dunn BK, Agurs-Collins T, Browne D, Lubet R, Johnson KA. Health disparities in breast cancer: biology meets socioeconomic status. Breast Cancer Res Treat 2010;121(2):281–92.

[15] Cross T, Racz G. U.S. Patent Application No. 10/025,112.

[16] DeSantis CE, Ma J, Jemal A. Trends in stage at diagnosis for young breast cancer patients in the United States. Breast Cancer Res Treat 2019;173(3):743–7.

[17] Enger SM, Ross RK, Paganini-Hill A, Bernstein L. Breastfeeding experience and breast cancer risk among postmenopausal women. Cancer 1998;7(5):365–9.

[18] Yaghjyan L, Colditz GA. Estrogens in the breast tissue: a systematic review. Cancer Causes Control 2011;22(4):529–40.

[19] Smalley M, Ashworth A. Stem cells and breast cancer: a field in transit. Nat Rev Cancer 2003; 3(11):832–44.

[20] Beral V, Reeves G, Bull D, Green J. Million Women Study Collaborators. Breast cancer risk in relation to the interval between menopause and starting hormone therapy. J Natl Cancer Inst 2011; 103(4):296–305.

[21] Faupel-Badger JM, Arcaro KF, Balkam JJ, Eliassen AH, Hassiotou F, Lebrilla CB, … Watson CJ. Postpartum remodeling, lactation, and breast cancer risk: summary of a National Cancer Institute–sponsored workshop. J Natl Cancer Inst 2013;105(3):166–74.

[22] Islami F, Liu Y, Jemal A, Zhou J, Weiderpass E, Colditz G, … Weiss M. Breastfeeding and breast cancer risk by receptor status—a systematic review and meta-analysis. Ann Oncol 2015;26(12): 2398–407.

[23] Ma H, Ursin G, Xu X, Lee E, Togawa K, Duan L, … Simon MS. Reproductive factors and the risk of triple-negative breast cancer in white women and African-American women: a pooled analysis. Breast Cancer Res 2017;19(1):6.

[24] Chlebowski RT, Manson JE, Anderson GL, Cauley JA, Aragaki AK, Stefanick ML, … Qi L. Estrogen plus progestin and breast cancer incidence and mortality in the Women's Health Initiative Observational Study. J Natl Cancer Inst 2013;105(8):526–35.

[25] Gaudet MM, Gierach GL, Carter BD, Luo J, Milne RL, Weiderpass E, … Wolk A. Pooled analysis of nine cohorts reveals breast cancer risk factors by tumor molecular subtype. Cancer Res 2018; 78(20):6011–21.

[26] Li K, Anderson G, Viallon V, Arveux P, Kvaskoff M, Fournier A, … Chirlaque MD. Risk prediction for estrogen receptor-specific breast cancers in two large prospective cohorts. Breast Cancer Res 2018;20(1):147.

[27] Manson JE, Chlebowski RT, Stefanick ML, Aragaki AK, Rossouw JE, Prentice RL, … Wactawski-Wende J. Menopausal hormone therapy and health outcomes during the intervention and extended poststopping phases of the Women's Health Initiative randomized trials. JAMA 2013;310 (13):1353–68.

[28] Chlebowski RT, Rohan TE, Manson JE, Aragaki AK, Kaunitz A, Stefanick ML, … Adams-Campbell LL. Breast cancer after use of estrogen plus progestin and estrogen alone: analyses of data from 2 women's health initiative randomized clinical trials. JAMA Oncol 2015;1(3):296–305.

[29] LaCroix AZ, Chlebowski RT, Manson JE, Aragaki AK, Johnson KC, Martin L, … Howard BV. Health outcomes after stopping conjugated equine estrogens among postmenopausal women with prior hysterectomy: a randomized controlled trial. JAMA 2011;305(13):1305–14.

[30] Bakken K, Fournier A, Lund E, Waaseth M, Dumeaux V, Clavel-Chapelon F, … Slimani N. Menopausal hormone therapy and breast cancer risk: impact of different treatments. The European Prospective Investigation Into Cancer and Nutrition. Int J Cancer 2011;128(1):144–56.

[31] Calle EE, Feigelson HS, Hildebrand JS, Teras LR, Thun MJ, Rodriguez C. Postmenopausal hormone use and breast cancer associations differ by hormone regimen and histologic subtype. Cancer 2009; 115(5):936–45.

[32] Pelucchi C, Levi F, La Vecchia C. The rise and fall in menopausal hormone therapy and breast cancer incidence. Breast 2010;19(3):198–201.

[33] Bassuk SS, Manson JE. Oral contraceptives and menopausal hormone therapy: relative and attributable risks of cardiovascular disease, cancer, and other health outcomes. Ann Epidemiol 2015; 25(3):193–200.

[34] Mørch LS, Skovlund CW, Hannaford PC, Iversen L, Fielding S, Lidegaard Ø. Contemporary hormonal contraception and the risk of breast cancer. N Engl J Med 2017;377(23):2228–39.

[35] Westhoff CL, Pike MC. Hormonal contraception and breast cancer. Contraception 2018;98(3): 171–3.

[36] Cao Y, Hou L, Wang W. Dietary total fat and fatty acids intake, serum fatty acids and risk of breast cancer: a meta-analysis of prospective cohort studies. Int J Cancer 2016;138(8):1894–904.

[37] Chen M, Rao Y, Zheng Y, Wei S, Li Y, Guo T, Yin P. Association between soy isoflavone intake and breast cancer risk for pre-and post-menopausal women: a meta-analysis of epidemiological studies. PLoS One 2014;9(2):1–10.

[38] Bakker MF, Peeters PH, Klaasen VM, Bueno-de-Mesquita HB, Jansen EH, Ros MM, … Rinaldi S. Plasma carotenoids, vitamin C, tocopherols, and retinol and the risk of breast cancer in the European Prospective Investigation into Cancer and Nutrition cohort, 2. Am J Clin Nutr 2016;103(2):454–64.

[39] Farvid MS, Chen WY, Rosner BA, Tamimi RM, Willett WC, Eliassen AH. Fruit and vegetable consumption and breast cancer incidence: repeated measures over 30 years of follow-up. Int J Cancer 2019;144(7):1496–510.

[40] Ogimoto I, Shibata A, Fukuda K. World Cancer Research Fund/American Institute of Cancer Research 1997 recommendations: applicability to digestive tract cancer in Japan. Cancer Causes Control 2000;11(1):9–23.

[41] Jiralerspong S, Goodwin PJ. Obesity and breast cancer prognosis: evidence, challenges, and opportunities. J Clin Oncol 2016;34(35):4203–16.

[42] Iyengar NM, Arthur R, Manson JE, Chlebowski RT, Kroenke CH, Peterson L, … Nassir R. Association of body fat and risk of breast cancer in postmenopausal women with normal body mass index: a

secondary analysis of a randomized clinical trial and observational study. JAMA Oncol 2019; 5(2):155–63.

[43] Picon-Ruiz M, Morata-Tarifa C, Valle-Goffin JJ, Friedman ER, Slingerland JM. Obesity and adverse breast cancer risk and outcome: mechanistic insights and strategies for intervention. CA Cancer J Clin 2017;67(5):378–97.

[44] Chlebowski RT, Luo J, Anderson GL, Barrington W, Reding K, Simon MS, … Strickler H. Weight loss and breast cancer incidence in postmenopausal women. Cancer 2019;125(2):205–12.

[45] Nelson HD, Zakher B, Cantor A, Fu R, Griffin J, O'Meara ES, … Mandelblatt JS. Risk factors for breast cancer for women aged 40 to 49 years: a systematic review and meta-analysis. Ann Intern Med 2012;156(9):635–48.

[46] Assi N, Rinaldi S, Viallon V, Dashti SG, Dossus L, Fournier A, … Boeing H. Mediation analysis of the alcohol-postmenopausal breast cancer relationship by sex hormones in the EPIC cohort. Int J Cancer 2020;146(3):759–68.

[47] Jung S, Wang M, Anderson K, Baglietto L, Bergkvist L, Bernstein L, … Falk R. Alcohol consumption and breast cancer risk by estrogen receptor status: in a pooled analysis of 20 studies. Int J Epidemiol 2016;45(3):916–28.

[48] Kerr J, Anderson C, Lippman SM. Physical activity, sedentary behaviour, diet, and cancer: an update and emerging new evidence. Lancet Oncol 2017;18(8):e457–71.

[49] McTiernan ANNE, Friedenreich CM, Katzmarzyk PT, Powell KE, Macko R, Buchner D, … George SM. Physical activity in cancer prevention and survival: a systematic review. Med Sci Sports Exerc 2019;51(6):1252–61.

[50] Moore SC, Lee IM, Weiderpass E, Campbell PT, Sampson JN, Kitahara CM, … Adami HO. Association of leisure-time physical activity with risk of 26 types of cancer in 1.44 million adults. JAMA Intern Med 2016;176(6):816–25.

[51] Pizot C, Boniol M, Mullie P, Koechlin A, Boniol M, Boyle P, Autier P. Physical activity, hormone replacement therapy and breast cancer risk: a meta-analysis of prospective studies. Eur J Cancer 2016;52:138–54.

[52] Beebe-Dimmer JL, Yee C, Cote ML, Petrucelli N, Palmer N, Bock C, … Simon MS. Familial clustering of breast and prostate cancer and risk of postmenopausal breast cancer in the Women's Health Initiative study. Cancer 2015;121(8):1265–72.

[53] Tung N, Lin NU, Kidd J, Allen BA, Singh N, Wenstrup RJ, … Garber JE. Frequency of germline mutations in 25 cancer susceptibility genes in a sequential series of patients with breast cancer. J Clin Oncol 2016;34(13):1460.

[54] Ehrhardt MJ, Howell CR, Hale K, Baassiri MJ, Rodriguez C, Wilson CL, … Wang Z. Subsequent breast cancer in female childhood cancer survivors in the St Jude Lifetime Cohort Study (SJLIFE). J Clin Oncol 2019;37(19):1647.

[55] Schaapveld M, Aleman BM, van Eggermond AM, Janus CP, Krol AD, van der Maazen RW, … Van Imhoff GW. Second cancer risk up to 40 years after treatment for Hodgkin's lymphoma. N Engl J Med 2015;373(26):2499–511.

[56] Ahern TP, Broe A, Lash TL, Cronin-Fenton DP, Ulrichsen SP, Christiansen PM, … Damkier P. Phthalate exposure and breast cancer incidence: a Danish nationwide cohort study. J Clin Oncol 2019;37(21):1800–9.

[57] Gaudet MM, Deubler EL, Kelly RS, Ryan Diver W, Teras LR, Hodge JM, … Palli D. Blood levels of cadmium and lead in relation to breast cancer risk in three prospective cohorts. Int J Cancer 2019; 144(5):1010–6.

[58] Rodgers KM, Udesky JO, Rudel RA, Brody JG. Environmental chemicals and breast cancer: an updated review of epidemiological literature informed by biological mechanisms. Environ Res 2018;160:152–82.

[59] Cordina-Duverger E, Menegaux F, Popa A, Rabstein S, Harth V, Pesch B, … Erren TC. Night shift work and breast cancer: a pooled analysis of population-based case–control studies with complete work history. Eur J Epidemiol 2018;33:369–79.

[60] Hansen J. Night shift work and risk of breast cancer. Curr Environ Health Rep 2017;4(3):325–39.

[61] Wegrzyn LR, Tamimi RM, Rosner BA, Brown SB, Stevens RG, Eliassen AH, … Schernhammer ES. Rotating night-shift work and the risk of breast cancer in the nurses' health studies. Am J Epidemiol 2017;186(5):532–40.

[62] Colquhoun A, Arnold M, Ferlay J, Goodman KJ, Forman D, Soerjomataram I. Global patterns of cardia and non-cardia gastric cancer incidence in 2012. Gut 2015;64(12):1881–8.

[63] Memon A, Bannister P. Epidemiology of cervical cancer. In: Uterine cervical cancer. Cham: Springer; 2019. p. 1–16.

[64] Huang FY, Kwok YK, Lau ET, Tang MH, Ng TY, Ngan HY. Genetic abnormalities and HPV status in cervical and vulvar squamous cell carcinomas. Cancer Genet Cytogenet 2005;157(1):42–8.

[65] Khazaei Z, Dehkordi AH, Amiri M, Adineh HA, Sohrabivafa M, Darvishi I, … Goodarzi E. The incidence and mortality of endometrial cancer and its association with body mass index and human development index in Asian population. World Cancer Res J 2018;5(4):11.

[66] Kumar R, Rai AK, Das D, Das R, Kumar RS, Sarma A, … Ramteke A. Alcohol and tobacco increases risk of high risk HPV infection in head and neck cancer patients: study from North-East Region of India. PLoS One 2015;10(10)e0140700 https://doi.org/10.1371/journal.pone.0140700.

[67] Merrill RM, Fugal S, Novilla LB, Raphael MC. Cancer risk associated with early and late maternal age at first birth. Gynecol Oncol 2005;96(3):583–93.

[68] Bidus MA, Elkas JC. Berek & Novak's gynecology. Philadelphia, PA: Lippincott Williams & Wilkins; 2007.

[69] Bruni L, Diaz M, Castellsagué M, Ferrer E, Bosch FX, de Sanjosé S. Cervical human papillomavirus prevalence in 5 continents: meta-analysis of 1 million women with normal cytological findings. J Infect Dis 2010;202(12):1789–99.

[70] Wardak S. Human papillomavirus (HPV) and cervical cancer. Med Dosw Mikrobiol 2016;68(1):73–84.

[71] Zur Hausen H. Papillomaviruses and cancer: from basic studies to clinical application. Nat Rev Cancer 2002;2(5):342–50.

[72] Crosbie EJ, Einstein MH, Franceschi S, Kitchener HC. Human papillomavirus and cervical cancer. Lancet 2013;382(9895):889–99.

[73] De Sanjosé S, Diaz M, Castellsagué X, Clifford G, Bruni L, Muñoz N, Bosch FX. Worldwide prevalence and genotype distribution of cervical human papillomavirus DNA in women with normal cytology: a meta-analysis. Lancet Infect Dis 2007;7(7):453–9.

[74] Galani E, Christodoulou C. Human papilloma viruses and cancer in the post-vaccine era. Clin Microbiol Infect 2009;15(11):977–81.

[75] Gunnell AS, Tran TN, Torrång A, Dickman PW, Sparén P, Palmgren J, Ylitalo N. Synergy between cigarette smoking and human papillomavirus type 16 in cervical cancer in situ development. Cancer 2006;15(11):2141–7.

[76] Ferenczy A, Coutlée F, Franco E, Hankins C. Human papillomavirus and HIV coinfection and the risk of neoplasias of the lower genital tract: a review of recent developments. CMAJ 2003; 169(5):431–4.

[77] Palefsky J. Human papillomavirus infection in HIV-infected persons. Top HIV Med 2007; 15(4):130–3.

[78] Mbulawa ZZ, Marais DJ, Johnson LF, Boulle A, Coetzee D, Williamson AL. Influence of human immunodeficiency virus and CD4 count on the prevalence of human papillomavirus in heterosexual couples. J Gen Virol 2010;91(12):3023–31.

[79] Pantanowitz L, Michelow P. Review of human immunodeficiency virus (HIV) and squamous lesions of the uterine cervix. Diagn Cytopathol 2011;39(1):65–72.

[80] Strickler HD, Palefsky JM, Shah KV, Anastos K, Klein RS, Minkoff H, … Fazzari M. Human papillomavirus type 16 and immune status in human immunodeficiency virus-seropositive women. J Natl Cancer Inst 2003;95(14):1062–71.

[81] Koskela P, Anttila T, Bjørge T, Brunsvig A, Dillner J, Hakama M, … Luostarinen T. Chlamydia trachomatis infection as a risk factor for invasive cervical cancer. Int J Cancer 2000;85(1):35–9.

[82] Smith JS, Bosetti C, Munoz N, Herrero R, Bosch FX, Eluf-Neto J, … Peeling RW. Chlamydia trachomatis and invasive cervical cancer: a pooled analysis of the IARC multicentric case-control study. Int J Cancer 2004;111(3):431–9.

[83] Silva J, Cerqueira F, Medeiros R. Chlamydia trachomatis infection: implications for HPV status and cervical cancer. Arch Gynecol Obstet 2014;289(4):715–23.

[84] Hildesheim A, Herrero R, Castle PE, Wacholder S, Bratti MC, Sherman ME, ... Helgesen K. HPV co-factors related to the development of cervical cancer: results from a population-based study in Costa Rica. Br J Cancer 2001;84(9):1219–26.

[85] Kim J, Kim BK, Lee CH, Seo SS, Park SY, Roh JW. Human papillomavirus genotypes and cofactors causing cervical intraepithelial neoplasia and cervical cancer in Korean women. Int J Gynecol Cancer 2012;22(9):1570–6.

[86] International Collaboration of Epidemiological Studies of Cervical Cancer. Carcinoma of the cervix and tobacco smoking: collaborative reanalysis of individual data on 13,541 women with carcinoma of the cervix and 23,017 women without carcinoma of the cervix from 23 epidemiological studies. Int J Cancer 2006;118(6):1481–95.

[87] Jensen KE, Schmiedel S, Norrild B, Frederiksen K, Iftner T, Kjaer SK. Parity as a cofactor for high-grade cervical disease among women with persistent human papillomavirus infection: a 13-year follow-up. Br J Cancer 2013;108(1):234–9.

[88] Castellsagué X, Bosch FX, Muñoz N. The male role in cervical cancer. Salud Publica Mex 2003;45:345–53.

[89] Liu ZC, Liu WD, Liu YH, Ye XH, Chen SD. Multiple sexual partners as a potential independent risk factor for cervical cancer: a meta-analysis of epidemiological studies. Asian Pac J Cancer Prev 2015; 16(9):3893–900.

[90] Silins I, Ryd W, Strand A, Wadell G, Törnberg S, Hansson BG, ... Persson K. Chlamydia trachomatis infection and persistence of human papillomavirus. Int J Cancer 2005;116(1):110–5.

[91] Winer RL, Hughes JP, Feng Q, O'Reilly S, Kiviat NB, Holmes KK, Koutsky LA. Condom use and the risk of genital human papillomavirus infection in young women. N Engl J Med 2006;354(25): 2645–54.

[92] Vanakankovit N, Taneepanichskul S. Effect of oral contraceptives on risk of cervical cancer. Med J Med Assoc Thailand 2008;91(1):7.

[93] Moreno V, Bosch FX, Muñoz N, Meijer CJ, Shah KV, Walboomers JM, ... International Agency for Research on Cancer (IARC) Multicentric Cervical Cancer Study Group. Effect of oral contraceptives on risk of cervical cancer in women with human papillomavirus infection: the IARC multicentric case-control study. Lancet 2002;359(9312):1085–92.

[94] Lacey Jr. JV, Swanson CA, Brinton LA, Altekruse SF, Barnes WA, Gravitt PE, ... Schwartz PE. Obesity as a potential risk factor for adenocarcinomas and squamous cell carcinomas of the uterine cervix. Cancer 2003;98(4):814–21.

[95] Lee JK, So KA, Piyathilake CJ, Kim MK. Mild obesity, physical activity, calorie intake, and the risks of cervical intraepithelial neoplasia and cervical cancer. PLoS One 2013;8(6) https://doi.org/10.1371/journal.pone.0066555.

[96] Poorolajal J, Jenabi E. The association between BMI and cervical cancer risk: a meta-analysis. Eur J Cancer Prev 2016;25(3):232–8.

[97] González CA, Travier N, Luján-Barroso L, Castellsagué X, Bosch FX, Roura E, ... Sacerdote C. Dietary factors and in situ and invasive cervical cancer risk in the European prospective investigation into cancer and nutrition study. Int J Cancer 2011;129(2):449–59.

[98] Tomita LY, Filho AL, Costa MC, Andreoli MAA, Villa LL, Franco EL, Cardoso MA. Diet and serum micronutrients in relation to cervical neoplasia and cancer among low-income Brazilian women. Int J Cancer 2010;126(3):703–14.

[99] Ghosh C, Baker JA, Moysich KB, Rivera R, Brasure JR, McCann SE. Dietary intakes of selected nutrients and food groups and risk of cervical cancer. Nutr Cancer 2008;60(3):331–41.

[100] Fonseca-Moutinho JA. Smoking and cervical cancer. ISRN Obstet Gynecol 2011;2011: https://doi.org/10.1371/journal.pone.0066555.

[101] Roura E, Castellsagué X, Pawlita M, Travier N, Waterboer T, Margall N, ... Tjønneland A. Smoking as a major risk factor for cervical cancer and pre-cancer: results from the EPIC cohort. Int J Cancer 2014;135(2):453–66.

[102] Plummer M, Herrero R, Franceschi S, Meijer CJ, Snijders P, Bosch FX, ... Muñoz N. Smoking and cervical cancer: pooled analysis of the IARC multi-centric case–control study. Cancer Causes Control 2003;14(9):805–14.

[103] Bray F, Ferlay J, Soerjomataram I, Siegel RL, Torre LA, Jemal A. Global cancer statistics 2018: GLO-BOCAN estimates of incidence and mortality worldwide for 36 cancers in 185 countries. CA Cancer J Clin 2018;68(6):394–424.

[104] Coburn SB, Bray F, Sherman ME, Trabert B. International patterns and trends in ovarian cancer incidence, overall and by histologic subtype. Int J Cancer 2017;140(11):2451–60.

[105] Yoneda A, Lendorf ME, Couchman JR, Multhaupt HA. Breast and ovarian cancers: a survey and possible roles for the cell surface heparan sulfate proteoglycans. J Histochem Cytochem 2012; 60(1):9–21.

[106] Chornokur G, Amankwah EK, Schildkraut JM, Phelan CM. Global ovarian cancer health disparities. Gynecol Oncol 2013;129(1):258–64.

[107] Mohammadian M, Ghafari M, Khosravi B, Salehiniya H, Aryaie M, Bakeshei FA, Mohammadian-Hafshejani A. Variations in the incidence and mortality of ovarian cancer and their relationship with the human development index in European Countries in 2012. Biomed Res Ther 2017;4(8): 1541–57.

[108] Chan JK, Urban R, Cheung MK, Osann K, Husain A, Teng NN, … Leiserowitz GS. Ovarian cancer in younger vs older women: a population-based analysis. Br J Cancer 2006;95(10):1314–20.

[109] Poole EM, Merritt MA, Jordan SJ, Yang HP, Hankinson SE, Park Y, … Terry KL. Hormonal and reproductive risk factors for epithelial ovarian cancer by tumor aggressiveness. Cancer 2013; 22(3):429–37.

[110] Walsh T, Casadei S, Lee MK, Pennil CC, Nord AS, Thornton AM, … Norquist B. Mutations in 12 genes for inherited ovarian, fallopian tube, and peritoneal carcinoma identified by massively parallel sequencing. Proc Natl Acad Sci 2011;108(44):18032–7.

[111] Toss A, Tomasello C, Razzaboni E, Contu G, Grandi G, Cagnacci A, … Cortesi L. Hereditary ovarian cancer: not only BRCA 1 and 2 genes. BioMed Res Int 2015; https://doi.org/10.1155/2015/341723.

[112] Nakamura K, Banno K, Yanokura M, Iida M, Adachi M, Masuda K, … Tominaga E. Features of ovarian cancer in Lynch syndrome. Mol Clin Oncol 2014;2(6):909–16.

[113] Helder-Woolderink JM, Blok EA, Vasen HFA, Hollema H, Mourits MJ, De Bock GH. Ovarian cancer in Lynch syndrome; a systematic review. Eur J Cancer 2016;55:65–73.

[114] Holschneider CH, Berek JS. Ovarian cancer: epidemiology, biology, and prognostic factors. Semin Surg Oncol 2000;19(1):3–10.

[115] Soegaard M, Jensen A, Høgdall E, Christensen L, Høgdall C, Blaakær J, Kjaer SK. Different risk factor profiles for mucinous and nonmucinous ovarian cancer: results from the Danish MALOVA study. Cancer Epidemiol Prev Biomarkers 2007;16(6):1160–6.

[116] Tsilidis KK, Allen NE, Key TJ, Dossus L, Lukanova A, Bakken K, … Tjønneland A. Oral contraceptive use and reproductive factors and risk of ovarian cancer in the European Prospective Investigation Into Cancer and Nutrition. Br J Cancer 2011;105(9):1436–42.

[117] Tung KH, Goodman MT, Wu AH, McDuffie K, Wilkens LR, Kolonel LN, … Sobin LH. Reproductive factors and epithelial ovarian cancer risk by histologic type: a multiethnic case-control study. Am J Epidemiol 2003;158(7):629–38.

[118] Riman T, Dickman PW, Nilsson S, Correia N, Nordlinder H, Magnusson CM, Persson IR. Risk factors for epithelial borderline ovarian tumors: results of a Swedish case–control study. Gynecol Oncol 2001;83(3):575–85.

[119] La Vecchia C, Franceschi S. Oral contraceptives and ovarian cancer. Eur J Cancer Prev 1999;8:297–304.

[120] Zhou J, Chng WJ. Roles of thioredoxin binding protein (TXNIP) in oxidative stress, apoptosis and cancer. Mitochondrion 2013;13(3):163–9.

[121] Lacey Jr. JV, Mink PJ, Lubin JH, Sherman ME, Troisi R, Hartge P, … Schairer C. Menopausal hormone replacement therapy and risk of ovarian cancer. JAMA 2002;288(3):334–41.

[122] Risch HA, Marrett LD, Jain M, Howe GR. Differences in risk factors for epithelial ovarian cancer by histologic type: results of a case-control study. Am J Epidemiol 1996;144(4):363–72.

[123] Mørch LS, Løkkegaard E, Andreasen AH, Krüger-Kjær S, Lidegaard Ø. Hormone therapy and ovarian cancer. JAMA 2009;302(3):298–305.

[124] Goodman MT, Wu AH, Tung KH, McDuffie K, Kolonel LN, Nomura AM, … Hankin JH. Association of dairy products, lactose, and calcium with the risk of ovarian cancer. Am J Epidemiol 2002;156(2):148–57.

[125] Mori M, Harabuchi I, Miyake H, Casagrande JT, Henderson BE, Ross RK. Reproductive, genetic, and dietary risk factors for ovarian cancer. Am J Epidemiol 1988;128(4):771–7.

[126] McCann SE, Freudenheim JL, Marshall JR, Graham S. Risk of human ovarian cancer is related to dietary intake of selected nutrients, phytochemicals and food groups. J Nutr 2003;133(6):1937–42.

[127] McCann SE, Moysich KB, Mettlin C. Intakes of selected nutrients and food groups and risk of ovarian cancer. Nutr Cancer 2001;39(1):19–28.

[128] Bandera EV, Lee VS, Qin B, Rodriguez-Rodriguez L, Powell CB, Kushi LH. Impact of body mass index on ovarian cancer survival varies by stage. Br J Cancer 2017;117(2):282–9.

[129] Delort L, Kwiatkowski F, Chalabi N, Satih S, Bignon YJ, Bernard-Gallon DJ. Central adiposity as a major risk factor of ovarian cancer. Anticancer Res 2009;29(12):5229–34.

[130] Anderson JP, Ross JA, Folsom AR. Anthropometric variables, physical activity, and incidence of ovarian cancer: the Iowa Women's Health Study. Cancer 2004;100(7):1515–21.

[131] Beehler GP, Sekhon M, Baker JA, Teter BE, McCann SE, Rodabaugh KJ, Moysich KB. Risk of ovarian cancer associated with BMI varies by menopausal status. J Nutr 2006;136(11):2881–6.

[132] Parkin DM, Bray FI, Devesa SS. Cancer burden in the year 2000. The global picture. Eur J Cancer 2001;37:4–66.

[133] Ferlay J, Soerjomataram I. GLOBOCAN: cancer incidence and mortality worldwide: IARC cancer base no. 11. International Agency for Research on Cancer; 2013.

[134] Fitzmaurice C, Akinyemiju TF, Al Lami FH, Alam T, Alizadeh-Navaei R, Allen C, … Aremu O. Global, regional, and national cancer incidence, mortality, years of life lost, years lived with disability, and disability-adjusted life-years for 29 cancer groups, 1990 to 2016: a systematic analysis for the global burden of disease study. JAMA Oncol 2018;4(11):1553–68.

[135] Bray F, dos Santos Silva I, Moller H, Weiderpass E. Endometrial cancer incidence trends in Europe: underlying determinants and prospects for prevention. Cancer 2005;14(5):1132–42.

[136] Purdie DM, Green AC. Epidemiology of endometrial cancer. Best Pract Res Clin Obstet Gynaecol 2001;15(3):341–54.

[137] Dossus L, Rinaldi S, Becker S, Lukanova A, Tjonneland A, Olsen A, … Clavel-Chapelon F. Obesity, inflammatory markers, and endometrial cancer risk: a prospective case–control study. Endocr Relat Cancer 2010;17(4):1007.

[138] Killackey MA. Endometrial adenocarcinoma in breast cancer patients receiving antiestrogens. Cancer Treat Rep 1985;69:237–8.

[139] Fearnley EJ, Marquart L, Spurdle AB, Weinstein P, Webb PM. Australian Ovarian Cancer Study Group and Australian National Endometrial Cancer Study Group: Polycystic ovary syndrome increases the risk of endometrial cancer in women aged less than 50 years: an Australian case-control study. Cancer Causes Control 2010;21(12):2303–8.

[140] Navaratnarajah R, Pillay OC, Hardiman P. Polycystic ovary syndrome and endometrial cancer. Semin Reprod Med 2008;26(01):062–71.

[141] Schumer ST, Cannistra SA. Granulosa cell tumor of the ovary. J Clin Oncol 2003;21(6):1180–9.

[142] Chang ET, Lee VS, Canchola AJ, Clarke CA, Purdie DM, Reynolds P, … Pinder R. Diet and risk of ovarian cancer in the California Teachers Study cohort. Am J Epidemiol 2007;165(7):802–13.

[143] Simpson ER, Brown KA. Obesity, aromatase and breast cancer. Expert Rev Endocrinol Metab 2011;6(3):383–95.

[144] Friberg E, Orsini N, Mantzoros CS, Wolk A. Diabetes mellitus and risk of endometrial cancer: a meta-analysis. 1365–74.

[145] Nagamani M, Stuart CA. Specific binding and growth-promoting activity of insulin in endometrial cancer cells in culture. Am J Obstet Gynecol 1998;179(1):6–12.

[146] Lucenteforte E, Bosetti C, Talamini R, Montella M, Zucchetto A, Pelucchi C, … La Vecchia C. Diabetes and endometrial cancer: effect modification by body weight, physical activity and hypertension. Br J Cancer 2007;97(7):995–8.

[147] Sjögren LL, Mørch LS, Løkkegaard E. Hormone replacement therapy and the risk of endometrial cancer: a systematic review. Maturitas 2016;91:25–35.

[148] Moore RG, MacLaughlan S, Bast Jr. RC. Current state of biomarker development for clinical application in epithelial ovarian cancer. Gynecol Oncol 2010;116(2):240–5.

[149] Lu KH, Dinh M, Kohlmann W, Watson P, Green J, Syngal S, … Terdiman J. Gynecologic cancer as a "sentinel cancer" for women with hereditary nonpolyposis colorectal cancer syndrome. Obstet Gynecol 2005;105(3):569–74.

[150] Zucchetto A, Serraino D, Polesel J, Negri E, De Paoli A, Dal Maso L, … Talamini R. Hormone-related factors and gynecological conditions in relation to endometrial cancer risk. Eur J Cancer Prev 2009;18(4):316–21.

CHAPTER 5

Quantum dot nanobody-based theranostics for triple negative breast cancer

Rama Rao Malla
Cancer Biology Lab, Department of Biochemistry and Bioinformatics, Institute of Science, GITAM (Deemed to be University), Visakhapatnam, Andhra Pradesh, India

Abstract

Triple negative breast cancer is a major health issue for women worldwide. Thus, there is a need for a novel and reliable therapy approach. With the advent of optical and chemical methods like quantum dots (QDs), QD-based nanotechnology has been given special attention in the field of cancer imaging. Here, we emphasize the applications of nanodots in imaging, targeting, and therapy of breast cancer as well as different nanobodies available in breast cancer therapy.

Keywords: Cancer imaging, Nanobody, Quantum dots, TNBC.

Abbreviations

BRET	bioluminescence resonance energy transfer
EGFR	epidermal growth factor receptor
FA	folic acid
HA	hyaluronic acid
MRI	magnetic resonance imaging
NIR	near–infrared
NPs	nanoparticles
PDT	photodynamic therapy
PEI	polyethylenimine polymers
QDs	quantum dots
SLN	sentinel lymph node
TNBC	triple negative breast cancer

1. Introduction

Triple negative breast cancer (TNBC) is a serious health issue for women worldwide due to it being highly aggressive and having a poor prognosis [1]. Complete control is highly difficult due to the lack of specific markers and treatments. However, early-stage diagnosis of TNBC can significantly reduce death rates in the long term. The stages of TNBC, such as invasion and metastasis, are complex and must be explored for better

G.P. Nagaraju, R.R. Malla (eds.)
A Theranostic and Precision Medicine Approach for Female-Specific Cancers
ISBN 978-0-12-822009-2
https://doi.org/10.1016/B978-0-12-822009-2.00005-4

diagnosis and prognosis. The major complexity is to understand the crosstalk between tumor cells and the microenvironment. Therefore, the development of typical cancer detection and a decisive approach is required to translate complex molecular data into imaging signals [2]. The most critical part of prognosis is to diagnose early-stage cancer cells.

Worldwide, researchers have investigated various diagnostic approaches for diagnosis of TNBC, including mammography, magnetic resonance imaging (MRI), ultrasound, computed tomography, positron emission tomography, and biopsy. These noninvasive imaging tools are effective for visualization of macroscopic tumors, confirmation of the presence of tumors, and location of primary as well as secondary sites of TNBC in combination with biopsies. However, these techniques have several limitations including poor sensitivity and specificity of detecting abnormalities at the microscopic level, trouble identifying tumors at incipient stage and detecting precancerous lesions at the molecular level, and nonsuitability for young women [3]. Therefore, a search for novel methods of detection is crucial, as is real-time monitoring for understanding the mechanisms of TNBC.

2. Quantum dots with biomedical applications

Optical and chemical methods like quantum dots (QDs) have been given special attention in the field of cancer imaging. QDs, having distinctive size and surface characters, have potential for biomedical application including in situ multiplex imaging [4]. Exploring unique properties of QDs, in situ QD-based multiplex imaging devices have been developed for quantitative analysis in tumor tissue specimens. QDs are engineered fluorescent nanoparticles (NPs) with exclusive optical and chemical properties. These properties make QDs promising platforms for biomedical applications. Early diagnosis and marker identification are most important in targeted therapy. Investigation of axillary lymph nodes to examine metastasis is a preliminary step in TNBC prognosis, using optical methods. This chapter summarizes biomedical applications of QDs in clinical studies of TNBC (Fig. 1).

3. Construction of QDs

QDs are nanocrystals in the size range of 2–10 nm, surface coated with particles to enhance solubility and reduce toxicity, and selectively target highly recognized tumor markers. The core of QDs mainly contain group II/IV metals, including cadmium/selenium or cadmium/technetium, while the shells are mainly made of zinc sulfide (ZnS) or cadmium sulfide (CdS). Currently used QDs are cadmium free and only made of indium/palladium owing to high biocompatibility.

QDs can be constructed by embedding NPs into nanospheres, incorporating NPs during the process of nanosphere formation, assembling NPs onto the surfaces of nanospheres, or in situ synthesis of NPs within the pores of nanospheres (Fig. 2).

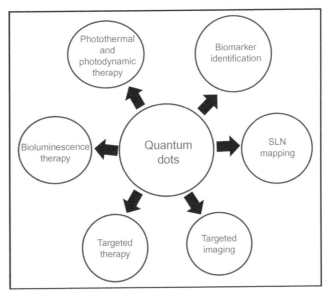

Fig. 1 Biomedical applications of quantum dots in clinical studies of TNBC. Quantum dots can be used in targeted therapy, targeted imaging, sentinel lymph node mapping, biomarker identification, photothermal and photodynamic therapy, and bioluminescence therapy.

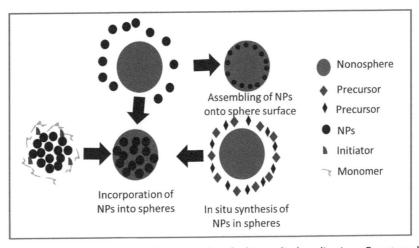

Fig. 2 Methods for the construction of quantum dots for biomedical applications. Quantum dots can be constructed by embedding NPs into nanospheres, incorporating NPs during the process of nanosphere formation, assembling NPs onto the surfaces of nanospheres, or in situ synthesis of NPs within the pores of nanospheres.

Table 1 Summary of quantum dot construction methods.

Item	Embedding method	Incorporation method	Assembly method	In situ synthesis
Operation process	Very convenient	Convenient	Very tedious	Convenient
Operation condition	Ultrasonication	Vigorously stirring	Gentle shaking	High temperature
Operation time	1 h	More than 1 h	More than 1 h	More than 1 h
NP distribution	Nonuniform	Uniform	Uniform	Uniform
NP aggregation	Some	Some	Little	Very little
Load capacity	Relatively low	Medium	Very high	High
Encoding type	Rich	Medium	Very rich	Poor
Controllability	Relatively random	Relatively controllable	Very controllable	Controllable
Stability	Stable in polar solvent	Stable	Stable	Stable
Composite size	Nanoscale	Nanoscale	Nanoscale	Microscale

The nanospheres and NPs are prepared separately using methods involving embedding or incorporation, and then combined together. In the other two methods, combination occurs during the formation process of nanospheres or NPs [5]. The comparison methods are summarized in Table 1.

4. Unique characteristics of QDs

Quantum and size effect make QDs exhibit specific optical and electrical properties. Nanometer particles exhibit different optical properties like size, dielectric and quantum confinement as well as surface effects. These specific properties allow for QDs to be used in biological application as fluorescent probes as well as functional materials [5]. QDs have exclusive fluorescence properties, owing to their electron–gap and the interaction with the surrounding configuration. If photon excitation is beyond the band gap, the QDs absorb photons and transfer electrons from the valence shell to the conduction band and exhibit luminescence, which varies with configuration and size of the QDs. On the contrary, to obtain color range, diverse fluorescent dyes can be used to obtain diverse excitation light, which is costly and complex for analysis [6]. Moreover, QDs exhibit high stokes shift compared to organic dyes, thus spectral analysis can be done at a low signal intensity [7]. As it is well known that tissues have autofluorescence ability and organic fluorescent dyes exhibit less stokes shift, a filter is required at the detector, which decreases the quality of signal intensity [8]. QDs have 20 times stronger fluorescence and 100 times the stability of rhodamine 6G, an organic fluorescent dye. Moreover, the modified CdS/ZnS QDs emitted intense fluorescence even at 500 mW with excitation at 488 nm for 14 h due to high photochemical stability and exhibited long time cellular interactions between biological molecules unlike organic fluorescent dyes [9].

5. QDs for biomarker identification

There is no test for characterization of TNBC subtypes (i.e., highly invasive vs less invasive and metastatic vs nonmetastatic). High throughput genomic and proteomic studies have identified some TNBC subtypes, which can be distinguished by specific biomarkers. Biomarkers are gaining importance in diagnosis, prognosis, and treatment and management. A cancer biomarker is a cellular indicator of a biological state that changes during cancer progression. It is well recognized that carcinogenic transformation causes the secretion of biomolecules or biomarkers at elevated and abnormal levels into body fluids [10]. Therefore, biomarkers could provide knowledge about the incidence and advancement of cancer. Several proteins have been recognized as unique biomarkers for several types of cancers. For example, prostate-specific antigen for prostate cancer, cancer antigen-125 for ovarian cancer [11], carcinoembryonic antigen for intestinal cancers, and HER-2 for breast cancer [12]. However, recognition of such biomarkers for TNBC is essential for early diagnosis. Advances in high-throughput genomics, proteomics, and metabolomics technologies have identified altered expression of mRNAs or proteins as well as different levels of metabolites [13]. These can be used as biomarkers for diagnosis, prognosis, assessment of risk, prediction of recurrence as well as therapeutic responses.

In quest of sensitive detection of biomarkers, nanotechnology has promise. QDs detect 10–100 cancer biomarkers simultaneously in blood and tissue biopsy samples [14]. In fact, QDs expanded cellular, tissue and whole-body multiplexed biomarkers for imaging of cancer. QDs also showed promise in personalized treatment of patients with distinct cancer phenotypes. Further, QDs possess potential for enhancing diagnostic ability and treatment of cancer [15].

6. Sentinel lymph node mapping

The sentinel lymph node (SLN) is the primary tumor drainage region and first site for metastasis. SLN mapping or lymphoscintigraphy is vital for surgical procedures of various cancers including breast cancer. The contemporary method for SLN mapping includes introducing an acidic sulfur-radiocolloid along with blue dye monitored by intraoperative photon probes sporadically using X-rays to find the SLN. Numerous reports have recommended mapping as the gold standard. Organic dyes as well as radiocolloids are the current choice for SLN mapping, however, SLN has limitations such as exposure to ionizing radiation and local tissue damage.

QDs can be used for SLN mapping without biopsy. The primary tumor resection by surgery is ideal for early diagnosis of breast cancer. However, due to complications in differentiating tumor boundaries and residual cancer cells, the tumor may recur frequently. QDs with functional groups are useful for image-guided surgery of tumors. Photostable near-infrared (NIR) QDs have excellent tissue penetration. Delayed clearance of QDs may lead to accumulation of QDs. However, cadmium-free QDs are promising and clinical trials prove their potential for cancer detection and therapy [16]. Indium-based QDs were used in a murine breast cancer model and were shown not to spread via the lymphatic system.

A PEGylated QD migrated to SLN from primary tumors and was noninvasively localized via the epithelium [17]. Conjugated QDs with hyaluronic acid (HA) and anionic glycosaminoglycans can be employed for lymphatic circulation imaging. Reports suggested they have less toxicity with high efficacy in lymphatic circulation. Conjugates of HA have application in enhancing permeability into tumors. Molecules with size less than 10 nm are suitable for SLN-based mapping, due to diffusion into lymph nodes, whereas particles with size in the range of 50–100 nm exhibited mixed action and poor penetration into lymphatic vessels. PEGylated QDs of only 10 nm size are not useful for the SLN mapping [18].

7. QDs in targeted imaging of TNBC

Fluorescence-based target imaging of tumors is a promising method for diagnosis of solid cancers localized in the body. This is a very convenient as well as inexpensive method compared to traditional imaging techniques. This method uses imaging agents with bright stable NIR fluorescence for imaging tumors with high efficiency. For example, NIR-QDs are attached to antibodies for high fluorescent quantum yield as well as photostability. Aptamer-coupled lipid nanocarriers encapsulated with QDs and siRNAs have potential theranostic delivery of TNBCs.

The indium phosphate core/zinc sulfide shell QDs (InP/ZnS QDs) with NIR fluorescence are used for biodistribution studies. The conjugation of the anti-EGFR (epidermal growth factor receptor) nanobody enhances the cellular uptake as well as cytotoxicity of the QD-based micelles. Analog to aminoflavone (AF)-encapsulated nontargeted micelles, the AF-encapsulated nanobody (Nb)-conjugated micelles were deposited in tumors at greater levels in TNBC models without systemic toxicity. Thus, a QD-based Nb-conjugated micelle can be used as a novel theranostic nanoplatform for TNBCs [19].

8. Self-illuminating QDs for bioluminescence imaging of TNBC

NIR-based fluorescent imaging has various advantages including excellent biological safety, sensitivity, easy operation, and real-time imaging. However, optical imaging based on bioluminescence resonance energy transfer (BRET) has several advantages over NIR fluorescent imaging, especially for deep tissue detection [20]. QD-BRET systems conjugated with mutant variants of *Rotylenchulus reniformis* luciferase (Luc8) emit blue luminescence by interacting with the substrate coelenterazine. This bioluminescence-based imaging can be applied for detection of deep tissue without interference from autofluorescence of tumor tissue [21]. Using multimodal fluorescence and bioluminescence imaging, TNBC lung metastases were detected [22].

9. QDs for detection of TNBC micro-metastasis

To prevent cancer relapse after surgery, it should be ensured that all cancer cells are eliminated. Recent studies have shown that intraoperative fluorescence imaging can be used for imaging of breast tumors during surgery. Real-time intraoperative fluorescence imaging gives an enhanced imaging of breast tumors [20]. The most common imaging modality for cancers is MRI, but it has a limitation as it fails to detect early stage cancer cells as well as occult cancer cells. Immunohistochemistry together with reverse transcription-polymerase chain reaction is normally used for imaging of micro-metastasis, but it has limitations of high time conception and failure to provide correct results [21]. Thus, novel theranostics are suggested for the management of optimal metastatic diseases. Intravenous injections of QDs bio-conjugated with cRGD to tumor sites have already been used for imaging using NIR fluorescence, and has been successful during tumor surgery. QDs tagged with cyclo-Arg-Gly-Asp-Tyr (c-RGDY) can target integrin tumor cells and help detect micrometastasis [23].

10. QDs for immunotherapy of TNBC

EGFR is highly expressed in most of TNBC patients. It is a prognostic and predictive marker of breast cancer [24]. HER2$^+$ breast cancer is aggressive and has a poor prognosis. EGFR-targeted therapy characterizes one of the most successful steps toward personalized precision medicine (Fig. 3). Humanized monoclonal antibodies (mAb) against EGFR have demonstrated improved survival rates with chemotherapeutics. A combination of anti-EGFR bioconjugated QD and immunohistochemistry has been reported to visualize single cells with enhanced sensitivity and efficacy with reduced cost in comparison to traditional methods. Conjugates of QDs against cancer are used to quantify markers at the cellular level. Microbeads functionalized with QD conjugated antibodies are used to target cancer antigens. Coupling of QDs to specific antibodies can serve for active targeting as well as passive targeting of drugs with enhanced permeability and retention [25].

11. QDs for gene therapy

Folic acid (FA)-QD complexes have been used in breast cancer therapy, as FA receptors are highly expressed in breast cancer and act as a low molecular weight targeting molecule

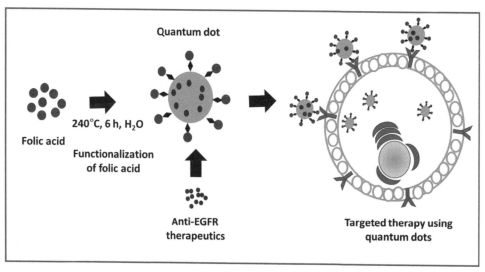

Fig. 3 Quantum dots in EGFR-targeted therapy of triple negative breast cancer. Coupling of quantum dots to specific antibodies can serve for active targeting as well as passive targeting of drugs with enhanced permeability and retention.

in breast cancer. FA complexes altered by carboxyl deuterogenic have durable binding ability to folate receptor. Essentially, small molecules, toxins, chemotherapeutics, radio-therapeutics, polymer-wrapped therapeutics, gene delivery agents, inhibitory prodrugs as well as immunotherapeutics can form complexes with FA via covalent binding [26].

The use of a d-dot/polymer nanocomplex for transfection of siRNAs to target the mutant oncogenic K-Ras gene is gaining interest in cancer therapy as the d-dot/PAH nano-complex is highly biocompatible at greater concentration. Recently, QD-Forster resonance energy transfer was used in an immunoassay as well as to design nanosensors for the release of nanodrugs [27]. Nanobody mediated activation immunotherapeutic in breast cancer is best explained by the selective anti-dinitrophenyl (anti-DNP) antibodies to the surface of TNBC cells. Moreover, SBT-100 (anti-STAT3 B VHH13) was developed against human STAT3, which can suppress the function of p-STAT3 [28].

12. QDs for photodynamic and photothermal TNBC therapy

Photodynamic therapy (PDT) with photosensitizers are used as an adjuvant therapy in medical practice. High–quality photosensitizers are required for improving the efficiency of PDT. Photosensitizers should have NIR absorption capacity. The absorption of QDs can be changed to NIR by enhancing their size. Such QDs are the best choice

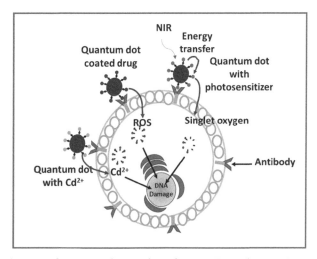

Fig. 4 Possible mechanism of quantum dots in photodynamic TNBC therapy. Quantum dots efficiently generate singlet oxygen and release toxic ions, especially heavy metal cations. Both singlet oxygen and heavy metal ions can damage DNA and induce cell apoptosis.

for deep-tissue imaging with PDT. Further, QDs release singlet oxygen and toxic heavy metals to induce DNA damage and apoptosis (Fig. 4).

New combinatory therapy of cannabinoid-based therapeutics as well as PDT upregulated the cannabinoid CB2 receptor as well as translocator protein (TSPO). The TSPO-PDT treatment showed synergistic reduction of TNBC tumors [29]. Further, CB2R-targeted photosensitizer IR700DX-mbc94 induced necrosis upon NIR irradiation, while PDT with TSPO-IR700DX-6T induced apoptosis. However, photosensitizers can significantly reduce tumor growth by targeting specific markers.

The cCB2R-TSPO-PDT caused increased cell death as well as reduction of tumor size at low doses of the drug [30]. Mechanistically, QDs eliminate tumor cells via photothermal effect. Brown or black QDs were more efficient for photothermal therapy, since they strongly absorb in the red and NIR. NIR penetrates deeply into tissues. Therefore, dark-colored QDs efficiently damage cells by transferring NIR light into heat energy. Cd- and Cu-based QDs have advantages and disadvantages in PTT of TNBC as well as their future prospects in biomedicine.

13. Nanobodies for targeted TNBC therapy

In breast cancer therapy, NPs of polymeric nanoparticles may be used as carriers of cytotoxic agents and deliver to the targets (Fig. 5). One of the best examples of such a system is delivery of saporin, which has specific targets in the cell but does not internalize.

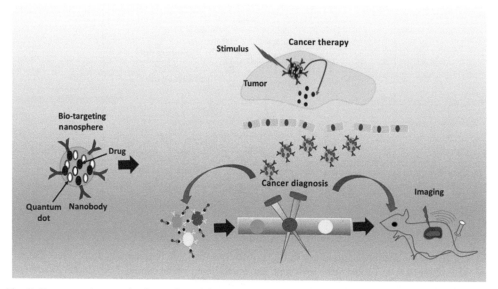

Fig. 5 Quantum dot nanobody-mediated diagnosis and therapy. The nanobody-conjugated quantum dots loaded with drugs internalize and release drugs and are also used for cancer diagnosis.

The carrier used against HER2+ breast cancer cells is PEGylated poly(lactic acid-*co*-glycolic acid-*co*-hydroxymethyl glycolic acid) NPs coated by 11A4 nanobody specific to HER2 receptors not to HER2− TNBC cells. This specific uptake was prevented by using the nanobody. Studies have shown that combination of saporin-11A4 NPs and photochemical internalization significantly decreased the proliferation and viability rate of the cell via inducing apoptosis [31].

Also, nanobodies are used for delivery of genes as well as targeting transcription. Commonly, polyethylenimine polymers (PEI) and poly(ethylene glycol) are covalents attached to anti-HER2 variable domains of nanobodies to the distal terminals of NHS-PEG3500 in PEI-PEG NPs [32]. In cancer stem cell (CSC) therapy, a modified TNBC cell line with HER2 overexpression was targeted with anti-HER2 Nb-coupled polyamidoamine polyplexes, which exhibited high efficiency against CSC [33]. Anti-PD-L1 mAbs for targeting PD-1/PD-L1 immune checkpoints have anticancer activity, where PD-L1 Nbs exhibited competitive binding inhibition to form a PD-1/PD-L1 complex and give simulations for molecular dynamics [34].

Recombinant immunotoxin is gaining attention in cancer therapy due to its specificity. The split inteins-fused heavy chain of trastuzumab (HER2 antibody) with Nb-Fc fusion targets HER2+ breast cancer. Intein, a fused exotoxin A, expressed fusion to maltose as a binding protein [35]. Nb-fused magnetic oleosomes deliver oil-based lipophilic drugs for breast cancer therapy [36]. Moreover, At-labeled sdAb conjugates, particularly iso-[211At] SAGMB-5F7, are efficient as a targeted α-particle radiotherapy of HER2-expressing breast cancer [37].

A QD-based micelle conjugated with EGFR Nband AF has been engineered as theranostics for TNBC [38]. Moreover, neutralization of TNFα in tumor microenvironment is a novel approach using nanobodies that can enhance paclitaxel therapy and inhibit metastasis in breast cancer. One such approach is the use of a TNFα-specific nanobody from *Pichia pastoris*, which inhibited proliferation and invasion of MDA-MB-231 [39]. Another study has proven that the conjugate tCAIX-specific nanobody-IRDye800CW could be used as a rapid imaging agent for invasive breast cancers by using hypoxia-targeting fluorescent Nbs [5]. Nbs can be used in therapy by activating immune response. For example, anti-DNP antibodies tagged to HER2-type breast cancers caused target destruction by antibody-mediated cellular cytotoxicity [40].

14. Conclusion

Though the studies on QD-based nanocarriers for delivery of drugs have accomplished considerable progress, the use of QD nanocarriers must be developed for disease screening and gene sequencing as well. Humanized antibodies can be progressed to effectively and selectively bind virtually any disease-relevant cell surface receptor. Further research on nanotechnology-based detection will expand the scope of its use in the field of cancer biology.

Acknowledgment

The authors thank the DST-EMR (EMR/2016/002694, dt. 21st August 2017), New Delhi, India for supporting the project.

Conflict of Interest

The author declared that there is no conflict of interest.

References

[1] Fang M, Peng CW, Pang DW, Li Y. Quantum dots for cancer research: current status, remaining issues, and future perspectives. Cancer Biol Med 2012;9(3):151–63.

[2] Sanz-Moreno V, Marshall CJ. The plasticity of cytoskeletal dynamics underlying neoplastic cell migration. Curr Opin Cell Biol 2010;22(5):690–6.

[3] Zhao MX, Zeng EZ. Application of functional quantum dot nanoparticles as fluorescence probes in cell labeling and tumor diagnostic imaging. Nanoscale Res Lett 2015;10:171.

[4] He X, Ma N. An overview of recent advances in quantum dots for biomedical applications. Colloids Surf B Biointerfaces 2014;124:118–31.

[5] Wen CY, Xie HY, Zhang ZL, Wu LL, Hu J, Tang M, Wu M, Pang DW. Fluorescent/magnetic micro/nano-spheres based on quantum dots and/or magnetic nanoparticles: preparation, properties, and their applications in cancer studies. Nanoscale 2016;8(25):12406–29.

[6] Bai M, Bornhop DJ. Recent advances in receptor-targeted fluorescent probes for in vivo cancer imaging. Curr Med Chem 2012;19(28):4742–58.

[7] Jaiswal JK, Mattoussi H, Mauro JM, Simon SM. Long-term multiple color imaging of live cells using quantum dot bioconjugates. Nat Biotechnol 2003;21(1):47–51.

[8] Viswanath AK. From clusters to semiconductor nanostructures. J Nanosci Nanotechnol 2014;14 (2):1253–81.

[9] Smith AM, Nie S. Semiconductor nanocrystals: structure, properties, and band gap engineering. Acc Chem Res 2010;43(2):190–200.

[10] Mason JN, Farmer H, Tomlinson ID, Schwartz JW, Savchenko V, DeFelice LJ, Rosenthal SJ, Blakely RD. Novel fluorescence-based approaches for the study of biogenic amine transporter localization, activity, and regulation. J Neurosci Methods 2005;143(1):3–25.

[11] Rosenblum LT, Kosaka N, Mitsunaga M, Choyke PL, Kobayashi H. Optimizing quantitative in vivo fluorescence imaging with near-infrared quantum dots. Contrast Media Mol Imaging 2011;6 (3):148–52.

[12] Doré-Savard L, Barrière DA, Midavaine É, Bélanger D, Beaudet N, Tremblay L, Beaudoin JF, Turcotte EE, Lecomte R, Lepage M, Sarret P. Mammary cancer bone metastasis follow-up using multimodal small-animal MR and PET imaging. J Nucl Med 2013;54(6):944–52.

[13] Fisher B, Bauer M, Wickerham DL, Redmond CK, Fisher ER, Cruz AB, Foster R, Gardner B, Lerner H, Margolese R, et al. Relation of number of positive axillary nodes to the prognosis of patients with primary breast cancer. An NSABP update. Cancer 1983;52(9):1551–7.

[14] Hulvat M, Rajan P, Rajan E, Sarker S, Schermer C, Aranha G, Yao K. Histopathologic characteristics of the primary tumor in breast cancer patients with isolated tumor cells of the sentinel node. Surgery 2008;144(4):518–24 discussion 524.

[15] Wu X, Wu M, Zhao JX. Recent development of silica nanoparticles as delivery vectors for cancer imaging and therapy. Nanomedicine 2014;10(2):297–312.

[16] Bakalova R, Zhelev Z, Kokuryo D, Spasov L, Aoki I, Saga T. Chemical nature and structure of organic coating of quantum dots is crucial for their application in imaging diagnostics. Int J Nanomedicine 2011;6:1719–32.

[17] Chowbay B, Jada SR, Wan Teck DL. Correspondence re: Cecchin et al., Carboxylesterase isoform 2 mRNA expression in peripheral blood mononuclear cells is a predictive marker of the irinotecan to SN38 activation step in colorectal cancer patients. Clin Cancer Res 2005;11:6901–7. Clin Cancer Res 2006;12(6):1942 author reply 1942–3.

[18] Meyer JS. Sentinel lymph node biopsy: strategies for pathologic examination of the specimen. J Surg Oncol 1998;69(4):212–8.

[19] Lin G, Ouyang Q, Hu R, Ding Z, Tian J, Yin F, Xu G, Chen Q, Wang X, Yong KT. In vivo toxicity assessment of non-cadmium quantum dots in BALB/c mice. Nanomedicine 2015;11(2):341–50.

[20] Tsoi KM, Dai Q, Alman BA, Chan WC. Are quantum dots toxic? Exploring the discrepancy between cell culture and animal studies. Acc Chem Res 2013;46(3):662–71.

[21] Shukur A, Rizvi SB, Whitehead D, Seifalian A, Azzawi M. Altered sensitivity to nitric oxide donors, induced by intravascular infusion of quantum dots, in murine mesenteric arteries. Nanomed Nanotechnol Biol Med 2013;9(4):532–9.

[22] Miao P, Han K, Tang Y, Wang B, Lin T, Cheng W. Recent advances in carbon nanodots: synthesis, properties and biomedical applications. Nanoscale 2015;7(5):1586–95.

[23] Huang X, Zhang F, Zhu L, Choi KY, Guo N, Guo J, Tackett K, Anilkumar P, Liu G, Quan Q, Choi HS, Niu G, Sun YP, Lee S, Chen X. Effect of injection routes on the biodistribution, clearance, and tumor uptake of carbon dots. ACS Nano 2013;7(7):5684–93.

[24] Abdullah Al N, Lee JE, In I, Lee H, Lee KD, Jeong JH, Park SY. Target delivery and cell imaging using hyaluronic acid-functionalized graphene quantum dots. Mol Pharm 2013;10(10):3736–44.

[25] Zheng XT, Ananthanarayanan A, Luo KQ, Chen P. Glowing graphene quantum dots and carbon dots: properties, syntheses, and biological applications. Small 2015;11(14):1620–36.

[26] Schneider R, Schmitt F, Frochot C, Fort Y, Lourette N, Guillemin F, Müller JF, Barberi-Heyob M. Design, synthesis, and biological evaluation of folic acid targeted tetraphenylporphyrin as novel photosensitizers for selective photodynamic therapy. Bioorg Med Chem 2005;13(8):2799–808.

[27] Zhang G, Gao J, Qian J, Zhang L, Zheng K, Zhong K, Cai D, Zhang X, Wu Z. Hydroxylated mesoporous nanosilica coated by polyethylenimine coupled with gadolinium and folic acid: a tumor-targeted T(1) magnetic resonance contrast agent and drug delivery system. ACS Appl Mater Interfaces 2015;7(26):14192–200.

[28] Lee SJ, Shim YH, Oh JS, Jeong YI, Park IK, Lee HC. Folic-acid-conjugated pullulan/poly(DL-lactide-co-glycolide) graft copolymer nanoparticles for folate-receptor-mediated drug delivery. Nanoscale Res Lett 2015;10:43.

[29] Shao D, Li J, Pan Y, Zhang X, Zheng X, Wang Z, Zhang M, Zhang H, Chen L. Noninvasive theranostic imaging of HSV-TK/GCV suicide gene therapy in liver cancer by folate-targeted quantum dot-based liposomes. Biomater Sci 2015;3(6):833–41.

[30] Yuan Y, Zhang J, An L, Cao Q, Deng Y, Liang G. Oligomeric nanoparticles functionalized with NIR-emitting CdTe/CdS QDs and folate for tumor-targeted imaging. Biomaterials 2014;35(27):7881–6.

[31] Yoo HS, Park TG. Folate-receptor-targeted delivery of doxorubicin nano-aggregates stabilized by doxorubicin-PEG-folate conjugate. J Control Release 2004;100(2):247–56.

[32] Martínez-Jothar L, Beztsinna N, van Nostrum CF. Selective cytotoxicity to HER2 positive breast cancer cells by saporin-loaded nanobody-targeted polymeric nanoparticles in combination with photochemical internalization. Mol Pharm 2019;16(4):1633–47.

[33] Saqafi B, Rahbarizadeh F. Polyethyleneimine-polyethylene glycol copolymer targeted by anti-HER2 nanobody for specific delivery of transcriptionally targeted tBid containing construct. Artif Cells Nanomed Biotechnol 2019;47(1):501–11.

[34] Reshadmanesh A, Rahbarizadeh F, Ahmadvand D, Jafari Iri Sofla F. Evaluation of cellular and transcriptional targeting of breast cancer stem cells via anti-HER2 nanobody conjugated PAMAM dendrimers. Artif Cells Nanomed Biotechnol 2018;46(Suppl 3):S105–15.

[35] Sun X, Yan X, Zhuo W, Gu J, Zuo K, Liu W, Liang L, Gan Y, He G, Wan H, Gou X, Shi H, Hu J. PD-L1 nanobody competitively inhibits the formation of the PD-1/PD-L1 complex: comparative molecular dynamics simulations. Int J Mol Sci 2018;19(7):1984–94.

[36] Pirzer T, Becher KS, Rieker M, Meckel T. Generation of potent anti-HER1/2 immunotoxins by protein ligation using split inteins. ACS Chem Biol 2018;13(8):2058–66.

[37] (a) Mazzega E, de Marco A. Engineered cross-reacting nanobodies simplify comparative oncology between humans and dogs. Vet Comp Oncol 2018;16(1):E202–6; (b) Choi J, Vaidyanathan G, Koumarianou E, Kang CM, Zalutsky MR. Astatine-211 labeled anti-HER2 5F7 single domain antibody fragment conjugates: radiolabeling and preliminary evaluation. Nucl Med Biol 2018;56:10–20.

[38] (a) Wang Y, Wang Y, Chen G, Li Y, Xu W, Gong S. Quantum-dot-based theranostic micelles conjugated with an anti-EGFR nanobody for triple-negative breast cancer therapy. ACS Appl Mater Interfaces 2017;9(36):30297–305; (b) Ji X, Peng Z, Li X, Yan Z, Yang Y, Qiao Z, Liu Y. Neutralization of TNFα in tumor with a novel nanobody potentiates paclitaxel-therapy and inhibits metastasis in breast cancer. Cancer Lett 2017;386:24–34.

[39] (a) van Brussel AS, Adams A, Oliveira S, Dorresteijn B, El Khattabi M, Vermeulen JF, van der Wall E, Mali WP, Derksen PW, van Diest PJ, Van Bergen En Henegouwen PM. Hypoxia-targeting fluorescent nanobodies for optical molecular imaging of pre-invasive breast cancer. Mol Imaging Biol 2016;18 (4):535–44; (b) Gray MA, Tao RN, DePorter SM, Spiegel DA, McNaughton BR. A nanobody activation immunotherapeutic that selectively destroys HER2-positive breast cancer cells. Chembiochem 2016;17(2):155–8.

[40] Gray MA, Tao RN, DePorter SM, Spiegel DA, McNaughton BR. A nanobody activation immunotherapeutic that selectively destroys HER2-positive breast cancer cells. Chembiochem 2016;17 (2):155–8.

CHAPTER 6

Nanotechnology advances in ovarian cancer

Kiranmayi Patnala[a], Rama Rao Malla[b], and Soumya Vishwas[a]

[a]Department of Biotechnology, Institute of Science, GITAM (Deemed to be University), Visakhapatnam, Andhra Pradesh, India
[b]Cancer Biology Lab, Department of Biochemistry and Bioinformatics, Institute of Science, GITAM (Deemed to be University), Visakhapatnam, Andhra Pradesh, India

Abstract

Ovarian cancer is the pervasive and asymptomatic type of female-specific cancer. Since it exhibits no symptoms during its early stages, detection is severely constrained until it reaches the advanced stages. Survival of such patients primarily depends on the stage of the cancer. Though patients may exhibit initial recovery after being treated with various drugs, prolonged exposure to the drugs may lead to development of drug resistance in most cases. Detected cancers can be treated by traditional cytoreductive therapy that involves abscission of tumor cells or by techniques based on nanotechnology to deliver chemotherapeutics at the specific tumor sites. Though removal of cancer can be achieved at stage I, administration of chemotherapeutics becomes a necessity at all stages of ovarian cancer, enhancing the scope of developing nanoscale formulations with high efficacy against cancer. The current chapter summarizes nanotechnology-based novel therapeutic strategies and formulations that show encouraging results in ovarian cancer imaging, detection, and treatment.

Keywords: Nanoparticles, Ovarian cancer, Diagnostics, Imaging, Treatment, Preclinical development, Chemotherapeutics, Novel therapies.

Abbreviations

ADCT	amphiphilic dendritic copolymer telodendrimer
Apo	apolipoprotein
Au-NP	gold nanoparticle
CA	cancer antigen
CE	carcino embryonic
DNA	deoxyribonucleic acid
DOPC	dioleoyl phosphatidylcholine
EGFR	epidermal growth factor receptor
EGP	ethylene glycol-glutamic acid-phenylalanine
ELISA	enzyme-linked immunosorbent assay
EOC	epithelial ovarian cancers
EPR	enhanced permeability and retention
FDA	food and drug administration
FILIP1L	filamin A interacting protein 1-like
Gd-B-dendrimers	Biotin conjugated dendrimers with Gd-chelates
HA-E1A	human adenovirus type 5 early region 1A
HE4	human epididymis

G.P. Nagaraju, R.R. Malla (eds.)
A Theranostic and Precision Medicine Approach for Female-Specific Cancers
ISBN 978-0-12-822009-2
https://doi.org/10.1016/B978-0-12-822009-2.00006-6

105

Heparin-PE	heparin-polyethyleneimine
HER	human epidermal growth factor receptor
HPMA	N-(2-hydroxypropyl) methacrylamide
HSulf-1	heparin sulfate 6-O-endosulfatase 1
ID4	inhibitor of DNA binding 4
IL-12	interleukin 12
IO-NPs	iron oxide nanoparticles
IP	intraperitoneal
IV	intravenous
LHR	luteinizing hormone receptor
LHRHan	luteinizing hormone-releasing hormone analogue
MEMS	microelectromechanical systems
MRI	magnetic resonance imaging
MUC	mucin
NIR	near infrared
NPs	nanoparticles
PA	photoacoustic
PAMAM	polyamidoamine
PDT	photodynamic therapy
PEG	polyethylene glycol
PEI	poly(ethylene imine)
PLA	polylactic acid
PLGA	poly(D,L-lactic-*co*-glycolic acid)
QDs	quantum dots
RES	reticuloendothelial system
ROMA	risk of ovarian malignancy algorithm
RONDEL	RNAi/oligonucleotide nanoparticle delivery
ROS	reactive oxygen species
RRM2	M2 subunit of ribonucleotide reductase
scFV	single-chain variable fragment
SELDI-TOF-MS	surface-enhanced laser desorption/ionization time of flight mass spectrometry
shRNA	short hairpin RNA
siRNA	small interfering
SPECT	single-photon emission computerized tomography
SR-B1	scavenger receptor type B1
STAT3	signal transducer and activator of transcription 3
TP	tumor protein
VEGF	vascular endothelial growth factor

1. Introduction

Ovarian cancer is a common cause of death in women worldwide. According to a study, 184,799 ovarian cancer patients out of the registered 295,414 cases died in 2018 [1]. Ovarian cancer can be mitigated if it is detected before it can spread to the peritoneum or any other organs surrounding the affected area. However, there is a 20% drop in the survival rate calculated over 5 years in metastatic patients. Over the last 30 years, there has been no improvement in the survival rate of patients with ovarian cancer. This is because more than 60% of the cases are detected only after reaching stage III [2]. More than 90% of the detected ovarian cancers are of epithelial origin and are referred to as epithelial ovarian cancers (EOCs). These EOCs, which are categorized into mucinous, clear cell, low-grade serous, high-grade serous, and endometrioid, all show similarity in their intraperitoneal (IP) localization holding various diseases [2]. Based on the origin of the tissue and analysis of the genome, these diseases are known to be recognizably different. Table 1 shows genetic mutations and their association with different types of ovarian cancer [2–7].

Currently, the serous type of ovarian cancer shows frequent occurrence in females with a large group of patients. Many endeavors to develop novel therapies for the serous type of ovarian cancer are therefore ongoing in the field of research. The repair of broken DNA generally requires BRCA1 and BRCA2 genes. Mutations caused in their germ lines result in 22% of ovarian cancer of high-grade serous type [8]. While endometriosis eventually leads to clear cell carcinomas in patients, the origin of the mucinous subtype is still unexplored [6]. However, conception, hysterectomy, and use of contraceptives have shown reduced risk of endometrioid carcinomas [9]. During early studies, the exterior epithelium of the ovary in an ovulating female was misinterpreted as progenitor site for ovarian cancer. Later studies demonstrated that it was the epithelium of the fallopian tube that leads to serious ovarian cancers. Most of the cases occurred due to genomic coalition of alteration caused in BRCA1/BRCA2 [10]. During ovulation when the ovarian epithelium ruptures, the fallopian tube fimbriae is believed to approach the ruptured site. This leads to formation of lesions that act as precursors for ovarian cancer. Ovarian cancer generally spreads via the body cavities and is frequently diagnosed after its extensive spread in the intraperitoneal cavity like in cases of serous type II cancers [7, 11].

Table 1 EOCs and associated genetic mutations.

Cancer subtype	EOC cases reported	Genetic mutations reported	Origin tissue
Type I: Mucinous	<5%	KRAS (75%)	–
Type I: Clear cell	10%	ARIK1A (50%) PIK3CA (50%) PTEN (20%)	Endometrial
Type I: Serous, low grade	<5%	BRAF/KRAS (50%) ERBB2 (9%)	Fimbriae of fallopian tube
Type II: Serous, high grade	70%	TP53 (96%) BRCA1/ BRCA2 (22%)	Fimbriae of fallopian tube
Type I—low grade and Type II—high grade: Endometrioid	10%	CTNNB1 (40%) PIK3CA (20%) PTEN (20%)	Endometrial

Cancer patients of stage III and IV exhibit excessive accumulation of fluids in the peritoneal cavity, leading to pain in the abdomen or its distension. Such cases showcase weak results of the patients [12].

2. Current diagnosis and clinical treatment

Ovarian cancers are often termed as "the silent killers." The reason for this is that ovarian cancer does not manifest any symptoms at its early stage. Three-fourths of ovarian cancer cases are diagnosed only after reaching stage III or IV [9]. In other words, ovarian cancers are diagnosed only after they spread far off from the ovaries. There are no efficient biomarkers to detect ovarian cancers at an early stage. This open up new research areas in hunting for alternatives. Proteomics studies revealed many new markers for ovarian tumors that were recognized from patients' serum [13]. MUC16, a cancer antigen, was derived from epithelial cell lines of ovarian cancer by Bast et al. [14]. This glycoprotein was also termed as CA125. Unfortunately, CA125 showed enormous false-positive and false-negative results when used as a marker for screening. Also, this marker is expressed during liver diseases and pregnancy, leading to incorrect disease diagnosis. It was observed that serum collected from cancer patients showed increased glycosylation of proteins (11 N-glycans), which showed better sensitivity and specificity when compared to CA125 [15–18]. However, these were not effective in diagnosing the cancer. In one study, secretory glycoprotein HE 4 that is generally found in half the population of ovarian cancer patients exhibited trivial expression in CA125. This protein therefore may minimize false negative results in patients [19]. In menopausal females, ROMA (combining HE4 and CA125) showed diversified yet improved results compared to lone CA125 screening. OVA1 is the test that was approved by FDA for ovarian cancer patients. This test uses a combination of serum markers like transferrin, ApoA1, transthyretin, CA125, and beta-2 microglobulin, all of which were recognized by surface-enhanced laser desorption/ionization time of flight mass spectrometry (SELDI-TOF-MS). However, there were no significant results using this test over using CA125 alone while screening for ovarian cancer [20, 21]. Ideal treatment pattern can be determined only when tumors are screened for changes in the expression levels of specific proteins. This is turn will aid in development of tumor-targeted therapies.

Nowadays, cytoreductive surgery is the standard for ovarian cancer treatment. Removal of the tumor, also known as tumor debulking, is a standard therapy used during the last 50 years. After debulking, the patients are generally kept on platinum-based IV chemotherapy for a period of 6 to 8 weeks. During this period, patients generally show good response to chemotherapeutic drugs like carboplatin or cisplatin. Unfortunately, most patients relapse after a certain period of improving health and develop resistance to platinum. Various approved drug combinations like bleomycin-etoposide-cisplatin, gemcitabine and taxol, and Doxil and carboplatin show good results in ovarian cancer patients. Combination chemotherapies that have been approved for ovarian cancer include carboplatin and Doxil, gemcitabine and taxol, and bleomycin-etoposide-cisplatin. Owing to the internment of initial lesions and ovarian tumors, oncologists have been administering these drugs through IP routes for several years [22]. Administration of drugs is done by placing an abdominal catheter in the patients, thereby making this type of therapy incommodious to patients. Also, IP therapies have unintentional effects like gastrointestinal problems or infections within the abdomen [23]. However, IP therapies took precedence over other drug delivery routes as they reduce systemic toxicity and enhance tumor targeting. The IP route can therefore be explored for imaging and treating cancers using nanotechnology.

3. Nanotechnology in diagnosis and imaging

Recently, technologies based on microelectrical systems (MEMS) and various nanoparticle-based techniques have shown advances in the fields of diagnostics. Most of these advances involve trials to improve contrast agents for imaging and biomarker detection methods. Recently, it was reported that CA125, mesothelin, and HE4 showed promising use in multiplex assays. They exhibited improved performance and good reliability in diagnostic methods [24, 25]. While it is known that a diagnostic test depends on availability of fine biomarkers, nanotechnological techniques can improve diagnostics. Assays and sensors based on nanotechnology are expeditious, dependable, cost-effective, and can be used in cancer diagnostics for multiplex screening [26–29].

In 2009, an assay based on antigen–antibody reaction was demonstrated by Jokerst et al., that integrated quantum dots [30]. The system uses quantum dot-labeled CA125 antibodies on a microfluidic nanoplatform. It comprises an array of wells, each of which is capable of clasping a solitary agarose particle. The agarose particles have static antibodies that capture the antigens. Once the antigen is captured, another quantum dot-labeled antibody permits fluorescent visualization. These agarose beads were used in multiplex assays for HER2 and CA125. This system showed enhanced responsiveness for CE antigen when compared to ELISA method. Upon further developments on this "programmable bio-nano-chip," it was recognized as a suitable screening method for ovarian cancer [31].

For any passable diagnosis, the sera level should extend to nearly 35 U/mL. Sensing CA125 without any labels can be achieved by using screen-printed Au-NP electrodes with serum levels of just 6.7 U/mL [32].

Primary antibody-labeled iron oxide NPs along with a dendrimer or luminol-altered secondary antibody also acts as a light-emitting sensing platform. This platform could detect concentrations of CA125 as low as 0.03 µU/mL [33, 34]. Systems based on various NPs, chips, MEMS, and other platforms are expected to improvise screening of ovarian cancer and play a crucial role in its diagnostics.

Among numerous imaging modalities, optical imaging, ultrasound, and MRI are the three most frequently used imaging methods for ovarian cancer. However, most clinical diagnostics favor transvaginal ultrasound. Table 2 summarizes the NPs used in imaging ovarian cancer cases.

In ultrasound imaging, high-frequency sound waves are targeted to the tissues where a transducer is placed. The reflected sound waves from the internal organs are used for image construction. For ultrasonography, the audile characteristics of the internal tissue have to be altered. Contrast agents like particles with perfluorocarbon-filled or air-filled cores are used commonly in ultrasound. The two commercially used imaging agents for ultrasound are OPTISON and Definity, both of which have been used for examining and recording microvasculature changes in the ovarian tumor. OPTISON is the terminology for perflutren particles stabilized with albumin, and Definity for perflutren–containing lipid microspheres [35, 36]. Color Doppler and 3D sonography along with microtubule contrast are promising ultrasound advancements that are currently being used for imaging or diagnostics of ovarian cancer [37].

Table 2 NPs used for imaging ovarian cancer.

Imaging modality	Nanoparticle	Targeting	Drug	Specimen	Cell line
Optical	Quantum dot	CA125-antibody	N	Xenograft	HO8910
	Quantum dot	EPR	N	Xenograft	HEYA8
	Quantum dot	Her2-antibody	N	In vitro	SKOV-3
	Quantum dot	MUC1-aptamer	Y	Xenograft	A2780
	Polymethacrylic acid (VIS)	Her2-antibody		In vitro	SKOV-3
	Lipoprotein (NIR)	Folic acid		Xenograft	–
	PEG–PLA (NIR)	EPR		Xenograft	A2780
	PEG-lipid (NIR)	EPR		Xenograft	SKOV-3
	$NaYF_4:Yb^{3+}/Er^{3+} @ SiO_2$	–		In vitro	SKOV-3
Magnetic resonance	Dendrimer	Biotin		Xenograft	SHIN3
	Liposome	Folic acid		Xenograft	IGROV-1, OVCAR-3
	Iron oxide	Folic acid		In vitro	SKOV-3
	Iron oxide/Gd	EPR		Xenograft	SKOV-3
Photo acoustic	Gold nanorods	EPR		Xenograft	2008, HEY, SKOV-3
Ultrasound	Liposome microbubbles	–		In vitro	N/A

The futuristic role of several nanoparticle assemblies can be applied to provide improved contrast in cancer imaging using MRI. While T1 imaging uses various NPs like dendrimers, micelles, and liposomes, T2 imaging frequently uses IO-NPs. Gd-B-dendrimers were initially used in MRI for in vivo imaging of cancer xenografts [38]. Gd3+ contained liposome with modified folate is also used for in vivo imaging [39]. The in vitro T2 imaging of SKOV-3 ovarian cells can be achieved by modifying the surface of IO-NPs using polyglycerol grafted with folic acid [40, 41], whereas T1 imaging can be achieved by gadolinium-embedded IO-NPs with zwitterionic surface. The accumulation of these NPs in circulatory system took place via EPR effect having half-life of approximately 1 h [42]. In the dual system using MRI and SPECT, it was recently observed that dextran-coated IO-NPs in combination with labeled mesothelin antibodies yielded effective imaging results in A431-K5 cell xenografts [43]. Similar particles can be used in imaging and diagnosis if the same receptors are highly expressed in ovarian cancer cells.

Probes designed for NIR optical imaging should possess photostability, fine quantum yield, high coefficient of molar adsorption, and huge Stokes shift. A large number of dendrimers, liposomes, and polymeric NPs integrate these NIR dyes. Tumor distribution and accumulation in mice xenograft HT29 (colon) and A2780 (ovarian) cell lines were evaluated using DiR (NIR dye). The dye was initially loaded into PEG–PLA copolymer NPs. After injecting, it was feasible to image up to 2 days followed by RES clearance. The 111-nm particles showed moderate RES clearance rate and elevated tumor accumulation when compared to 166-nm particles [44]. In a study, ICG (NIR dye) was encapsulated in PEGylated micelles along with tanespimycin and paclitaxel. These encapsulated micelles when introduced into mice with SKOV-3 xenografts, exhibited enhanced therapeutic effect when compared to free drug. These NPs showed accumulation in tumors through EPR effect and enabled imaging up to 2 days [45].

Quantum dots are inorganic NPs. In comparison to several dyes, QDs have better optical qualities like large Stokes shift, limited window for emission, broad range of excitation, and diminished photobleaching [46–48]. QDs, like polymeric NPs, have also been used for treatment of A2780 cancer xenografts. They exhibit emission at around 500 nm. QDs, when in conjugation with MUC1 (targeting aptamer) and drug doxorubicin, are not only used for imaging but also used for ovarian cancer treatment [49].

PA imaging, like ultrasound, uses NIR to create an image based on sound waves via local heating. Here, local heating can be accomplished by administering nanomaterials or tiny molecules. For example, gold nanorods on being prompted with NIR radiations not only produce local heating effect but also impart contrast. They are used on mice for PA imaging of ovarian cancer xenografts. On injecting these nanorods, they showed greatest signal after 180 min and extended up to 48 h. Studies revealed that nanorods with a 3.5 aspect ratio were ideal for in vivo imaging [50].

4. Nanotechnology in chemotherapeutics

Nanotechnology-based formulations that incorporate liposomes have tremendous advantages of increased drug accumulation, retentivity time, and compatibility. For instance, Doxil is an FDA-approved NP formulation for patients with recurrent cancer. This approved formulation is based on a PEGylated liposome with doxorubicin drug and is currently used in gemcitabine combination therapy for cancer patients who have developed resistance to platinum. Patients treated with NP liposomal formulations showed enhanced plasma half-life and improved accumulation in tumors when compared to patients who were treated with unfettered doxorubicin. These liposomes were capable of staying unscathed for several days after their administration [51]. Phase II trials of Doxil demonstrated low toxicity in comparison to free doxorubicin [52]. Recurrent or intractable ovarian cancer patients who are treated via standard therapies using topotecan and gemcitabine showed benefit when treated with Doxil [53–56]. It also showed an increase in the survival rate of the platinum–resistant patient pool, displaying no further cancer progress [55, 56]. In another trial study on platinum–sensitive cancer patients, it was observed that therapy combining Doxil with carboplatin yielded parallel results to standard therapy using paclitaxel with carboplatin. However, Doxil–carboplatin treatment on patients with partial resistance to platinum showed no cancer progression [57]. Also, cancer patients with partial or complete platinum resistance showed improved survival rate when Doxil was used in combination with trabectedin rather than using Doxil alone [58, 59]. The success of Doxil in treating ovarian cancer has opened up prospects for new clinical trials using various nanoformulations. However, another drug, topotecan, showed parallel efficacy to Doxil as well as paclitaxel in treating relapsed or recurrent ovarian metastatic cancer [60]. Liposomes with topotecan are under phase I clinical trials. PEGylating the liposomal topotecan, though considered for use, did not show considerable improvement with topotecan [61]. Moreover, it showed detrimental effects in rats by causing excessive blood clearance [62]. Lurtotecan, yet another formulation based on liposomes, showed increased exposure when compared to free drug. In spite of this feature, the formulation could show optimum activity only after being given a daily inhibitor dose. This resulted in several side effects, especially in populations that were already on prescribed treatment. Since it could not show significant activity in such patient population, it could not gain approval for clinical use [63].

Abraxane is an FDA-approved drug for treating phase I cancers of the breast, lung, and pancreas. It is a commercial name given to nab-paclitaxel or albumin paclitaxel bound to NPs [64]. Nab-paclitaxel is more advantageous than other paclitaxel formulations. The albumin can stabilize paclitaxel particles of nearly 130 nm in size. General paclitaxel formulations involve synthetic solvents for carrying and delivering drugs to the target cells and cause hypersensitive reactions [65]. The first ever usage of Abraxane for ovarian cancer was set forth in 2006 in patients aged 60 years to curtail hypersensitivity induced by paclitaxel. The formulation showed no incidence while undergoing three treatment rounds in these patients [66]. Since then, nab-paclitaxel has been in phase II clinical trials for ovarian cancer patients with platinum sensitivity as well as resistivity. Infusing Abraxane into a patient pool with platinum resistivity once a week resulted in a notable survival rate of 135 days without progression [67]. Similarly, the drug showed 64% response rate in patients who were sensitive to platinum. These studies dealing with two types of patient population indicated that Abraxane showed impressive activity and can therefore be summoned for phase III trials [68]. Xyotax, a paclitaxel drug formulation, can enter the cellular metabolism after degrading itself into L–glutamic acid from its paclitaxel conjugate. It is basically a nanoscale solvent-free formulation that showed inspiring efficacy in phase I and II clinical trials, in the absence of any prior medications [69]. Xyotax when administered in patients with repetitive EOC, showed fair activity in phase II trials. The drug was given at an interval of 3 weeks for such EOC patients as a second- or third-line treatment. Though it showed modest activity, it has future prospects of being used as a maintaining drug [70]. Xyotax when used in conjugation with carboplatin exhibited workable results in first-line therapy. The other drug formulations include Nanotax, a paclitaxel NP suspension that is under phase I clinical trials, and Etirinotecan pegol, a polymer conjugate (NKTR–102, Nektar) that is considered a probable drug for treating ovarian cancer patients with platinum resistance. In Etirinotecan pegol, a decomposable linking element binds PEG-NP to irinotecan, which is a topoisomerase I inhibitor. This formulation gets activated once it breaks down to release the inhibitor. During phase I clinical studies, this drug showed anti-cancerous activity in a group of patients who were on prior treatments [71]. The formulation under phase II trials showed high activity in ovarian cancer patients who developed resistance to platinum. When administered once a month, it increased the survival rate of these patients up to an average of 5.4 months. The results further led to phase III trials for breast and ovarian cancer [72]. Camptothecin (topoisomerase inhibitor) when incorporated into cyclodextrin-PEG-NPs and tested in phase I clinical trials showed greater efficiency in patients than free camptothecin [73]. Like PEG, HPMA also has uses in drug delivery. It is a compatible polymer to which a carboplatinate analogue or an oxaliplatin analogue can be bound using linker molecules that are sensitive to acids. ProLindac (an analogue of oxaliplatin bound to HPMA) showed better efficacy than cisplatin in the preclinical trials. It showed 16 times more accumulation of platinum in cancer cells than cisplatin. In spite of exhibiting high tolerance, it could not however express efficacy over oxaliplatin in phase I trials [74]. In phase II trials for single-agent efficacy, two thirds of late-stage heavily pretreated ovarian cancer patients exhibited disease stabilization, opening avenues for further combination clinical studies. The use of carboplatinate analogue bound to HPMA also showed encouraging toxicity in phase I and II clinical trials [75, 76]. Table 3 shows nanoformulations that are under clinical trials for ovarian cancer.

Table 3 Clinical trials of the nanoformulations for ovarian cancer.

Compound name	Formulation	Active agent	Status for ovarian cancer	Trial no., Refs.
Doxil	PEGylated liposomal doxorubicin	Doxorubicin	FDA approved	NCT00945139, NCT00862355, NCT00248248
–	Liposomal topotecan	Topotecan	Phase I	NCT00765973
OSI–211	Liposomal lurotecan	Lurotecan	Phase II	NCT00010179
Abraxane, ABI–007	Nanoparticle-bound albumin paclitaxel	Paclitaxel	Phase II	Approved for breast, small cell and pancreatic cancers NCT00466986, NCT00407563
Xyotax	Paclitaxel poliglumex	Paclitaxel	Phase II	NCT00060359, NCT00017017
Nanotax	Nanoparticle suspension	Paclitaxel	Phase I	NCT00666991
NKTR–102	Etirinotecan pegol	Topoisomerase I inhibitor	Phase II	NCT00806156103
CRLX101	Cyclodextrin–PEG Camptothecin	Camptothecin	Phase I	NCT00333502, NCT01652079
ProLindac (AP5346)	HPMA-oxaliplatin analogue	Oxaliplatin analogue	Phase II	[74, 75]
AP5280	HPMA-carboplatinate analogue	Carboplatin analogue	Phase II	[75, 76]

5. Preclinical development

Polymeric NP structures aid in targeted drug delivery to tumor sites. A study showed that micelles loaded with paclitaxel as well as cisplatin exhibited toxicity on cells. The micelle core made of EGP is enclosed within a PEG shell. This NP polymer biodegrades in the presence of proteolytic enzymes like cathepsin B to release both drugs at the targeted site. Ovarian cancers, on being treated for a long time, gradually become resistant to multiple drugs. However, the combined NP formulation of paclitaxel and cisplatin yielded excellent results in such cancers when compared to free drugs or micelles carrying only one drug [77]. Another copolymer micelle was also developed for improving pharmacokinetics. This new NP system consists of PEG-poly(ε-caprolactone)–PEG polymer aggregates that form a gel. This gel is designed in such a way that it degrades gradually releasing the drug doxorubicin via IP region, further leading to total dismissal from the cavity. Though this is a biodegradable copolymer with high therapeutic efficiency, it still has to enter clinical trials [78]. A similar type of micelle that uses gossypol and cylopamine inhibitors for IP delivery of loaded paclitaxel showed reduction in growth and magnitude of the tumor with enhanced survival rate when tested in a mouse xenograft [79]. Dendrimers of various radii (1.3–2.9 nm) like PAMAM were also used traditionally for drug delivery. Cisplatin-loaded PAMAM reduced the tumor and its toxicity, exhibiting high tolerated dose [80]. Factors like steric hinderance, usage of linkers that are sensitive to pH, or chemistry involved in linking all affect the controlled drug delivery to the targeted site in various types of cancer [81–83]. A group from the University of California Davis Cancer Center introduced an ADCT that can be used for ovarian cancer therapy. Either the core or the shell of this system is made of PEG, lysin, and cholic acid that have the capacity to load paclitaxel in its aqueous state. The in vitro activity of this formulation against the tumor was similar to that of Abraxane or Taxol. However, it showed deeper penetration when tested in a mouse, showing predominant antitumor effects than most of the FDA-approved formulations [84]. When this formulation was radiolabeled before biodistribution, it showed slower pharmacokinetics when compared to Taxol, exhibiting preferential uptake at the sites of tumor. With these advantages, it can further be used as a propitious drug carrier in clinical trials [85, 86]. A polymeric NP system was designed for controlled release of drugs at the targeted site. This system incorporates two drug components: a slow-releasing PLGA containing ceramide and paclitaxel that is enclosed in a pH-responsive poly (beta-amino ester). Though it was not used intrinsically in clinical trials for ovarian cancer, it showed effective tumor reduction in xenografts [87].

In ovarian cancer cells, surface proteins like integrins, mucins, and claudins get highly expressed. The expression of receptor proteins (LHR, EGFR, and HER2) also get elevated in ovarian cancer cases. The most general techniques for tumor targeting and localization include binding of these surface proteins with antibodies and peptides. However, prospects of a new targeting system based on magnetic NPs is encouraging. The drug cisplatin, for instance, is loaded onto iron oxide particles, which is then localized to cancer cells magnetically. This magnetic localization system resulted in almost 110-fold toxicity of the cells [88]. One of the main disadvantages of this system for clinical trials is the difficulty in manipulating the magnetic field, which should be powerful and capable of precisely targeting the tumor cells. Liposomal formulations also showed inspiring results in the in vivo ovarian cancer studies. Several in vivo xenograft studies reveal promising results using targeted liposomal formulations for ovarian cancer. When cholesterol liposomes loaded with Docetaxel were targeted to the tumor site using LHRHa, it showed a nine-fold increase in drug accumulation in tumor cells when compared to free drug within 1 h of its administration. It also decreased undesirable drug accumulation in the spleen and liver [89].

A-3 integrin receptors, which are highly expressed on surfaces of tumor cells, have high affinity for "OA02" peptides. These elements are used in micelle carriers and aid in cellular uptake and localization of PEG micelles. The micelle complex when loaded with paclitaxel exhibited decreased systemic toxicity and enhanced efficiency against the tumor when compared to the Taxol formulation and untargeted PEG complex. Such features might be useful in decreasing the ill effects of chemo drugs [85, 86].

Nukolova et al. designed a nanogel using di-block copolymers that can target folate during ovarian cancer therapy. The effects of this nanogel on ovarian cancer were studied on a murine model. The complex, when loaded with doxorubicin or cisplatin, displayed specified and targeted drug delivery to tumor cells with stirring activity against them [90]. Similar results were seen when nanogel loaded with cisplatin was targeted to LHRH in ovarian cancer cases. This gel has further prospective applications in other cancer types, as this marker is seen to be highly expressed in prostrate and breast cancer as well [91].

In doxorubicin-loaded poly(L-histidine) micelles, folate was again used as a targeting ligand. This system showed interesting results of controlled drug release even in acidic pH. When tested on mouse xenografts of human ovarian carcinoma, the micelles showed greater accumulation at the site of the tumor, releasing high doses of drug. These micelles lead to higher treatment efficiency by exhibiting a five-fold increase in the plasma half-life when compared to free doxorubicin treatment [92]. Again, a micelle modified with folate was devised by the same group for enhanced uptake of drug via receptor-mediated endocytosis. These micelles displayed growth inhibition of tumors that were resistant to multi-drugs with negligible loss of weight in the tested animal [93]. When the pH sensitivity was turned to 6.0, it showed greater suppression of the tumor in mice for nearly 2 months [94].

6. Nanotechnology in novel therapies

Drug encapsulation and drug targeting, though they result in improved tumor response rate to the drug, might gradually lead to the development of drug resistance by cancer patients. However, many NP-mediated novel therapies are gaining traction. Photodynamic therapy (PDT) is one such novel therapy that is being used for ovarian cancers. During the initial trial, polylactic-acid NPs loaded with hypericin were used for PDT of ovarian cancer cells in vitro [95], followed by in vivo study on rats. The in vivo study conducted on the epithelial ovarian cancer cell line in rats using NP particles showed improved accumulation compared to treatment with free drug [96]. A PEG nanoformulation with chlorine photosensitizer (Ce6) tested on a murine model showed efficacy for ovarian cancer [33, 34]. By using effective photosensitizers and NIR radiations, penetration depth in ovarian cancers can be increased. However, successful irradiation using PDT within the IP cavity is still a challenge. A combination of NIR with NPs along with

Ce6 co-administration in a mouse model with breast cancer showed improved penetration and less toxicity [40, 41].

A porphyrin photosensitizer enclosed in a virus capsid and altered with an aptamer that targets nucleolin was tested in vitro and was found to target breast cancer cells specifically [97]. This nucleolin receptor is seen to be highly expressed in ovarian cancer cells, thereby increasing the scope of using similar systems for targeting. When illuminated by visible light, functionalized fullerene shows ability to produce reactive oxygen species (ROS). These particles were tested in a mouse model with colon adenocarcinoma for IP PDT [98]. Fullerenes proved to be effective photosensitizers either while administering them intravenously or via exposure against bacterial infections and fibrosarcoma [99, 100]. With improved photosensitizers and further advancements in targeted NP delivery systems, there exists a great scope for PDT implementation in near future.

The NP system for plasmid and oligonucleotide delivery is chosen based on its biocompatibility. Their charge is generally modified to enable efficient binding of the nucleotide for delivery of drug [101–103]. Several gene delivery methods use functionalized NPs to reach ovarian cancer cells [102].

The DNA plasmid can be delivered to the cancerous cells by forcefully expressing tumor suppressor genes. Since TP53 mutations were observed to be high in most of the cancer cases, methods to target this repressor gene using NPs are under consideration for clinical trials. There were recent clinical trials on "SGT-53." SGT-53 is an antitransferrin receptor scFV antibody that is bound to liposomal nanocarriers for delivery of the TP53 gene. The study depicted impressive output by stabilizing tumors of different origins in patients without much toxicity [104]. Almost all the high-grade serous ovarian cancer cases have mutations in TP53. Therefore, aiming at this pathway by gene therapy will show beneficial results in such patients. In the 1990s, delivery of the TP53 gene (Ad-p53) during clinical trials was attempted via adenoviral elements. However, this system did not prove to be of much utility to patients as the antibodies present within the body could recognize the incorporated virus as a foreign particle, rendering it incapable to reach the tumor sites [105]. NPs can be used for delivery of genetic material to ovarian cancer sites. In later phase I and II studies, liposomal HA-E1A in combination with paclitaxel was administered to ovarian cancer patients who developed platinum resistance with low toxicity profile. However, effects of the long-term usage of this therapy has not been reported [106]. A phase I clinical trial conducted in 2013 used a PEG-PEI-cholesterol lipopolymer for delivering a plasmid intraperitoneally. This plasmid encoded for IL-12 ("EGEN-001"). The purpose of the IP delivery of plasmid, besides carboplatin treatment, was to evoke a local immune response against the cancer cells in patients. The patients responded positively by enhancing the production of IL-12–stimulated cytokines (interferon-g and tumor necrosis factor-a) and showed bearable toxicity in the abdomen [107]. Patients who suffer from repeated or tenacious ovarian cancer are being engaged in phase II clinical trials for treatment with IP EGEN-001 or in combination with IP Doxil.

With a hope to elicit antitumor action via successful uptake and expression, NPs have been used for delivering expression vectors to ovarian cancer cells. The DNA plasmids were delivered by nanogel Heparin-PE for expressing HSulf-1 [108] and FILIP1L [109]. These proteins are usually negatively regulated in ovarian cancers. NP-containing expression plasmids of these genes when injected intraperitoneally into nude mice xenografts showed significant cell death with reduced cell proliferation and terminated angiogenesis. Treatment with cisplatin further induced enhanced Hsulf-1 expression [108].

In order to silence expression of oncogenes, NP formulations can be considered as potent methods for siRNA or shRNA delivery. The first trial on NP-based siRNA exhibited knockdown of RRM2 in tumor patients following IV administration of RONDEL [110]. This NP polymer formulation based on PEGylated cyclodextrin is embedded with transferrin protein that targets its corresponding receptor. This revealed that siRNA could affect the expression of RRM2 by targeting tumor tissue. Likewise, liposomal siRNA nanocarriers toward kinesin spindle protein (ALN-VSP) and VEGF showed encouraging outcomes in phase II trials with low toxicity and high tumor targeting. The study also reports retrogression of liver metastases in endometrial cancer patients [111]. Examples of promising preclinical studies on NP-based gene therapy are listed in Table 4.

Tyrosine kinase ephrin receptor EphA2 is known to be related to a high rate of angiogenesis and metastatic invasion when it is highly expressed in ovarian cancer [119, 121, 128–130].

Table 4 Preclinical studies on NP-based gene therapy.

Gene target tumor suppressors		NP composition	Target	References
Human adenovirus type 5 – E1A Trial drug:	Plasmid vector	Liposome	–	Phase I and II: NCT00102622, [106, 112]
"tgDCC-E1"	Plasmid vector	PEG-PEI-cholesterol Lipopolymer	–	Phase I: NCT01489371 Phase I: NCT00473954 Phase II: NCT01118052, [107]
HSulf-1	Plasmid vector	Heparin-polyethyleneimine nanogels	–	Xenograft, [108]

Table 4 Preclinical studies on NP-based gene therapy—cont'd

Gene target tumor suppressors		NP composition	Target	References
FILIP1L	Plasmid vector	Heparin-polyethyleneimine nanogels	–	Xenograft, [109]
Diphtheria toxin	Plasmid vector (ovarian-specific HE4 and MSLN promoters)	Poly(β-amino ester) polymer–DNA complex	–	Xenograft, [113]
HIF1α	siRNA	Liposome	–	Xenograft, [114]
PARP1	siRNA	Liposome	–	Xenograft, [115]
Src	siRNA	Chitosan	–	Xenograft, [116]
Claudin-3	Codelivery of both shRNA plasmid vectors	PLGA nanoparticles	–	Xenograft, [117]
CD44 and FAK	Codelivery of both shRNA plasmid vectors	PLGA nanoparticles	–	Xenograft, [118]
EphA2 Trial Drug: "siRNAEphA2-DOPC"	siRNA	Mesoporous silicon loaded with DOPC liposomes	–	Xenograft, [119, 120] Phase I: NCT01591356
EphA2 miR-520d-3p (targets EphB2)	siRNA with codelivery of miRNA	Mesoporous silicon loaded with DOPC liposomes	–	Xenograft, [121]
Jagged1	siRNA		–	Xenograft, [122]
EGFR	siRNA		EphA2	Cell lines, [123]
STAT3 and FAK	siRNA	HDL nanoparticles	SR–B1	Xenograft, [124]
ID4	siRNA	Peptide nanocomplex	Neuropilin-1	Xenograft, [108]
PLXDC1	siRNA	Chitosan	Integrin	Xenograft, [125]
–	Nontargeting siRNA	Polyethylenimine nanoparticles	TLR5	Xenograft, [126, 127]

When EphA2 expression was downregulated via NP-mediated siRNA delivery, there was tumor reduction observed in ovarian cancer mice xenografts [119–121]. EphA2 siRNA containing DOPC liposomes loaded on silicon particles [131] enables sustained and slow release of liposomes, thereby leading to EphA2 knockdown. This knockdown further increased docetaxel sensitivity in HeyA8 ovarian tumors that are resistant to chemotherapeutics. It also inhibited growth of tumors in docetaxel- and siRNA-treated animals [119].

A group working on cancer reported that codelivery of microRNA (miR-520d-3p) and EphA2 siRNA loaded in DOPC liposomes/silicon NPs showed improved antitumor activity in xenografts. miR 520d-3p expression generally targets EphB2, another ephrin receptor, and is linked with positive outcome in ovarian cancer patients [121].

A hydrogel NP developed by Dickerson et al. to improve ovarian cancer loads EGFR siRNA. It is functionalized with a peptide that binds to EphA2 receptor. These particles in the hydrogel reduced the expression of EGFR and enhanced doxetacel sensitivity in EpHA2-positive cells [123]. Though EGFR targeting via small molecules was not beneficial in clinical trials, NPs functionalized with EphA2-specific peptides may improve NP localization in ovarian cancer cells.

SR–B1 expresses highly on tumor cells and binds to a lipoprotein for maintaining fast growth. When siRNA-encapsulated liposome NPs against STAT3 and focal adhesion kinase were tested in mouse models with colon and ovarian cancer, there was enhancement in tumor targeting and protein silencing [124].

ID4 was identified as a transcriptional regulator necessary for the proliferation of ovarian cancer [108]. In order to reduce ID4 expression, siRNA was bound electrostatically to it via tandem peptide sequence. The peptide consists of an arginine domain that binds siRNA and a tumor-penetrating domain, cyclic nanopeptide LyP-1. This technique is known to enhance tumor targeting and vascular penetration [132]. When an ID4–siRNA nanocomplex was injected into a xenograft, it decreased the subcutaneous tumor. Similarly, IP treatment could terminate growth of the tumor in orthotopic ovarian cancer xenograft [108].

In the same way, chitosan NPs with arginyl glycyl aspartic acid peptide linkages have been developed in order to target integrin receptors for specific delivery of siRNA against many genes that are responsible for ovarian cancer cell proliferation. When tested in a mouse xenograft model, it showed reduced tumor growth [125].

NP gene therapy was also used for influencing tumor microenvironment. Notch ligand Jagged1 is highly expressed in some ovarian cancer cells and the microenvironment. siRNA against this ligand was delivered using chitosan. Investigations revealed that administered chitosan NPs could downregulate Jagged-1 expression derived from either a human tumor cell or murine stromal cell. It also effectively reduced microenvironment-derived angiogenesis and proliferation of the xenograft [122].

The dendritic cells showed preferential uptake of polyethylenimine-based (PEI-based) NPs in the microenvironment of the tumor. These particles could elicit immune response through toll-like receptor 5 when loaded with nontargeting siRNA [126, 127]. Most strategies for NP gene or RNAi delivery are still under phase I or II clinical trials. However, their use as novel strategies in the near future is highly expected.

7. Conclusion

To summarize, NPs are gaining grip in current therapeutics for ovarian cancer. Owing to their properties, NPs have an excellent ability to deliver therapeutic drugs specifically to targeted sites, in addition to having tremendous applications in tumor detection and imaging. However, further developments in targeted and nontargeted NP systems are anticipated to enter clinical trials shortly along with NP-based novel strategies that announce their utility for ovarian cancer therapies in the near future.

Acknowledgments

The authors are grateful to the Department of Biotechnology, Institute of Science, Gandhi Institute of Technology and Management, Visakhapatnam, Andhra Pradesh, India.

Conflict of interest

The authors declared that there is no conflict of interest.

References

[1] Bray F, Ferlay J, Soerjomataram I, Siegel RL, Torre LA, Jemal A. Global Cancer Statistics 2018: GLO-BOCAN estimates of incidence and mortality worldwide for 36 cancers in 185 countries. CA Cancer J Clin 2018;68:394–424.

[2] Vaughan S, Coward JI, Bast Jr. RC, Berchuck A, Berek JS, Brenton JD, Coukos G, Crum CC, Drapkin R, Etemadmoghadam D, Friedlander M, Gabra H, Kaye SB, Lord CJ, Lengyel E, Levine DA, McNeish IA, Menon U, Mills GB, Nephew KP, Oza AM, Sood AK, Stonach EA, Walczak H, Bowtell DD, Balkwill FR. Rethinking ovarian cancer: recommendations for improving outcomes. Nat Rev Cancer 2011;11:719–25.

[3] Banerjee S, Kaye SB. New strategies in the treatment of ovarian cancer: current clinical perspectives and future potential. Clin Cancer Res 2013;19:961–8.

[4] Prat J. Ovarian carcinomas: five distinct diseases with different origins, genetic alterations, and clinicopathological features. Virchows Arch 2012;460:237–49.

[5] Berns EM, Bowtell DD. The changing view of high-grade serous ovarian cancer. Cancer Res 2012;72:2701–4.

[6] Kurman RJ, Shih IM. Molecular pathogenesis and extraovarian origin of epithelial ovarian cancer—shifting the paradigm. Hum Pathol 2011;42:918–31.

[7] Landen Jr. CN, Birrer MJ, Sood AK. Early events in the pathogenesis of epithelial ovarian cancer. J Clin Oncol 2008;26:995–1005.

[8] Cancer Genome Atlas Research Network. Integrated genomic analyses of ovarian carcinoma. Nature 2011;474:609–15.

[9] Jelovac D, Armstrong DK. Recent progress in the diagnosis and treatment of ovarian cancer. CA Cancer J Clin 2011;61:183–203.

[10] Piek JM, Verheijen RH, Kenemans P, Massuger LF, Bulten H, van Diest PJ. BRCA1/2-related ovarian cancers are of tubal origin: a hypothesis. Gynecol Oncol 2003;90:491.

[11] Tan DS, Agarwal R, Kaye SB. Mechanisms of transcoelomic metastasis in ovarian cancer. Lancet Oncol 2006;7:925–34.

[12] Davidson B. Ovarian carcinoma and serous effusions. Changing views regarding tumor progression and review of current literature. Anal Cell Pathol 2001;23:107–28.

[13] Husseinzadeh N. Status of tumor markers in epithelial ovarian cancer has there been any progress? A review. Gynecol Oncol 2011;120:152–7.

[14] Bast Jr. RC, Feeney M, Lazarus H, Nadler LM, Colvin RB, Knapp RC. Reactivity of a monoclonal antibody with human ovarian carcinoma. J Clin Invest 1981;68:1331–7.

[15] Qian Y, Wang Y, Zhang X, Zhou L, Zhang Z, Xu J, Ruan Y, Ren S, Xu C, Gu J. Quantitative analysis of serum IgG galactosylation assists differential diagnosis of ovarian cancer. J Proteome Res 2013;12:4046–55.

[16] Biskup K, Braicu EI, Sehouli J, Fotopoulou C, Tauber R, Berger M, Blanchard V. Serum glycome profiling: a biomarker for diagnosis of ovarian cancer. J Proteome Res 2013;12:4056–63.

[17] Li B, An HJ, Kirmiz C, Lebrilla CB, Lam KS, Miyamoto S. Glycoproteomic analyses of ovarian cancer cell lines and sera from ovarian cancer patients show distinct glycosylation changes in individual proteins. J Proteome Res 2008;7:3776–88.

[18] Saldova R, Royle L, Radcliffe CM, Abd Hamid UM, Evans R, Arnold JN, Banks RE, Hutson R, Harvey DJ, Antrobus R, Petrescu SM, Dwek RA, Rudd PM. Ovarian cancer is associated with changes in glycosylation in both acute-phase proteins and IgG. Glycobiology 2007;17:1344–56.

[19] Moore RG, McMeekin DS, Brown AK, DiSilvestro P, Miller MC, Allard WJ, Gajewski W, Kurman R, Bast Jr. RC, Skates SJ. A novel multiple marker bioassay utilizing HE4 and CA125 for the prediction of ovarian cancer in patients with a pelvic mass. Gynecol Oncol 2009;112:40–6.

[20] Moore LE, Pfeiffer RM, Zhang Z, Lu KH, Fung ET, Bast Jr. RC. Proteomic biomarkers in combination with CA 125 for detection of epithelial ovarian cancer using prediagnostic serum samples from the Prostate, Lung, Colorectal, and Ovarian (PLCO) Cancer Screening Trial. Cancer 2012;118:91–100.

[21] Zhang Z, Chan DW. The road from discovery to clinical diagnostics: lessons learned from the first FDA-cleared in vitro diagnostic multivariate index assay of proteomic biomarkers. Cancer Epidemiol Biomarkers Prev 2010;19:2995–9.

[22] Armstrong DK, Bundy B, Wenzel L, Huang HQ, Baergen R, Lele S, Copeland LJ, Walker JL, Burger RA, Gyencologic Oncology Group. Intraperitoneal cisplatin and paclitaxel in ovarian cancer. N Engl J Med 2006;354:34–43.

[23] Jaaback K, Johnson N, Lawrie TA. Intraperitoneal chemotherapy for the initial management of primary epithelial ovarian cancer. Cochrane Database Syst Rev 2011; CD005340.

[24] Nolen BM, Lokshin AE. Biomarker testing for ovarian cancer: clinical utility of multiplex assays. Mol Diagn Ther 2013;17:139–46.

[25] Sarojini S, Tamir A, Lim H, Li S, Zhang S, Goy A, Pecora A, Suh KS. Early detection biomarkers for ovarian cancer. J Oncol 2012;2012:709049.

[26] Swierczewska M, Liu G, Lee S, Chen X. Highsensitivity nanosensors for biomarker detection. Chem Soc Rev 2012;41:2641–55.

[27] Perfezou M, Turner A, Merkoci A. Cancer detection using nanoparticle-based sensors. Chem Soc Rev 2012;41:2606–22.

[28] Bellan LM, Wu D, Langer RS. Current trends in nanobiosensor technology. Rev Nanomed Nanobiotechnol 2011;3:229–46.

[29] Kimmel DW, LeBlanc G, Meschievitz ME, Cliffel DE. Electrochemical sensors and biosensors. Anal Chem 2011;84:685–707.

[30] Jokerst JV, Raamanathan A, Christodoulides N, Floriano PN, Pollard AA, Simmons GW, Wong J, Gage C, Furmaga WB, Redding SW, McDevitt JT. Nano-bio-chips for high performance multiplexed protein detection: determinations of cancer biomarkers in serum and saliva using quantum dot bioconjugate labels. Biosens Bioelectron 2009;24:3622–9.

[31] Raamanathan A, Simmons GW, Christodoulides N, Floriano PN, Furmaga WB, Redding SW, Lu KH, Bast Jr. RC, McDevitt JT. Programmable bio-nano-chip systems for serum CA125 quantification: toward ovarian cancer diagnostics at the point-of-care. Cancer Prev Res (Phila) 2012;5:706–16.

[32] Ravalli A, dos Santos GP, Ferroni M, Faglia G, Yamanaka H, Marrazza G. New label free CA125 detection based on gold nanostructured screen-printed electrode. Sens Actuators B 2013;179: 194–200.

[33] Li J, Xu Q, Fu C, Zhang Y. A dramatically enhanced electrochemiluminescence assay for CA125 based on dendrimer multiply labeled luminol on Fe_3O_4 nanoparticles. Sens Actuators B 2013;185: 146–53.

[34] Li Z, Wang C, Cheng L, Gong H, Yin S, Gong Q, Li Y, Liu Z. PEG-functionalized iron oxide nanoclusters loaded with chlorin e6 for targeted, NIR light induced, photodynamic therapy. Biomaterials 2013;34:9160–70.

[35] Fleischer AC, Lyshchik A, Andreotti RF, Hwang M, Jones HW, Fishman DA. Advances in sonographic detection of ovarian cancer: depiction of tumor neovascularity with microbubbles. Am J Roentgenol 2010;194:343–8.

[36] Fleischer AC, Lyshchik A, Jones Jr. HW, Crispens M, Loveless M, Andreotti RF, Williams PK, Fishman DA. Contrast-enhanced transvaginal sonography of benign versus malignant ovarian masses: preliminary findings. J Ultrasound Med 2008;27:1011–8.

[37] Fleischer AC, Lyshchik A, Hirari M, Moore RD, Abramson RG, Fishman DA. Early detection of ovarian cancer with conventional and contrast-enhanced transvaginal sonography: recent advances and potential improvements. J Oncol 2012;2012:11.

[38] Xu H, Regino CAS, Koyama Y, Hama Y, Gunn AJ, Bernardo M, Kobayashi H, Choyke PL, Brechbiel MW. Preparation and preliminary evaluation of a biotin-targeted, lectin-targeted dendrimer-based probe for dual-modality magnetic resonance and fluorescence imaging. Bioconjug Chem 2007;18:1474–82.

[39] Kamaly N, Kalber T, Thanou M, Bell JD, Miller AD. Folate receptor targeted bimodal liposomes for tumor magnetic resonance imaging. Bioconjug Chem 2009;20:648–55.

[40] Wang L, Neoh KG, Kang E-T, Shuter B. Multifunctional polyglycerol-grafted $Fe_3O_4@SiO_2$ nanoparticles for targeting ovarian cancer cells. Biomaterials 2011;32:2166–73.

[41] Wang C, Tao H, Cheng L, Liu Z. Near-infrared light induced in vivo photodynamic therapy of cancer based on upconversion nanoparticles. Biomaterials 2011;32:6145–54.

[42] Zhou Z, Wang L, Chi X, Bao J, Yang L, Zhao W, Chen Z, Wang X, Chen X, Gao J. Engineered iron-oxide-based nanoparticles as enhanced T1 contrast agents for efficient tumor imaging. ACS Nano 2013;7:3287–96.

[43] Misri R, Meier D, Yung AC, Kozlowski P, Häfeli UO. Development and evaluation of a dual-modality (MRI/SPECT) molecular imaging bioprobe. Nanomedicine 2012;8:1007–16.

[44] Schädlich A, Caysa H, Mueller T, Tenambergen F, Rose C, Göpferich A, Kuntsche J, Mäder K. Tumor accumulation of NIR fluorescent PEG–PLA nanoparticles: impact of particle size and human xenograft tumor model. ACS Nano 2011;5:8710–20.

[45] Katragadda U, Fan W, Wang Y, Teng Q, Tan C. Combined delivery of paclitaxel and tanespimycin via micellar nanocarriers: pharmacokinetics, efficacy and metabolomic analysis. PLoS One 2013;8: e58619.

[46] Zdobnova TA, Dorofeev SG, Tananaev PN, Zlomanov VP, Stremovskiy OA, Lebedenko EN, Balalaeva IV, Deyev SM, Petrov RV. Imaging of human ovarian cancer SKOV-3 cells by quantum dot bioconjugates. Dokl Biochem Biophys 2010;430:41–4.

[47] Nathwani BB, Jaffari M, Juriani AR, Mathur AB, Meissner KE. Fabrication and characterization of silk-fibroin-coated quantum dots. IEEE Trans Nanobioscience 2009;8:72–7.

[48] Wang H-Z, Wang H-Y, Liang R-Q, Ruan K-C. Detection of tumor marker CA125 in ovarian carcinoma using quantum dots. Acta Biochim Biophys Sin 2004;36:681–6.

[49] Savla R, Taratula O, Garbuzenko O, Minko T. Tumor targeted quantum dot-mucin 1 aptamer-doxorubicin conjugate for imaging and treatment of cancer. J Control Release 2011;153:16–22.

[50] Jokerst JV, Cole AJ, Van de Sompel D, Gambhir SS. Gold nanorods for ovarian cancer detection with photoacoustic imaging and resection guidance via raman imaging in living mice. ACS Nano 2012;6:10366–77.

[51] Gabizon A, Catane R, Uziely B, Kaufman B, Safra T, Cohen R, Martin F, Huang A, Barenholz Y. Prolonged circulation time and enhanced accumulation in malignant exudates of doxorubicin encapsulated in polyethylene-glycol coated liposomes. Cancer Res 1994;54:987–92.

[52] Muggia FM, Hainsworth JD, Jeffers S, Miller P, Groshen S, Tan M, Roman L, Uziely B, Muderspach L, Garcia A, Burnett A, Greco FA, Morrow CP, Paradiso LJ, Liang LJ. Phase II study of liposomal doxorubicin in refractory ovarian cancer: antitumor activity and toxicity modification by liposomal encapsulation. J Clin Oncol 1997;15:987–93.

[53] Ferrandina G, Ludovisi M, Lorusso D, Pignata S, Breda E, Savarese A, Del Medico P, Scaltriti L, Katsaros D, Priolo D, Scambia G. Phase III trial of gemcitabine compared with pegylated liposomal doxorubicin in progressive or recurrent ovarian cancer. J Clin Oncol 2008;26:890–6.

[54] Mutch DG, Orlando M, Goss T, Teneriello MG, Gordon AN, McMeekin SD, Wang Y, Scribner Jr. DR, Marciniack M, Naumann RW, Secord AA. Randomized phase III trial of gemcitabine compared with pegylated liposomal doxorubicin in patients with platinum-resistant ovarian cancer. J Clin Oncol 2007;25:2811–8.

[55] Gordon AN, Tonda M, Sun S, Rackoff W. Long-term survival advantage for women treated with pegylated liposomal doxorubicin compared with topotecan in a phase 3 randomized study of recurrent and refractory epithelial ovarian cancer. Gynecol Oncol 2004;95:1–8.

[56] Gordon AN, Fleagle JT, Guthrie D, Parkin DE, Gore ME, Lacave AJ. Recurrent epithelial ovarian carcinoma: a randomized phase III study of pegylated liposomal doxorubicin versus topotecan. J Clin Oncol 2001;19:3312–22.

[57] Wagner U, Marth C, Largillier R, Kaern J, Brown C, Heywood M, Bonaventura T, Vergote I, Piccirillo MC, Fossati R, Gebski V, Lauraine EP. Final overall survival results of phase III GCIG CALYPSO trial of pegylated liposomal doxorubicin and carboplatin vs paclitaxel and carboplatin in platinum-sensitive ovarian cancer patients. Br J Cancer 2012;107:588–91.

[58] Poveda A, Vergote I, Tjulandin S, Kong B, Roy M, Chan S, Filipczyk-Cisarz E, Hagberg H, Kaye SB, Colombo N, Lebedinsky C, Parekh T, Gómez J, Park YC, Alfaro V, Monk BJ. Trabectedin plus pegylated liposomal doxorubicin in relapsed ovarian cancer: outcomes in the partially platinumsensitive (platinum-free interval 6-12 months) subpopulation of OVA-301 phase III randomized trial. Ann Oncol 2011;22:39–48.

[59] Monk BJ, Herzog TJ, Kaye SB, Krasner CN, Vermorken JB, Muggia FM, PujadeLauranine E, Lisyanskaya AS, Makhson AN, Rolski J, Gorbounova VA, Ghatage P, Bidzinski M, Shen K, Ngan HY, Vergote IB, Nam JH, Park YC, Lebedinsky CA, Poveda AM. Trabectedin plus pegylated liposomal doxorubicin in recurrent ovarian cancer. J Clin Oncol 2010;28:3107–14.

[60] Herzog TJ. Update on the role of topotecan in the treatment of recurrent ovarian cancer. Oncologist 2002;7(Suppl 5):3–10.

[61] Dadashzadeh S, Vali AM, Rezaie M. The effect of PEG coating on in vitro cytotoxicity and in vivo disposition of topotecan loaded liposomes in rats. Int J Pharm 2008;353:251–9.

[62] Ma Y, Yang Q, Wang L, Zhou X, Zhao Y, Deng Y. Repeated injections of PEGylated liposomal topotecan induces accelerated blood clearance phenomenon in rats. Eur J Pharm Sci 2012; 45:539–45.

[63] Dark GG, Calvert AH, Grimshaw R, Poole C, Swenerton K, Kaye S, Coleman R, Jayson G, Le T, Ellard S, Trudeau M, Vasey P, Hamilton M, Cameron T, Barrett E, Walsh W, Mcintosh L, Eisenhauer EA. Randomized trial of two intravenous schedules of the topoisomerase I inhibitor liposomal lurtotecan in women with relapsed epithelial ovarian cancer: a trial of the National Cancer Institute of Canada Clinical Trials Group. J Clin Oncol 2005;23:1859–66.

[64] Ma P, Russell J. Mumper, paclitaxel nano-delivery systems: a comprehensive review. J Nanomed Nanotechnol 2013;18:1000164–99.

[65] Mirtsching B, Cosgriff T, Harker G, Keaton M, Chidiac T, Min M. A phase II study of weekly nanoparticle albumin-bound paclitaxel with or without trastuzumab in metastatic breast cancer. Clin Breast Cancer 2011;11:121–8.

[66] Micha JP, Goldstein BH, Birk CL, Rettenmaier MA, Brown 3rd. JV. Abraxane in the treatment of ovarian cancer: the absence of hypersensitivity reactions. Gynecol Oncol 2006;100:437–8.

[67] Coleman RL, Brady WE, McMeekin DS, Rose PG, Soper JT, Lentz SS, Hoffman JS, Shahin MS. A phase II evaluation of nanoparticle, albumin-bound (nab) paclitaxel in the treatment of recurrent or persistent platinum-resistant ovarian, fallopian tube, or primary peritoneal cancer: a Gynecologic Oncology Group study. Gynecol Oncol 2011;122:111–5.

[68] Teneriello MG, Tseng PC, Crozier M, Encarnacion C, Hancock K, Messing MJ, Boehm KA, Williams A, Asmar L. Phase II evaluation of nanoparticle albumin-bound paclitaxel in platinum-sensitive patients with recurrent ovarian, peritoneal, or fallopian tube cancer. J Clin Oncol 2009;27:1426–31.

[69] Singer JW. Paclitaxel poliglumex (XYOTAX, CT-2103): a macromolecular taxane. J Control Release 2005;109:120–6.

[70] Sabbatini P, Sill MW, O'Malley D, Adler L, Secord AA, Gynecologic Oncology Group S. A phase II trial of paclitaxel poliglumex in recurrent or persistent ovarian or primary peritoneal cancer (EOC): a Gynecologic Oncology Group study. Gynecol Oncol 2008;111:455–60.

[71] Awada A, Garcia AA, Chan S, Jerusalem GH, Coleman RE, Huizing MT, Mehdi A, O'Reilly SM, Hamm JT, Barrett-Lee PJ, Cocquyt V, Sideras K, Young DE, Zhao C, Chia YL, Hoch U, Hannah AL, Perez EA; NKTR-102 Study Group. Two schedules of etirinotecan pegol (NKTR-102) in patients with previously treated metastatic breast cancer: a randomised phase 2 study. Lancet Oncol 2013;14:1216–25.

[72] Iqbal S, Tsao-Wei DD, Quinn DI, Gitlitz BJ, Groshen S, Aparicio A, Lenz HJ, El-Khoueiry A, Pinski J, Garcia AA. Phase I clinical trial of pegylated liposomal doxorubicin and docetaxel in patients with advanced solid tumors. Am J Clin Oncol 2011;34:27–31.

[73] Svenson S, Wolfgang M, Hwang J, Ryan J, Eliasof S. Preclinical to clinical development of the novel camptothecin nanopharmaceutical CRLX101. J Control Release 2011;153:49–55.

[74] Nowotnik DP, Cvitkovic E. ProLindac (AP5346): a review of the development of an HPMA DACH platinum polymer therapeutic. Adv Drug Deliv Rev 2009;61:1214–9.

[75] Duncan R. Development of HPMA copolymer-anticancer conjugates: clinical experience and lessons learnt. Adv Drug Deliv Rev 2009;61:1131–48.

[76] Lin X, Zhang Q, Rice JR, Stewart DR, Nowotnik DP, Howell SB. Improved targeting of platinum chemotherapeutics. The antitumour activity of the HPMA copolymer platinum agent AP5280 in murine tumour models. Eur J Cancer 2004;40:291–7.

[77] Desale SS, Cohen SM, Zhao Y, Kabanov AV, Bronich TK. Biodegradable hybrid polymer micelles for combination drug therapy in ovarian cancer. J Control Release 2013;171:339–48.

[78] Gong C, Yang B, Qian Z, Zhao X, Wu Q, Qi X, Wang Y, Guo G, Kan B, Luo F, Wei Y. Improving intraperitoneal chemotherapeutic effect and preventing postsurgical adhesions simultaneously with biodegradable micelles. Nanomedicine 2012;8:963–73.

[79] Cho H, Lai TC, Kwon GS. Poly(ethylene glycol)-block-poly(epsilon-caprolactone) micelles for combination drug delivery: evaluation of paclitaxel, cyclopamine and gossypol in intraperitoneal xenograft models of ovarian cancer. J Control Release 2013;166:1–9.

[80] Kirkpatrick GJ, Plumb JA, Sutcliffe OB, Flint DJ, Wheate NJ. Evaluation of anionic half generation 3.5-6.5 poly(amidoamine) dendrimers as delivery vehicles for the active component of the anticancer drug cisplatin. J Inorg Biochem 2011;105:1115–22.

[81] Zhu S, Hong M, Tang G, Qian L, Lin J, Jiang Y, Pei Y. Partly PEGylated polyamidoamine dendrimer for tumor-selective targeting of doxorubicin: the effects of PEGylation degree and drug conjugation style. Biomaterials 2010;31:1360–71.

[82] Zhu S, Hong M, Zhang L, Tang G, Jiang Y, Pei Y. PEGylated PAMAM dendrimerdoxorubicin conjugates: in vitro evaluation and in vivo tumor accumulation. Pharm Res 2010;27:161–74.

[83] Kurtoglu YE, Mishra MK, Kannan S, Kannan RM. Drug release characteristics of PAMAM dendrimer-drug conjugates with different linkers. Int J Pharm 2010;384:189–94.

[84] Xiao K, Luo J, Fowler WL, Li Y, Lee JS, Xing L, Cheng RH, Wang L, Lam KS. A self-assembling nanoparticle for paclitaxel delivery in ovarian cancer. Biomaterials 2009;30:6006–16.

[85] Xiao W, Luo J, Jain T, Riggs JW, Tseng HP, Henderson PT, Cherry SR, Rowland D, Lam KS. Biodistribution and pharmacokinetics of a telodendrimer micellar paclitaxel nanoformulation in a mouse xenograft model of ovarian cancer. Int J Nanomedicine 2012;7:1587–97.

[86] Xiao K, Li Y, Lee JS, Gonik AM, Dong T, Fung G, Sanchez E, Xing L, Cheng HR, Luo J, Lam KS. "OA02" peptide facilitates the precise targeting of paclitaxel-loaded micellar nanoparticles to ovarian cancer in vivo. Cancer Res 2012;72:2100–10.

[87] Van Vlerken LE, Duan Z, Little SR, Seiden MV, Amiji MM. Augmentation of therapeutic efficacy in drug-resistant tumor models using ceramide coadministration in temporalcontrolled polymer-blend nanoparticle delivery systems. AAPS J 2010;12:171–80.

[88] Wagstaff AJ, Brown SD, Holden MR, Craig GE, Plumb JA, Brown RE, Schreiter N, Chrzanowski W, Wheate NJ. Cisplatin drug delivery using gold-coated iron oxide nanoparticles for enhanced tumour targeting with external magnetic fields. Inorg Chim Acta 2012;393:328–33.

[89] Qin Y, Song Q-G, Zhang Z-R, Liu J, Fu Y, He Q, Liu J. Ovarian tumor targeting of docetaxel-loaded liposomes mediated by luteinizing hormone-releasing hormone analogues. Arzneimittelforschung 2008;58:529–34.

[90] Nukolova NV, Oberoi HS, Cohen SM, Kabanov AV, Bronich TK. Folate-decorated nanogels for targeted therapy of ovarian cancer. Biomaterials 2011;32:5417–26.

[91] Nukolova NV, Oberoi HS, Zhao Y, Chekhonin VP, Kabanov AV, Bronich TK. LHRH-targeted nanogels as a delivery system for cisplatin to ovarian cancer. Mol Pharm 2013;10:3913–21.

[92] Gao ZG, Lee DH, Kim DI, Bae YH. Doxorubicin loaded pH-sensitive micelle targeting acidic extracellular pH of human ovarian A2780 tumor in mice. J Drug Target 2005;13:391–7.

[93] Kim D, Lee ES, Park K, Kwon IC, Bae YH. Doxorubicin loaded pH-sensitive micelle: antitumoral efficacy against ovarian A2780/DOXR tumor. Pharm Res 2008;25:2074–82.

[94] Kim D, Gao ZG, Lee ES, Bae YH. In vivo evaluation of doxorubicin-loaded polymeric micelles targeting folate receptors and early endosomal pH in drug-resistant ovarian cancer. Mol Pharm 2009;6:1353–62.

[95] Zeisser-Labouebe M, Lange N, Gurny R, Delie F. Hypericin-loaded nanoparticles for the photodynamic treatment of ovarian cancer. Int J Pharm 2006;326:174–81.

[96] Zeisser-Labouèbe M, Delie F, Gurny R, Lange N. Benefits of nanoencapsulation for the hypercin-mediated photodetection of ovarian micrometastases. Eur J Pharm Biopharma 2009;71:207–13.

[97] Cohen BA, Bergkvist M. Targeted in vitro photodynamic therapy via aptamer-labeled, porphyrin-loaded virus capsids. J Photochem Photobiol B 2013;121:67–74.

[98] Mroz P, Xia Y, Asanuma D, Konopko A, Zhiyentayev T, Huang YY, Sharma SK, Dai T, Khan UJ, Wharton T, Hamblin MR. Intraperitoneal photodynamic therapy mediated by a fullerene in a mouse model of abdominal dissemination of colon adenocarcinoma. Nanomedicine 2011;7:965–74.

[99] Liu J, Ohta S, Sonoda A, Yamada M, Yamamoto M, Nitta N, Murata K, Tabata Y. Preparation of PEG-conjugated fullerene containing Gd3+ ions for photodynamic therapy. J Control Release 2007;117:104–10.

[100] Tabata Y, Murakami Y, Ikada Y. Photodynamic effect of polyethylene glycol-modified fullerene on tumor. Jpn J Cancer Res 1997;88:1108–16.

[101] Heidel JD, Schluep T. Cyclodextrin-containing polymers: versatile platforms of drug delivery materials. J Drug Deliv 2012;2012:262731.

[102] Kim DH, Rossi JJ. Strategies for silencing human disease using RNA interference. Nat Rev Genet 2007;8:173–84.

[103] Mansouri S, Cuie Y, Winnik F, Shi Q, Lavigne P, Benderdour M, Beaumont E, Fernandes JC. Characterization of folate-chitosan-DNA nanoparticles for gene therapy. Biomaterials 2006;27:2060–5.

[104] Senzer N, Nemunaitis J, Nemunaitis D, Bedell C, Edelman G, Barve M, Nunan R, Pirollo KF, Rait A, Chang EH. Phase I study of a systemically delivered p53 nanoparticle in advanced solid tumors. Mol Ther 2013;21:1096–103.

[105] Zeimet AG, Marth C. Why did p53 gene therapy fail in ovarian cancer? Lancet Oncol 2003;4:415–22.

[106] Madhusudan S, Tamir A, Bates N, Flanagan E, Gore ME, Barton DP, Harper P, Secki M, Thomas H, Lemoine NR, Charnock M, Habib NA, Lechler R, Nicholls J, Pignatelli M, Ganesan TS. A multicenter phase I gene therapy clinical trial involving intraperitoneal administration of E1A-lipid complex in patients with recurrent epithelial ovarian cancer overexpressing HER-2/neu oncogene. Clin Cancer Res 2004;10:2986–96.

[107] Anwer K, Kelly FJ, Chu C, Fewell JG, Lewis D, Alvarez RD. Phase I trial of a formulated IL-12 plasmid in combination with carboplatin and docetaxel chemotherapy in the treatment of platinum-sensitive recurrent ovarian cancer. Gynecol Oncol 2013;131:169–73.

[108] Ren Y, Cheung HW, von Maltzhan G, Agrawal A, Cowley GS, Weir BA, Boehm JS, Tamayo P, Karst AM, Liu JF, Hirsch MS, Mesirov JP, Drapkin R, Root DE, Lo J, Fogal V, Ruoslahti E, Hahn WC, Bhatia SN. Targeted tumor-penetrating siRNA nanocomplexes for credentialing the ovarian cancer oncogene ID4. Sci Transl Med 2012;4: 147ra12.

[109] Xie C, Gou ML, Yi T, Deng H, Li ZY, Liu P, Qi XR, He X, Wei Y, Zhao X. Efficient inhibition of ovarian cancer by truncation mutant of FILIP1L gene delivered by novel biodegradable cationic heparin-polyethyleneimine nanogels. Human Gene Ther 2011;22:1413–22.

[110] Davis ME, Zuckerman JE, Choi CH, Seligson D, Tolcher A, Alabi CA, Yen Y, Heidel JD, Ribas A. Evidence of RNAi in humans from systemically administered siRNA via targeted nanoparticles. Nature 2010;464:1067–70.

[111] Tabernero J, Shapiro GI, LoRusso PM, Cervantes A, Schwartz GK, Weiss GJ, PazAres L, Cho DC, Infante JR, Alsina M, Gounder MM, Falzone R, Harrop J, White AC, Toudjarska I, Bumcrot D, Meyers RE, Hinkle G, Svrzikapa N, Hutabarat RM, Clausen VA, Cehelsky J, Nochur SV, Gamba-Vitalo C, Vaishnaw AK, Sah DW, Gollob JA, Burris 3rd HA. First-in-humans trial of an RNA interference therapeutic targeting VEGF and KSP in cancer patients with liver involvement. Cancer Discov 2013;3:406–17.

[112] Xing X, Zhang S, Chang J, Tucker SD, Chen H, Huang L, Hung MC. Safety study and characterization of E1A–liposome complex gene-delivery protocol in an ovarian cancer model. Gene Ther 1998;5:1538–44.

[113] Huang YH, Zugates GT, Peng W, Holtz D, Dunton C, Green JJ, Hossain N, Chernick MR, Padera Jr RF, Langer R, Anderson DG, Sawicki JA. Nanoparticle-delivered suicide gene therapy effectively reduces ovarian tumor burden in mice. Cancer Res 2009;69:6184–91.

[114] Wang Y, Liu Y, Malek SN, Zheng P, Liu Y. Targeting HIF1α eliminates cancer stem cells in hematological malignancies. Cell Stem Cell 2011;8:399–411.

[115] Goldberg MS, Xing D, Ren Y, Orsulic S, Bhatia SN, Sharp PA. Nanoparticle-mediated delivery of siRNA targeting Parp1 extends survival of mice bearing tumors derived from Brca1-deficient ovarian cancer cells. Proc Natl Acad Sci U S A 2011;108:745–50.

[116] Kim HS, Han HD, Armaiz-Pena GN, Stone RL, Nam EJ, Lee JW, Shahzad MM, Nick AM, Lee SJ, Roh JW, Nishimura M, Mangala LS, Bottsford-Miller J, Gallick GE, LopezBerestein G, Sood AK. Functional roles of Src and Fgr in ovarian carcinoma. Clin Cancer Res 2011;17:1713–21.

[117] Sun C, Yi T, Song X, Li S, Qi X, Chen X, Lin H, He X, Li Z, Wei Y, Zhao X. Efficient inhibition of ovarian cancer by short hairpin RNA targeting claudin-3. Oncol Rep 2011;26:193–200.

[118] Zou L, Song X, Yi T, Li S, Deng H, Chen X, Li Z, Bai Y, Zhong Q, Wei Y, Zhao X. Administration of PLGA nanoparticles carrying shRNA against focal adhesion kinase and CD44 results in enhanced antitumor effects against ovarian cancer. Cancer Gene Ther 2013;20:242–50.

[119] Shen H, Rodriguez-Aguayo C, Xu R, Gonzalez Villasana V, Mai J, Huang Y, Zhang G, Guo X, Bai L, Qin G, Deng X, Li Q, Erm DR, Aslan B, Liu X, Sakamoto J, Chavez-Reyes A, Han HD, Sood AK, Ferrari M, Lopez-Berestein G. Enhancing chemotherapy response with sustained EphA2 silencing using multistage vector delivery. Clin Cancer Res 2013;19:1806–15.

[120] Hasan N, Mann A, Ferrari M, Tanaka T. Mesoporous silicon particles for sustained gene silencing. Methods Mol Biol 2013;1049:481–93.

[121] Nishimura M, Jung EJ, Shah MY, Lu C, Spizzo R, Shimizu M, Han HD, Ivan C, Rossi S, Zhang X, Nicoloso MS, Wu SY, Almeida MI, Bottsford-Miller J, Pecot CV, Zand B, Matsuo K, Shahzad MM, Jennings NB, Rodriguez Aguayo C, Lopez-Berestein G, Sood AK, Calin GA. Therapeutic synergy between microRNA and siRNA in ovarian cancer treatment. Cancer Discov 2013;3:1302–15.

[122] Steg AD, Katre AA, Goodman B, Han HD, Nick AM, Stone RL, Coleman RL, Alvarez RD, Lopez-Berestein G, Sood AK, Landen CN. Targeting the notch ligand JAGGED1 in both tumor cells and stroma in ovarian cancer. Clin Cancer Res 2011;17:5674–85.

[123] Dickerson EB, Blackburn WH, Smith MH, Kapa LB, Lyon LA, McDonald JF. Chemosensitization of cancer cells by siRNA using targeted nanogel delivery. BMC Cancer 2010;10:10.

[124] Shahzad MM, Mangala LS, Han HD, Lu C, Bottsford-Miller J, Nishimura M, Mora EM, Lee JW, Stone RL, Pecot CV, Thanapprapasr D, Roh JW, Gaur P, Nair MP, Park YY, Sabnis N, Deavers MT, Lee JS, Ellis LM, LopezBerestein G, McConathy WJ, Prokai L, Lacko AG, Sood AK. Targeted delivery of small interfering RNA using reconstituted highdensity lipoprotein nanoparticles. Neoplasia 2011;13:309–19.

[125] Han HD, Mangala LS, Lee JW, Shahzad MM, Kim HS, Shen D, Nam EJ, Mora EM, Stone RL, Lu C, Lee SJ, Roh JW, Nick AM, LopezBerestein G, Sood AK. Targeted gene silencing using RGD-labeled chitosan nanoparticles. Clin Cancer Res 2010;16:3910–22.

[126] Cubillos-Ruiz JR, Engle X, Scarlett UK, Martinez D, Barber A, Elgueta R, Wang L, Nesbeth Y, Durant Y, Gewirtz AT, Sentman CL, Kedl R, Conejo-Garcia JR. Polyethylenimine-based siRNA nanocomplexes reprogram tumor-associated dendritic cells via TLR5 to elicit therapeutic antitumor immunity. J Clin Invest 2009;119:2231–44.

[127] Cubillos-Ruiz JR, Baird JR, Tesone AJ, Rutkowski MR, Scarlett UK, Camposeco Jacobs AL, Anadon-Arnillas J, Harwood NM, Korc M, Fiering SN, Sempere LF, Conejo Garcia JR. Reprogramming tumor-associated dendritic cells in vivo using miRNA mimetics triggers protective immunity against ovarian cancer. Cancer Res 2012;72:1683–93.

[128] Pasquale EB. Eph receptor signalling casts a wide net on cell behaviour. Nat Rev Mol Cell Biol 2005;6:462–75.

[129] Thaker PH, Deavers M, Celestino J, Thornton A, Fletcher MS, Landen CN, Kinch MS, Kiener PA, Sood AK. EphA2 expression is associated with aggressive features in ovarian carcinoma. Clin Cancer Res 2004;10:5145–50.

[130] Lin YG, Han LY, Kamat AA, Merritt WM, Landen CN, Deavers MT, Fletcher MS, Urbauer DL, Kinch MS, Sood AK. EphA2 overexpression is associated with angiogenesis in ovarian cancer. Cancer 2007;109:332–40.

[131] Tanaka T, Mangala LS, Vivas-Mejia PE, Nieves-Alicea R, Mann AP, Mora E, Han HD, Shahzad MM, Liu X, Bhavane R, Gu J, Fakhoury JR, Chiappini C, Lu C, Matsuo K, Godin B, Stone RL, Nick AM, Lopez-Berestein G, Sood AK, Ferrari M. Sustained small interfering RNA delivery by mesoporous silicon particles. Cancer Res 2010;70:3687–96.

[132] Sugahara KN, Teesalu T, Karmali PP, Kotamraju VR, Agemy L, Greenwald DR, Ruoslahti E. Coadministration of a tumor-penetrating peptide enhances the efficacy of cancer drugs. Science 2010;328:1031–5.

CHAPTER 7

Phytotherapy for breast cancer

Phaniendra Alugoju[a], Nyshadham S.N. Chaitanya[b], V.K.D. Krishna Swamy[a], and Pavan Kumar Kancharla[c]

[a]Department of Biochemistry and Molecular Biology, School of Life Sciences, Pondicherry University, Puducherry, India
[b]Department of Animal Biology, School of Life Sciences, University of Hyderabad, Gachibowli, Hyderabad, India
[c]Department of Biotechnology, School of Life Sciences, Pondicherry University, Puducherry, India

Abstract

Breast cancer is one of the leading causes of death in humans worldwide and is the most common type of cancer in women. Several factors are responsible for the development of breast cancer in women. Disturbances in various molecular pathways and the concomitant altered expression of several molecular markers have been well reported. Despite the potent beneficial effects of various therapeutic strategies such as surgery, chemotherapy, and radiotherapy, the associated detrimental side effects of these strategies has spurred the scientific community to search for alternatives. Phytotherapy is one such potential therapeutic strategy involving the usage of traditional medicinal plants to target various molecular markers altered during cancer and thereby to protect from cancer without inducing side effects. In this chapter we explain the possible beneficial effects of a few traditional Indian medical plants against breast cancer.

Keywords: Breast cancer, Phytotherapy, Risk factors, Molecular mechanisms, Medicinal plants.

Abbreviations

ROS	reactive oxygen species
MAPK	mitogen–activated protein kinase
mTOR	mammalian target of rapamycin
PI3K	phosphoinositide 3-kinase
ER	estrogen receptor
PR	progesterone receptor
DMBA	Dimethylbenz(a)anthracene
MPTP	mitochondrial permeability transition pore
TNBC	triple negative breast cancer

1. Introduction

Cancer is characterized by uncontrolled division and abnormal growth of cells that can metastasize to other parts of the body by invading other tissues. Cancer develops progressively and it may be caused by internal as well as external factors. The type of cancer is named for the type of cell from which it forms. For example, carcinoma (from epithelial cells), sarcoma (formed in bone and soft tissues), leukemia (blood

G.P. Nagaraju, R.R. Malla (eds.)
A Theranostic and Precision Medicine Approach for Female-Specific Cancers
ISBN 978-0-12-822009-2
https://doi.org/10.1016/B978-0-12-822009-2.00007-8

tissue), lymphoma (formed in B or T lymphocytes), multiple Myeloma (plasma cells, myeloma cells), and melanoma (melanocytes). Cancer type is also based on the organ(s) or tissue(s) from which it originates. For example, breast, blood, brain, liver, lung, and skin cancer. A 2018 report on the burden of cancer worldwide reveals that there will be an estimated 18.1 million new cancer cases and 9.6 million cancer deaths globally. Among both the sexes, lung cancer is the most prevalent. The second most prevalent cancer is breast cancer, which is a common cause of death in women [1]. Breast cancer incidence is highly prevalent in more developed regions (North America, Australia, Europe, and Japan) compared to less developed regions (Central America, Africa, South America, Micronesia, and Polynesia) [2]. The incidence of breast cancer varies widely due to differences in race, health resources, and lifestyle patterns [2–6]. In this chapter, we discuss the prevalence of breast cancer, factors promoting breast cancer, breast cancer-associated molecular mechanisms, and phytotherapy as a means to treat breast cancer.

2. Breast cancer prevalence in India

Breast cancer is the second most common cancer among Indian women, and is diagnosed with a case fatality ratio of 40% [7, 8]. India has the highest incidence of cancer deaths worldwide [8, 9], with triple negative breast cancer (TNBC) being most lethal accounting for 20%–43% of deaths [7, 10–12]. The peak age for TNBC is 40–55 years in India, whereas it is 50–70 years in other countries. Reports reveal that TNBC is more prevalent in women ≤50 years in Bangalore [13], Hyderabad [14], Srinagar [15], and Pune [16]. Delhi is the exception in that TNBC is found more among postmenopausal than premenopausal women [17]. The prevalence of TNBC is highest in Mumbai at 32.1% [18], followed by Delhi at 24.2% [17], and Hyderabad at 22.8% [14]. Recently, it was found that TNBC was prevalent among 31.57% of younger women, which is a higher percentage than that found in older women, and highest among all Indian, Hispanic, Chinese, and non-Hispanic women [19]. The Luminal Androgen Receptor Subtype population was the smallest and the Human epidermal growth factor receptor 2 coded by proto-oncogene neu (HER2/neu) subtype was also observed to occur more frequently in younger Indian women [20].

3. Risk factors for the development of breast cancer

Cellular events in breast cancer explain epidemiology, but its specific cause is largely unknown. Even though there are defects in endogenous steroid sex hormones that act as promoters, the only known initiator is ionizing radiation (IR). Female sex, age, age at menopause, age at first child, nulliparity, lactation, and exogenous estrogens are some of the factors that implicate hormones as tumor promoters, whereas IR, family history of breast cancer, geography, nutritional factors, weight, height, body size, and benign breast disease are probable factors with unknown mechanisms of action [21]. Cancer causation is multifactorial and etiology includes (1) unmodifiable intrinsic risk factors that involve errors in DNA replication and (2) nonintrinsic risk factors. The latter group includes modifiable exogenous factors such as radiation, chemical carcinogens, oncovirus, smoking, lack of exercise, nutrient imbalance, as well as partially modifiable endogenous factors such as biological aging, DNA repair machinery, hormones, and inflammation [22].

4. Breast cancer-associated molecular mechanisms

In breast cancer the cells in the breast grow out of control, which is diagnosed using a mammogram in the early stages. In the later stages, the cells may invade axillary lymph nodes or distant sites in the body. Breast cancer can be detected by changes in the skin or nipple of the breast and is categorized into stages I–IV based on occurrence [23]. Stage IV is metastatic, which means cancer has spread from the breast to axillary lymph nodes [24]. Breast cancer cells are divided based on three different types of proteins: estrogen receptor (ER), progesterone (PgR), and HER2 (human epidermal growth factor receptor 2). The majority of cancer cells have ER or PgR proteins. Up to 15%–20% of breast cancers are ERBB2-positive (formerly known as HER2-positive). TNBC, in which none of these proteins are expressed, accounts for up to 15% of breast cancers [25]. The prognosis and treatment depend on stage and type of cancer. The goal of treatment for stages I–III is curation, and the goal for stage IV treatment is control [26]. Various types of medicine are used for various types of breast cancers, as they all respond differently. In the case of ER/PgR-positive breast cancer, antiestrogen pills are effective. In ERBB2-positive breast cancer, intravenous medicines are given. In TNBC, intravenous chemotherapy is used.

   Cancer complexity is multifactorial and multidimensional. The disruption of redox balance is one major cause of cancer and its progression and metastasis [27]. ROS production is initiated through intracellular ROS and environmental factors lead to oxidative stress. This in turn leads to DNA/protein/lipid degradation [28] that results in autophagy, apoptosis, and inflammation. High levels of ROS inside the cell leads to metastasis through the stimulation of PI3K/Akt, mammalian target of the rapamycin (mTOR), mitogen-activated protein kinase (MAPK) signaling pathways that activate downstream proteins, the snail transcription factor, and matrix metalloproteinases (MMP2 and MMP9) that cause epithelial-mesenchymal transition (EMT) and lead to metastasis [29]. On the other hand, hypoxia induces expression of vascular endothelial growth factor (VEGF) via PI3K/Akt/mTOR, phosphatase and tensin homolog (PTEN), MAPK signaling cascades through HIF-1α and p70S6K1, which release various cytokines, growth factors, and upregulate MMPs leading to angiogenesis [30]. Independent of hypoxia, oxidative lipid ligand activates NF-κB through TLRs. ROS produced exogenously or endogenously activates apoptosis pathways [31] and are mediated by the activation of MAPK, Bcl-2, and Bax [32] (Fig. 1).

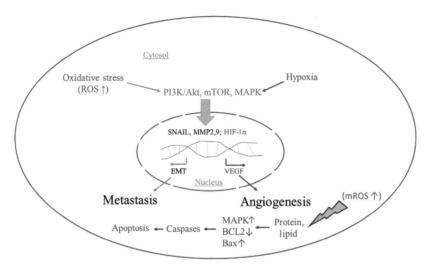

Fig. 1 Overview of molecular mechanisms involved in breast cancer.

4.1 Oxidative stress

Aerobic respiration produces several other compounds essential for certain subcellular events, including gene expression, signal transduction, disulfide bond formation, and the control of caspase activity. Enzymes such as xanthine oxidase and NADPH oxidase complexes act as internal sources of oxidative stress, whereas UV radiation and chemical compounds act as external sources. Reactive species can be classified into ROS, reactive nitrogen species (RNS), and so on [33] based on the atom involved. These mechanisms and pathways are conserved in mammalian cells [33].

4.2 Apoptosis

Pathogenesis of a disease can be best studied by apoptosis, which also gives clues to the treatment of the disease. In cancer, the balance between cell division and cell death is lost. Caspase activation involves intrinsic and extrinsic pathways as well as the intrinsic endoplasmic reticulum pathway, which is less well known [34]. The reduction of apoptosis or resistance to it plays a role in carcinogenesis due to which there is a disturbance in pro-apoptotic and antiapoptotic proteins that provide mechanistic insights [35]. Apoptosis is reduced in breast cancer cells due to the imbalance between cell division and cell death [36].

4.3 MAPK signaling

MAPK signaling links extracellular signals to the intracellular process that controls growth, proliferation, migration, and apoptosis. MAPK pathways that have been characterized in mammals include MAPK/ERK and MAPK/JNK [37]. Tumorigenesis in cancer cells is acquired by independent proliferative signals, high replication potential, evading apoptosis, ability to invade other tissues, and angiogenesis for nutrient supply [38]. Abnormalities in MAPK signaling play a role in progression and development of cancer [39]. The MAPK pathway is activated in breast cancer [40].

4.4 PI3K/AKT pathway

Protein signaling pathways regulating cell growth, differentiation, and development such as the PI3K/Akt pathway undergo oncogenic changes in cancerous tissue due to frequent disturbances [41]. PI3K/Akt pathway activation in breast cancer plays a role in endocrine resistance as well as leads to cell growth and tumor proliferation [42].

4.5 mTOR pathway

The mTOR signaling pathway is necessary for cell growth, survival, and proliferation under lack of nutrients and growth factors [43]. In vitro and in vivo studies reveal that the mTOR pathway is involved in tumor growth, angiogenesis, and metastasis [44], PI3K/AKT signaling is deregulated by upstream proteins such as HER-2 and IGFR through mutations in PI3K or amplification of AKT [45]. PTEN is the negative regulator of PI3K signaling and gets downregulated in many cancers due to expression of regulatory miRNA, mutation, methylation, and protein instability [46]. PI3K/mTOR pathway activation is seen in breast cancers [47].

In addition, the other molecular markers that are altered during breast cancer include cyclin-dependent kinase (CDK)/cyclin and poly (ADP-ribose) polymerase (PARP).

4.6 CDKs

CDK/cyclins play a role in dysregulating cell cycle, growth, and proliferation in several human cancers [38, 48]. CDK2 is found to be overexpressed in laryngeal carcinoma, melanoma, and breast cancer [49, 50]. Decreased proliferation of breast cancer is found through knockdown studies of CDK8, which suggests its role in breast cancer [51]. Partner cyclins such as cyclin A and E are found to be overexpressed in breast, lung, and thyroid cancer [52, 53]. High levels of cyclin D1 are found in breast cancers through gene amplification or overexpression [54], whereas cyclin E levels are found to be high in the nucleus [55].

4.7 PARP

PARPs catalyze the transfer of ADP-ribose to target proteins, which include 18 members that share sequence homology for conserved catalytic domain. PARPs play a role in DNA repair process, cell death, and proliferation [56]. Tankyrase 1 (PARP 5a) expression is high in gastric cancer and breast cancer [57].

5. Induction of mammary tumorigenesis

One of the most effective approaches for studying breast cancer is to create animal models for laboratory experiments. One of the methods used to induce tumors in breast tissue is applying chemical compounds that belong to the family of polycyclic aromatic hydro-carbons (PAHs) [58]. Compounds used to induce breast tumors include 7,12-dimethylbenz[a]anthracene (DMBA) [59], N-nitroso-N-methylurea (NMU) [59], 2-amino-1-methyl-6-phenylimidazo[4,5-b]pyridine (PhIP) [60], and 3-methylcholanthrene (MC) [61]. Metabolism of DMBA takes place in the liver by the cytochromes P450 system and produces free radicals thereby causing mutations in DNA and NF-κB activation. This eventually results in a change of the cell's genetic material by disrupting DNA repair through depurination thereby inducing cell death pathways and tumor development in tissues [62]. NMU effects on cellular proteins produced from methylated DNA, and ultimately leads to tumors. PhIP is a mutagenic and carcinogenic compound found especially in meat and fish and cigarette smoke, which directly affects on DNA thereby producing mammary, colon, and prostate tissues [63]. MC induces estrogen activity through aryl hydrocarbon receptor (AhR)–estrogen receptor alpha (ERα) interaction by acting as an agonist of AhR and increasing expression of C-fos and C-myc [64].

6. Types of cell lines used for breast cancer study

The first cell line used for breast cancer study was established in 1958. It is called BT-20 [65]. The nomenclature of cell lines was derived from its source, i.e., the lab or the patient or subculture isolated serially from the same initial population [66]. For example, HCC series cell lines are from the Hamon Cancer Center [67] and the MDA series is from the MD Anderson Hospital and Tumor Institute [68]. Cell lines for breast cancer study include LY2, MCF7, MDAMB134, MDAMB134VI, T47D, ZR751, EFM192A, IBEP1, MDAMB330, MDAMB361, UACC812, ZR7527, ZR7530, 21MT1, OCUB-F, SKBR3, HMT3522, KPL-3C, MA11, MDAMB435, MDAMB436, MDAMB468, MFM223, SUM185PE, MDAMB231, SKBR7, SUM149PT, and SUM159PT [66].

7. Therapeutic strategies to treat breast cancer

The most prevalent cancer among women is breast cancer, which accounts for about 25.5 percent of all cancers. This implies that there is a continuing need for development of new drugs, drug combinations, and potent therapeutic strategies for the successful treatment of breast cancer. Different therapeutic strategies including surgery (a procedure in which cancer tissue is removed by surgeon), radiation therapy (a type of treatment that requires high dosage of radiation to destroy and shrink tumors), chemotherapy (a type of treatment that involves usage of drugs to destroy cancer cells), hormone therapy (a treatment that involves usage of hormones to inhibit the growth of cancer cells), and precision medicine (a treatment that involves usage of medicine to help patients based on a genetic understanding of their disease). Nevertheless, these therapeutic strategies are associated with various side effects and are often ineffective for treating cancer. Thus there is a need to develop alternative approaches for treating cancer, such as herbal medicines.

Common treatment for cancer is unfavorable because it causes many detrimental side effects, and lately there has been a growing resistance toward anticancer drugs. Therefore, the focus has now shifted toward natural products, such as spices and plants.

8. Phytotherapy

Since ancient times, plants have been used to treat various ailments. Medicinal plants contribute to health care and more than 100 plant-derived chemical substances are currently used as drugs in modern medicine. It is also important to note that many other drugs are simple synthetic modifications of naturally available chemical substances. Phytotherapy is the study of utilizing plant extracts as medicine or health-promoting agents [69]. There has been increasing research in the past few decades to test the efficacy of herbal preparations and the associated active ingredient chemical substances against various human disease conditions [70]. Plant products have been used to treat cancer for several years, however, there is lack of information on potent active compounds that selectively target cancerous cells without harming normal cells. Resistance to anticancer drugs and modest tumor specificity hampers successful treatment of cancer patients. Regardless of recent advancements, chemotherapy for breast cancer is still faulty. In an effort to kill less-sensitive tumor subpopulations, implementing high doses of drugs becomes necessary and provokes severe side effects in cancer patients. On the other hand, suboptimal drug doses allow resistant tumor cells to survive and result in refractory tumors that don't respond to cytostatic therapy and are fatal to the patients. This makes it necessary to identify efficient natural agents such as herbs and other plant-based formulations that induce the least side effects. In the following sections, we discuss traditional Indian medicinal plants with antibreast cancer activity. In addition, we examine some other medicinal plants with potent antiproliferative activity against various breast cancers cell lines that are mentioned in Table 1.

8.1 Andrographis paniculata

Andrographis paniculata is an annual, branched, and herbaceous plant commonly known as King of Bitters or Mahatikta. It belongs to the family Acanthaceae [172]. It is native to India and Sri Lanka and is widely used as a traditional medicine in Bangladesh, China, and Indonesia [173]. It is traditionally used for the treatment of snake bite, diabetes, dysentery, and malaria. The leaf extract is a traditional remedy for the treatment of infectious disease,

Table 1 List of medicinal plants with antibreast cancer activity.

Plant name	Plant part	Active constituents	Cell line	Mechanism of action	References
Acacia nilotica (Fabaceae)	Leaves	γ–Sitosterol	MCF-7	• Cell cycle arrest at the G2/M phase • ↓c-Myc	[71]
Amoora rohituka (Meliaceae)	Stem Bark	Amooranin	MCF-7, MCF-7/TH, and MCF-10A	Apoptosis mediated cell death through • Induction of DNA ladder formation • ↑Total caspase and caspase–8 activities	[72]
Andrographis paniculata (Acanthaceae)	Whole plant	Andrographolide	MDA-MB-231	Induction of both intrinsic and extrinsic apoptosis pathway. • Cell cycle arrest at S and G2/M phase • ↑ROS production • ↓ΔΨm and externalization of phosphatidylserine • ↑Caspase-3 and 9 • ↑Bax and Apaf-1 • ↓Bcl-2 and Bcl-xL • Downregulation of PI3 kinase/Akt activation • Inhibition of pro-angiogenic molecules such as OPN and VEGF expressions	[73–76]
Annona reticulata Linn. (Annonaceae)	Leaves		T–47D	• ↓Bcl-2 • ↑Bax and Bak, and caspases' activation • Cell cycle arrest at G2/M phase	[77]
Artemisia absinthium (Asteraceae)	Methanol extract		MDA-MB–231 and MCF-7	• Nuclear condensation • Cell cycle arrest sub-G1 • Activation of caspase–7 • ↑Bad and Bcl-2 • ↑Both MEK1/2 and ERK1/2	[78]
Adiantum capillus veneris (Pteridaceae)			MCF7 and BT47	Crude extract has antiproliferative and apoptosis-inducing properties against MCF7 and BT47 cell lines	[79]
Aerva javanica (Amaranthaceae)	Leaf		MCF-7	Antiproliferative activity	[80]

Plant (Family)	Part used	Compound	Cell line	Activity / Mechanism	Ref.
Anogeissus latifolia (Combretaceae)	Whole plant		T47D and MCF-7	Whole plant extracts inhibit cell proliferation	[81]
Acacia catechu (fabaceae)	Whole plant		T47D & MCF-7	Whole plant extracts inhibit cell proliferatio	[81]
Annona muricata L. (Annonaceae)	Leaves		MDA–MB–435S	Cytotoxicity	[82]
Annona muricata (Annonaceae)	Leaves		MCF-7	Significant cell inhibition	[83]
Anisochilus carnosus (L.f.) wall (Lamiaceae)	Leaves		BT-549	Potent cytotoxic effect against BT–549	[84]
Artemisia nilagirica (Asteraceae)				Ethyl acetate (AR–03) and hexane (AR–04) fractions were found to be the most cytotoxic against breast cancer cell lines	[85]
Arisaema tortuosum (Wall.) Schott (Araceae)	Tuber and leaves			Chloroform fraction (leaves) showed significant reduction in the cell viability of MCF-7 cell line	[86]
Allium atroviolaceum (Amaryllidaceae)	Flower		MCF-7 and MDA-MB-231	• Cell cycle arrest at S, G2/M phase and sub-G0 • ↓Cdk1 (in a p53-independent pathway) • ↓Bcl-2 • Caspase-dependent apoptosis	[87]
Butea monosperma (Fabaceae)	Bark		MCF-7	Butanol fraction arrest cell cycles at sub-G1 ↑ROS ↓MMP indicating mitochondrial-dependent pathway of apoptosis	[88]
Phaseolus vulgaris L. (Fabaceae)	Seeds		MCF-7 and MDA-MB231	• Activation of caspase 3/7 • ↑Bax • ↓Bcl-2 and Bcl-xL • Cell cycle arrest in S and G2/M phase • ↓ΔΨm indicating mitochondrial-dependent pathway of apoptosis	[89]
Barleria buxifolia (Acanthaceae)		Barleriaquinone-I (BQ-I) and barleriaquinone-II (BQ-II)	MCF7	Cytotoxicity BQ-II to be more active than BQ-I	[90]

Continued

Table 1 List of medicinal plants with antibreast cancer activity—cont'd

Plant name	Plant part	Active constituents	Cell line	Mechanism of action	References
Brassica juncea (Brassicaceae)	Seeds	Glucosinolates	MCF-7 and MDA-MB-231	• ↑ROS • ↓ ΔΨm indicating mitochondrial-dependent pathway of apoptosis	[91]
Boerhaavia diffusa L. (Nyctaginaceae)	Whole plant		MCF-7	• ↓mRNA expression of pS2 • Cell cycle arrest at G0–G1 phase	[92]
Boswellia ovalifoliolata (Burseraceae)	Leaves		MDA-MB-231 and MDA-MB-453	• ↓Phospho-NF-κB (ser536), PCNA, Bcl-2 • ↑Bax	[93]
Betula utilis (Betulaceae)	Bark	Betulin, betulinic acid, lupeol, ursolic acid, oleanolic acid and β-amyrin		• ↑DR4, DR5, and PARP cleavage causing extrinsic apoptosis pathway • ↑ROS • ↓ ΔΨm	[94]
Cyperus rotundus (Cyperaceae)	Rhizomes		MCF-7, MDA-MB-231 and MDA-MB-468	Cytotoxicity thorough induction of apoptosis Induction of apoptosis was associated with ↑Death receptor 4 (DR4), DR5 and pro-apoptotic Bax ↓Antiapoptotic survivin and Bcl-2 ↓Bid Activation of caspase-8 and -9, respective initiator caspases of the extrinsic and intrinsic apoptotic pathways ↑Mitochondrial membrane depolarization was correlated with activation of caspase-3 and cleavage of PARP, a vital substrate of activated caspase-3	[95–97]
Chenopodium album (Amaranthaceae)	Leaves		MCF-7 and MDA-MB-468	Methanolic extract of *C. album* (leaves) exhibited maximum anticancer activity	[98]
Costus pictus D. Don (Costaceae)	Leaves and rhizome	Costunolide	MDA-MB-231 and MCF 10A	Moderate cytotoxicity	[99, 100]
Cassia auriculata (Caesalpiniaceae)	Leaves		MCF-7	• ↓NF-κB subunits—p65, 52 and 100 • Nuclear fragmentation and condensation • ↓Bcl-2 protein • ↑Bax protein	[101]

Plant (Family)	Plant material	Compound	Cell line	Mechanism	References
Centratherum anthelminticum (Asteraceae)	Seeds	Vernodalin and morpholinoethyl isothiocyanate	MCF-7, MDA-MB 231 and LA7	• ↑ROS • ↓Bcl-2, Bcl-Xl • ↓ΔΨm and release of cytochrome c • Release of cytochrome c from mitochondria to cytosol triggered activation of caspase cascade, PARP cleavage, DNA damage, and eventually cell death through apoptosis • Activation of FOXO transcription factors and its downstream targets (Bim, p27Kip1, p21Waf1/cip1, cyclin D1, cyclin E) • Akt kinase activity was downregulated and accumulation of FOXO3a • Upregulation of p27Kip1 and FOXO3a and decrease in the p-FOXO3a level • Inhibition of TNF-α release	[102–105]
Cephalotaxus griffithii (Taxaceae)			ZR751	↓hTERT, hTR, and c-Myc expression	[106]
Citrullus colocynth (L.) (Cucurbitaceae)	Fruit pulp		MCF-7 and MDA-MB-231	• ↓BCL2 and BCLXL • ↑BAX and caspase 3 • ↑Epithelial gene keratin 19 • ↓Mesenchymal genes, vimentin, N-cadherin, Zeb1 and Zeb2 • Suggesting a suppressive impact of these extracts in epithelial–mesenchymal transition (EMT) • ↓Stemness-associated genes BMI-1 and CD44	[107]
Caesalpinia sappan L. (Fabaceae)	Heartwood and leaves	Brazilin A	MCF-7	Induction of cell death in MCF-7 cells	[108]
Dioscorea deltoidea (Dioscoreaceae)	Plant material	Diosgenin		Significant antiproliferative effect against HBL-100 cell line	[109]

Continued

Table 1 List of medicinal plants with antibreast cancer activity—cont'd

Plant name	Plant part	Active constituents	Cell line	Mechanism of action	References
Drosera burmannii Vahl (Droseraceae)	Plant		MCF-7	• Cell cycle arrest at G2/M phase • ↓Cyclin A1, cyclin B1 and Cdk-1 • ↑p53, Bax/Bcl-2 ratio leading to activation of caspases and PARP degradation • ↓iNOS, COX-2, and TNF-α along with suppression on intracellular ROS confirms the antiinflammatory potential of extract	[110]
Dysoxylum Binectariferum Hook.f (Meliaceae)	Stem bark	Rohitukine		Flavopiridol is a potent inhibitor of several cyclin–dependent kinases (CDKs)	[111]
Eulophia nuda L. (Orchidaceae)	Crude methanolic extract	9,10–Dihydro-2, 5–dimethoxyphenan threne-1,7–diol	MCF-7 and MDA-MB-231	Significant antiproliferative activity	[112]
Fraxinus micrantha (Oleaceae)	Bark		MCF-7	↑NO and DNA fragmentation	[113]
Ficus religiosa (Moraceae)	Acetone extract			• Cell cycle arrest in G1 phase • Induction of chromatin condensation • ↓ΔΨm and induced Caspase activation • ↑ROS • ↑BAX and activation of Caspase 9	[114]
Glycyrrhiza glabra (Fabaceae)	Root	Quercetin and glabrol		Inhibition of CYP1B1	[115]
Glycine max (Fabaceae)	Seed		MCF-7	Soybean extract acts as a promoter of MCF-7 cell growth, the fenugreek extract induces apoptosis	[116]
Garcinia hombroniana (Clusiaceae)	Bark extract			Significant cytotoxic towards MCF-7	[117]
Glycosmis pentaphylla (Retz.) DC (Rutaceae)	Dried leaves	Lupeol, chrysin, quercetin, β-sitosterol and kaempferol	MCF-7 and MDA-MB-231	Apoptotic induction through the mitochondrial pathway by the activation of caspase-3/7	[118]
Juglans regia (Juglandaceae)	Root bark		MDA-MB-231	↑Bax, caspases, tp53, and TNF-alpha	[119]
Leptadenia reticulata (Asclepiadaceae)			MCF-7	Strong cytotoxicity of ethyl acetate extract	[120]

Plant name (Family)	Part	Compound	Cell line	Mechanism	Ref.
Launaea procumbens (Asteraceae)	Leaves		MCF-7	Potent cytotoxicity against breast (MCF-7)	[121]
Trigonella foenum-graecum (Fabaceae)	Seeds		MCF-7	Induction of apoptosis	[116]
Murraya koenigii Spreng (Rutaceae)	Leaf		MCF-7 and MDA-MB-231 & 4T1	• Cell cycle arrest in S phase • Decreased the activity of the 26S proteasome • Inhibition of trypsin-like, but not the chymotrypsin-like proteolytic activity of the proteasome • MK reduced the tumors' size and lung metastasis • Inhibited the cell viability of 4T1 cells • Decreased the level of NO and iNOS, iCAM, NF-kB, and c-MYC	[122–124]
Morinda citrifolia L. (Rubiaceae)	Fruits and whole plant		MCF-7, MDA-MB-231 and T47D	• Cell cycle arrested in G1/S and G0/G1 phase • Antiproliferation activity • Induction of DNA damage	[81, 125]
Mucuna pruriens (L.) DC (Fabaceae)	Seeds	L–Dopa	T47D, MCF-7, MDA-MB-468, and MDA-MB-231	• Cell cycles arrest in G1 phase • ↓PRL expression, further suppressing the JAK2/STAT5A/Cyclin D1 signaling pathway	[126]
Morus alba L. (Moraceae)	Leaves	Lectin	MCF-7	Lectin showed significant antiproliferative activity towards MCF-7	[127, 128]
Macrosolen parasiticus (L.) (Loranthaceaea)	Stem		MCF-7	• Fragmentation of DNA • Caspase-3 activation • ↓NO • ↑iNOS	[129]
Maesa macrophylla (Primulaceae)	Leaf		MCF7 c	Significant cytotoxic activity against MCF-7	[130]
Madhuca indica (Sapotaceae)	Flower		MCF-7 and MDA-MB-468	Potent cytotoxicity	[131]
				• ↑Caspase 3/7 • ↓COX-2 mRNA and COX-2 protein	

Continued

Table 1 List of medicinal plants with antibreast cancer activity—cont'd

Plant name	Plant part	Active constituents	Cell line	Mechanism of action	References
Nardostachys jatamansi DC (Caprifoliaceae)	Roots and rhizomes	Lupeol and β-sitosterol	MCF-7 and MDA-MB-231	Cell cycle arrest at G2/M and G0/G1 phase	[132]
Nyctanthes arbortristis Linn (Oleaceae)	Seeds	Arbortristoside-A (1) and 7-O-trans-cinnamoyl 6β-hydroxyloganin (2)	MCF-7	Exhibited moderate in vitro anticancer activity	[133]
Oenothera biennis L. (Onagraceae)		Oenotheralanosterol A and oenotheralanosterol B		Significant antiproliferative activity	[134]
Orthosiphon pallidus (Lamiaceae)			MCF-7 and MDA-MB-231	Significant cytotoxicity	[135]
Oroxylum indicum (L.) (Bignoniaceae)			MDA–MB-231	Induction of apoptosis and significant antimetastatic activity	[136]
Parmotrema reticulatum (Parmeliaceae)	Dried lichen		MCF-7	• Cell cycle arrest in S and G2/M phases • ↓Cyclin B1, Cdk-2 and Cdc25C as well as Cdk-1 and cyclin A1 • ↑Upregulation of p53 and p21 • ↑BAX and ↓Bcl-2 expression, which results in increasing Bax/Bcl-2 ratio and activation of caspase cascade, ultimately leads to PARP degradation thereby apoptosis	[137]
Psoralea corylifolia (Fabaceae)	Seeds		MCF7	• Induction of apoptotic death through the cleavage of caspase-9 and caspase-7 • ↑Upregulation of Bax, release of cytochrome-c • ↑ROS • ↓ΔΨm • Increased cleavage of PARP leading to apoptotic mode of cell death	[138]
Pithecellobium dulce (Fabaceae)	Leaves		MCF-7	• ↑mRNA expression of Bax, p21, p53, TNF and fas. • ↓mRNA expression of Bcl-2, NF-KB and Cdk	[139]

Plant (Family)	Part	Compound	Cell line	Mechanism/Effect	Reference
Pteris quadriureta (Pteridaceae)			MCF7 and BT47	Significant antiproliferative and apoptosis-inducing properties against MCF7 and BT47 cell lines	[79]
Pomegranate Punicaceae	Pericarp		MCF-7, MDA-MB-231	↓Estrogen response elements (ERE)–mediated transcription. PME can compete with 27HC (27-hydroxycholesterol) for ERα and reduce 27HC-induced proliferation of MCF-7 cells	[140]
Punica granatum (Punicaceae)	Fruit peel	Polysaccharide	MCF-7	Cytotoxicity	[141]
Picrorhizakurroa Royle ex Benth (Scrophulariaceae)	Rhizome		MDA-MB-435S	Cytotoxicity and induces apoptosis	[142]
Podophyllum hexandrum (Berberidaceae)	Rhizome		MCF-7	Methanol and 70% ethanolic extracts of the rhizome showed highest cytotoxic effect on MCF-7	[143]
Pteris vittata L. (Pteridaceae)	Methanolic		MCF-7	Dose-dependent decrease in viability of MCF-7 cells	[144]
Pongamia pinnata (L.) Pierre. (Fabaceae)	Seeds			Cell cycle arrest at the G0–G1 phase ↓Cyclin D1 levels	[145]
Polygonatum verticillatum (L.) (Ruscaceae)	Rhizomes			Chloroform extract exhibited the strongest cytotoxicity against the human breast cancer cell line, MCF-7	[146]
Pyrenacantha volubilis (Icacinaceae)	Seeds	Camptothecine		Exhibited the strongest cytotoxicity	[147]
Rheum emodi Wall. ex Meissn. (Polygonaceae)	Rhizome		MDA-MB-435S	Induction of apoptosis	[148]
Soyabean (fabaceae) and flaxseed (Linaceae)	Soyabean seed and flax seed		MCF-7 and MDA-MB-231	↓SOD and GPx ↑Intra-cellular ROS levels ↓MMP Cell cycle arrest at S and G2/M phase and Sub G1 phase	[149]

Continued

Table 1 List of medicinal plants with antibreast cancer activity—cont'd

Plant name	Plant part	Active constituents	Cell line	Mechanism of action	References
Semecarpus anacardium (Anacardiaceae)	Nut		T47D	Cytotoxicity is due to induction of apoptosis Induction of rapid Ca^{2+} mobilization from intracellular stores correlated with altered mitochondrial transmembrane potential $\downarrow bcl(2)$ $\uparrow bax$, cytochrome c, caspases and PARP cleavage	[150]
Semecarpus anacardium (Anacardiaceae)	Root	Total alkaloids and phenolics	MCF-7	Alkaloid fraction had maximum cytotoxicity and apoptotic induction	[151]
Semecarpus lehyam			MCF-7 and MDA 231	*n*-Hexane and chloroform fractions showed significantly cytotoxic	[152]
Syzygium aromaticum L. (Myrtaceae)			MCF-7 and MDA-MB-231	Maximal cytotoxicity	[153]
Saraca indica (Caesalpiniaceae)	Bark			Antibreast cancer activity	[154]
Salacia oblonga (Celastraceae)	Root and aerial parts		MCF7	Significant antiproliferative activity	[155]
Tinospora cordifolia (Menispermaceae)			MDA–MB–231 and MCF-7	$\uparrow ROS$	[156]
Tiliacora racemosa (Menispermaceae)	Root	Total alkaloids and phenolics	MCF-7	$\uparrow BAX$ and $\downarrow Bcl-2$ Maximum cytotoxicity	[151]
Thespesia populnea (Malvaceae)		Quinones	MCF-7	Cytotoxicity	[157]
Terminalia bellerica (Combretaceae)	Whole plant		T47D and MCF-7	Antiproliferation activity	[81]
Taxus wallichiana (Taxaceae)		Taxiresinol	MCF-7	Cytotoxicity	[158]
Vernonia cinerea Less. (Asteraceae)			MCF-7	• Induction of apoptosis • Inhibited functional activity of MDR transporters (ABC–B1 and ABC–G2), enhanced DNR-uptake in cancer cells	[159]

Plant (Family)	Part	Compound	Cell line/Model	Mechanism	Ref.
Wrightia tomentosa (Apocynaceae)		Oleanolic acid and ursolic acid	MCF-7 and MDA-MB-231	Cell cycle arrest in G1 ↑ROS ↓MMP and subsequent apoptosis ↑Bax/Bcl-2 ratio Enhanced Annexin-V positivity, caspase 8 activation and DNA fragmentation	[160]
Withania somnifera (Solanaceae)	Root		MDA-MB-231	Induced mitochondrial-mediated apoptosis via ↑ROS, ↑Bax/Bcl-2, ↓MMP and caspase-3 activation	[161]
Zanthoxylum zanthoxyloides (Rutaceae)		Zantholic acid		Cell cycle arrest at G2/M-phase, cleavage of nuclear lamin A/C proteins Cytotoxic activity	[162]

In vivo studies

Plant (Family)	Part	Compound	Cell line/Model	Mechanism	Ref.
Bacopa monnieri (Plantaginaceae)		Bacoside A, B and cucurbitacins	Ehrlich ascites carcinoma (EAC) tumor-bearing mice	Prominent reduction in tumor weight, packed cell volume, tumor volume, and viable tumor cell	[163]
Butea monosperma (Fabaceae)	Flower		Methylnitrosourea (MNU) induced mammary cancer in nulliparous Sprague-Dawley rats	Decreased expression of estrogen and progesterone, nucleic acid content and increased latency period Induction of apoptosis, inhibition of angiogenesis and metastasis	[164]
Cedrus deodara (Pinaceae)	Stem wood	(−)-wikstromal, (−)-matairesinol and benzylbutyrolactol		Induction of tumor regression in vivo Induces apoptosis as indicated by annexin V positive cells, induction of intracellular caspases, DNA fragmentation and DNA cell cycle analysis	[165]
Garcinia morella (Clusiaceae)		Garcinol	MCF7, MDAMB231 and SKBR3	apoptosis induction thorough P53 dependent ↑Bax ↓Bcl XL	[166]

Continued

Table 1 List of medicinal plants with antibreast cancer activity—cont'd

Plant name	Plant part	Active constituents	Cell line	Mechanism of action	References
Murraya koenigii (Rutaceae)	Leaves		Xenograft tumor mouse model	• ↓Release of nitrite and TNF-α level Significantly inhibited paw inflammation Inhibition of endogenous 26S proteasome activity. Reduction in tumor growth was associated with a decrease in proteasomal enzyme activities • ↑Caspase-3 activity and TUNEL-positive cells • ↓Angiogenic and antiapoptotic gene markers is indicative of inhibition of angiogenesis and promotion of apoptosis	[167]
Pyracantha fortuneana (Rosaceae)	Dry fruit	Se-containing polysaccharides	Mouse xenograft model	• Cell cycle arrest in G2 phase • Inhibition of CDC25C-CyclinB1/CDC2 pathway • ↑p53, Bax, Puma and Noxa • ↓Bcl2 • ↑Increased Bax/Bcl2 ratio • ↑Activities of caspases 3/9 • Reduction of tumor growth	[168]
Ricinus communis L. (Euphorbiaceae)	Dry fruit		MCF-7 and highly aggressive, triple negative MDA-MB-231 breast cancer cells	Inhibition of migration, adhesion, invasion and expression of MMPs 2 and 9 ↓Bcl-2 ↑Bax and caspase-7 expressions as well as PARP cleavage Reduction in tumor volume	[169]
Withania somnifera (Solanaceae)	Leaf extract	Withaferin-A, a withanolide	HCT116 cells in xenograft mouse tumor model	Significant decrease in volume and weight of tumors	[170]
Zizyphus nummularia (Rhamnaceae)	Root bark		female Swiss albino mice against Ehrlich ascites carcinoma (EAC)	Decreased tumor volume and viable tumor cell count and increased body weight and life span	[171]

↓—indicates decrease in; ↑—indicates increase in.

fever-causing diseases, colic pain, irregular stools, and diarrhea. It has been reported to possess a wide range of pharmacological effects such as anticancerous, antihepatitic, antihyperglycemic, antiinflammatory, antimalarial, antioxidant, cardiovascular, and sexual dysfunctions. The bioactive components of this plant include diterpenes, flavonoids, xanthones, and nocardioides [174]. Andrographolide, a diterpene lactone isolated from *A. paniculata* is the most important active component and has been reported to elicit a number of biological activities, including antioxidant, antiinflammatory, cytotoxic, immunomodulatory, cardioprotective, hepatoprotective, and neuromodulatory effects. Andrographolide was found to exert anticancer activity against a wide range of cell lines tested. The underlying anticancer mechanism associated with this compound involves oxidative stress, cell cycle arrest, antiinflammatory, apoptosis, necrosis, autophagy, inhibition of cell adhesion, proliferation, migration, invasion, and antiangiogenic activity [175].

Andrographolide inhibits breast cancer cell proliferation and migration and arrests the cell cycle at the G2/M phase and induces apoptosis through the caspase independent pathway. The antitumor activity of andrographolide is associated with repression of PI3 kinase/Akt and angiogenic molecules such as osteopontin (OPN) and vascular endothelial growth factor (VEGF). It also attenuates tumor-endothelial cell interaction. Therefore, breast tumor growth in orthotopic NOD/SCID mice was suppressed [73]. Andrographolide favored normal breast epithelial cell growth witnessed by an increase in the number of cells in the S as well as G2/M phase. However, an inhibitory effect was observed on MDA-MB-231 (metastatic breast cancer cells) with nonfunctional p53 and ER. The progressive mechanisms on MDA-MB-231 inhibition include elevated ROS production, decrease in mitochondrial membrane potential ($\Delta\Psi$m), externalization of phosphatidylserine, increase in pro-apoptotic (Bax/Apaf-1) and decrease in anti-apoptotic (Bcl2/BclxL) signals, and activation of caspase 3 and caspase 9, which finally results in apoptotic cell death [74]. Being cancer-cell specific, andrographolide and analogues need to be examined further in clinical and biomedical studies associated with cancer chemoprevention. Collectively, it can be suggested that andrographolide may be a potential anticancer agent in the near future.

8.2 Withania somnifera

Withania somnifera (Ashwagandha), also known as Indian ginseng, is a short perennial shrub of the Solanaceae family. This plant has been used in Indian medicine for a long time and its roots are used in more than 200 formulations [176]. Phytochemicals such as withanolides, withaferins, withanones, and withanosides were effective against cancer cell lines [177, 178]. Glycoproteins and lectin-like protein obtained from this plant were reported as antimicrobials and antisnake venom [179, 180]. Further, preclinical studies involving neuroprotective, cardioprotective, anti-inflammatory, and anti-diabetic properties are attributed to diverse phytoconstituents of this plant. A novel protein fraction from *W. somnifera* (WSPF) exhibited prominent apoptotic activity against MDA-MB-231 human breast cancer cells. Progression of apoptosis in MDA-MB-231 was mediated by this fraction with G2/M phase cell cycle arrest. The series of apoptotic events include increased ROS generation followed by loss of $\Delta\Psi$m, dysregulation of Bax/Bcl-2, caspase-3 activation, and cell death. Further, cleavage of nuclear lamin A/C proteins and nuclear morphological changes were also observed. This highlights the potential therapeutic properties of WSPF against TNBC [161]. Likewise, root extract inhibited the proliferation of MDA-MB-231 cells and cell cycle arrest in sub-G1 phase. It inhibited the proliferation of xenografted MDA-MB-231 by reduced expression of cytokine CCL2 (C-C motif Ligand 2) thereby reducing xenograft size [181].

Withaferin A (WA) hampers proteasomal degradation system and perturb autophagy. On suppressing the intracellular degradation systems, ubiquitinated proteins accumulate, which in turn unfolds protein response and ER stress-mediated proteotoxicity in human breast cancer cell-lines MCF-7 and MDA-MB-231. Therefore, WA targets breast cancer cells through cellular proteotoxicity upon simultaneous inhibition of proteasome system and induction of impaired autophagy [182].

8.3 Tinospora cordifolia

Tinospora cordifolia, also known as Amrita and Guduchi, belongs to the Menispermaceae family. It is commonly found in the Indian subcontinent and China. The plant is used alone or in combination with other plants in Ayurveda and folk medicine [183]. The plant phytochemicals (alkaloids, terpenoids, lignans, steroids, etc.) confer the plant with a broad spectrum of pharmacological properties such as antioxidant, antimicrobial, antidiabetic, antistress activity, anticancer, antiHIV, and immunomodulating activities [184].

Pharmacologically active compounds like rutin and quercetin are present in the Chloroform fraction *T. cordifolia* (TcCF), which confers the plant with anticancer properties against breast cancer cells (MDA-MB-231 and MCF 7). Elevated ROS, reduced colony-forming properties, increase in pro-apoptotic/antiapoptotic gene expression ratio, and finally apoptosis was observed in TcCF-treated breast cancer cells. However, the mechanism of apoptosis was abrogated upon ROS inhibition, emphasizing the role of ROS generation in TcCF-induced apoptosis in breast cancer [156].

A natural compound, Bis(2- ethyl hexyl) 1H-pyrrole-3,4–dicarboxylate (TCCP), isolated from butanol fraction of TcCF leaves was pro-apoptotic and arrested MDA-MB-231 cell cycle in the sub-G1 phase. The mechanism involves a series of events that begins with decreased mitochondrial membrane potential, elevated intracellular calcium, phosphorylation of p53, endogenous ROS generation, cardiolipin peroxidation, formation of mitochondrial permeability transition pores (MPTP), increase in Bax/Bcl2 ratio, cytochrome release in the cytosol, caspase activation, and apoptotic DNA fragmentation.

Further, in vivo studies with Ehrlich ascites tumors (EATs) in mice showed a twofold increase in mice survival with trivial hepato-renal toxicity. Therefore, in vitro and in vivo studies revealed the efficiency of TCCP against tumor proliferation involving MDA-MB-231 cells and EATs in mice [185].

8.4 Murraya koenigii

Murraya koenigii (MK), also known as curry leaves, belongs to the Rutaceae family and is widely distributed in eastern Asia. Its medicinal properties are well documented in Ayurveda, and pharmacologically they were reported to possess antiviral, antiinflammatory, antidiabetic antileishmanial, and antitumor activities [186–188]. Proteolytic inhibition of 26S proteasome is a promising approach for cancer therapy as this mechanism selectively kills cancer cells and enhances their sensitivity to chemotherapeutic agents. The total alkaloid extract and mahanine, a carbazole alkaloid of MK leaves inhibits trypsin-like but not chymotrypsin-like proteolytic activity of proteasome [122]. Hydromethanolic polyphenol-rich extract of MK leaves induced cell cycle arrest of cancer cells at the S phase, but not in normal lung fibroblasts. The mechanism involves inhibition of 26S proteasome in cancer cells and the resulting apoptotic cell death is suggested by Annexin V binding data [123]. Further in vivo studies were carried out with xenograft tumor mouse models and MK extract did not show any toxic effect in mice. Analysis of the extract revealed the presence of flavonoids (quercetin, apigenin, kaempferol, and rutin) effectively inhibiting endogenous 26S proteasome in MDA-MB-231 cells. Therefore, reduction in tumor growth was associated with a decrease in proteasomal enzyme activities. Further, increased caspase 3 activity and decreased antiapoptotic and angiogenic gene expression were indicative of inhibited tumor progression [167]. MK reduced tumor size and lung metastasis in mice with 4T1 breast cancer by reducing nitrate and inflammatory cytokines such as iNOS, iCAM, NF-kB, and c-MYC [124].

8.5 *Cyperus rotundus* L.

Cyperus rotundus (CR) L., also known as nutgrass, is a wild weed belonging to the Cyperaceae family. In Ayurveda it is widely used to treat diarrhea, diabetes, inflammation, malaria, and bowel disorders. Likewise, pharmacologically nutgrass was reported to have neuroprotective, cardioprotective, hepatoprotective, antidiabetic, antiuropathogenic, and anticonvulsant properties [189]. MCF7 apoptosis was induced by methanol extract of CR rhizome, whereas apoptotic activity of TNBC MDA-MB-231 was greater in ethanolic extract (EECR) compared to methanolic extract of CR rhizome (MECR). TNBC showed cell cycle arrest at G0/G1 phase with EECR. The mechanism of apoptosis progresses with upregulation of death receptors (DR4, DR5) and pro-apoptotic Bax along with the downregulation of antiapoptotic surviving and Bcl2. Activation of caspase 8 and 9, represents the initiation of both extrinsic and intrinsic apoptotic pathways. Likewise, activation of effector caspase 3 was correlated with mitochondrial membrane depolarization and cleavage of PARP, a substrate of activated caspase 3 [95, 96, 190]. Further, 3-methyladenine inhibited TNBC cells pro-survival autophagy and increased sensitivity to EECR [97].

HC9 is a polyherbal formulation used by Ayurvedic practitioners for cleansing and detoxifying breast milk in lactating mothers. CR is included in HC9 as one of the nine equal ratios of medicinal plants. HC9 arrested cell cycle of MCF7 at S phase and MDA-MB-231 at G1 phase. Upregulation of p53, p21 is responsible for cell cycle arrest, whereas MCF7 arrest at S phase is represented by upregulation of p16 and TNBC arrest, and G1 phase is represented by upregulation of pRb. HC9 significantly reduced migration in both cell lines by reduced expression of MMP-2/9, HIF1α, and VEGF. HC9 also suppressed inflammatory markers (NF-kB and COX-2) and altered the expression of chromatin modulators (SMAR1 and CDP/Cux) in both MCF7 and MDA-MB-231. Such potent anticancer activity against breast cancer cells warrants preclinical and clinical studies in the future [191].

8.6 Centratherum anthelminticum

Centratherum anthelminticum (L) Kuntze, also known as cumin or kalizeeri in Hindi, belongs to the Asteraceae family, normally found in India, Sri Lanka, and Afghanistan. Traditionally, the seeds were used to treat diabetes, and cumin is a major portion of Kaya-kalp, a preparation used for whole-body rejuvenation. Likewise, pharmacologically the seeds were reported to possess anticancer, antidiabetic, anti-inflammatory, antiviral, anti-filarial, and antimicrobial properties [192]. In 2004, Lambertini et al. reported that 80 percent ethanol seed extract of *C. anthelminticum* displayed antiproliferative effects on MCF7 and MDA-MB-231 cell lines. Additionally, seed extract treatment upon both the breast cancer cell lines induced ERα mRNA accumulation, which is a well-established response of breast tumors to endocrine therapy [102]. Chloroform fraction of *C. anthelminticum* (CACF) mediated inhibition of MCF7 witnessed by morphological changes, disrupted cytoskeletal structures, and DNA fragmentation. As a consequence of bio-assay-guided fractionation, vernodalin was identified as a cytotoxic agent in CACF [103]. Further, in vitro and in vivo reports showed that vernodalin mediated apoptosis in breast cancer cells by the PI3K-Akt/FOXO3a pathway [104].

8.7 Juglans regia

Juglans regia L., also known as walnut, belongs to the Juglandaceae family. It is native to south-eastern Europe, Asia Minor, India, and China [193]. Traditionally, leaves were used to treat venous insufficiency and hemorrhoidal symptomatology [194] *J. regia* flowers possess bioactive compounds such as azacyclo-indoles, phenolics, alkaloids, flavones, tetralones, and naphthoquinones [195, 196]. Juglanin from green walnut husks of

Juglans mandshurica displayed G2/M phase arrest and induced apoptosis and autophagy via the ROS/JNK signaling pathway in human breast cancer cells [197]. Bio-assay–guided isolation of 5,7-dihydroxy-3,4'-dimethoxyflavone and regiolone from leaf chloroform extract inducing caspase 3 independent apoptotic pathway and cell cycle arrest at the G0/G1 phase in MCF7 cells [198, 199]. Chloroform extract of *J. regia* root bark (RBJR) improved Bax, caspases, p53, and TNFα-mediated cytotoxicity in MDA-MB-231 [119]. Walnut oil is rich in alpha linolenic acid (ALA) and β-sitosterol. Therefore, extracts derived from walnut oil reduced MCF7 proliferation as did ALA and β-sitosterol. ALA-treated mouse breast cancer cell line TM2H showed multiple cellular targets with PPAR, LXR, and FXR target genes. Walnut oil increased FXR activity more extensively than other tested nuclear receptors [200].

8.8 Linum usitatissimum

Flax is a food and dietary supplement commonly used for menopausal symptoms. Flax-seeds (FS) rich in phytoestrogen lignans and ALA may be associated with reduced risk of breast cancer [201]. FS extract induced apoptosis-mediated cytotoxicity in MCF7 cells. The mechanism of cytotoxicity begins with elevated ROS and loss of mitochondrial membrane potential, and ends with caspase cascade-mediated apoptotic cell death [202]. In rodent models 2.5%–10% FS diet or equivalent amount of lignan or oil reduced tumor growth. Likewise, in clinical trials, FS (25 g/day with 50 mg lignans for 32 days) supplementation suppressed tumor growth in breast cancer patients, whereas 50 mg/day of lignans for a year reduced the risk in premenopausal women [203]. Further, FS lignans significantly sensitized SKBR3 and MDA-MB-231 breast cancer cell lines and enhanced the cytotoxic ability of chemotherapeutic agents such as docetaxel, doxorubicin, and carboplatin [204]. Secoisolariciresinol diglucoside (SDG) is an abundant lignan in FS with antiestrogenic activity in high estrogen environments. SDG treatment restored several biomarkers in mammary gland tissue. Further, it doesn't promote preneoplastic progression in the ovarian epithelium [205]. Flax straw flavonoid C-glucosides (vitexin, orientin, and isoorientin) and individual pure compounds are cytotoxic for MCF7 cells. Apoptotic cell death was associated with an increased Bax/Bcl2 ratio and caspases 7, 8, and 9 expression [206]. FS sprouts significantly suppressed MCF7 and MDA-MB-231 cells by upregulating p53 mRNA, whereas no effect was observed with MCF-10A normal mammalian epithelial cells. Therefore, these studies indicate the anticancer potential of phytoconstituents in FS and straw [207].

8.9 Syzygium aromaticum

Syzgium aromaticum, also known as cloves, are aromatic flower buds belonging to the Myrataceae family. Clove extract displayed antiproliferative and pro-apoptotic effects in MCF7 cells, which displayed cell cycle arrest at S phase followed by reduced mitochondrial membrane potential, suppressed Bcl2 expression, and increased caspase 7. In an N-nitroso-N-methylurea-induced mammary cancer model, dried flower buds of cloves were administered along with diet and resulted in reduced tumor frequency and decreased lipid peroxidation in rat carcinoma cells. It also suppressed antiapoptotic gene bcl–2 and various metastasis- and stemness-related genes including Ki67, VEGFA, CD24, and CD44. Further it increased pro-apoptotic Bax, effector caspase-3, and ALDH1 expression in the cancer cells of animals. Clove supplementation increased lysine trimethylations and acetylations (H4K20me3, H4K16ac) of histone proteins in carcinoma. Likewise, methylation levels on promoter islands of tumor suppressor TIMP3 and RASSF1A genes were altered in clove-treated cancer cells [208].

Eugenol is a major volatile constituent of clove oil. In H-ras oncogene transfected (MCF10A-ras) cells, intracellular ATP levels were significantly reduced on eugenol treatment by inhibition of oxidative phosphorylation complexes and expression of fatty acid oxidizing (FAO) proteins such as PPARα, MCAD, and CPT1C by downregulating the c-Myc/PGC-1β/ERRα pathway. This decreased oxidative stress in MCF10A-ras cells [209, 210]. Therefore, eugenol prevents breast cancer by regulating energy metabolism. However, the mechanism was not witnessed in normal MCF10A cells treated with eugenol. Further, clove oil also displayed greatest cytotoxic effect against MCF7 cells [211]. Combination of clove extracts with nanoparticles is also a promising potential cancer treatment [212].

8.10 Semecarpus anacardium

Semecarpus anacardium (SA) nut extract induced apoptosis in T47D cells by a series of events that initiate with increased intracellular Ca^{2+} from intracellular stores, reduced mitochondrial membrane potential, decreased Bcl2/Bax ratio, increased cytosolic cytochrome C, and caspase-induced PARP cleavage that ultimately ends with DNA fragmentation [150]. Polycyclic aromatic hydrocarbons (PAHs) are environmental risk factors with potential genotoxic effect. Xenobiotic metabolizing enzymes in breast tissue are responsible for both susceptibility to such chemical carcinogens and response to chemotherapy. However, treatment with SA nut extract in 7,12-dimethylbenz(a)anthracene (PAH family)-induced breast cancer model of rodents restored the altered Phase I (NADPH-cytochrome (P(450)) reductase, NADPH-cytochrome (b(5)) reductase, epoxide hydrolase) and Phase II (glutathione-S-transferase, gluthatione peroxidase, gluatathione reductase, UDP-glucuronyl transferease) biotransformation enzymes and achieved complete detoxification of the carcinogen. Likewise, it also delayed tumor growth, as glycolytic enzyme activities were restored with decreased expression of glucose transporter 1 and carbonic anhydrase IX. Further, it suppressed angiogenic activity by inhibiting iNOS, VEGF, and HIF-1, and prevented endothelial cell proliferation by suppressing the surviving cytokines [213, 214].

Kalpaamruthaa (KA) is a Siddha preparation that includes *Semecarpus anacardium* (SA) along with *Emblica officinalis* and honey. In 7,12-dimethylbenz(a)anthracene-induced breast cancer model of rodents, altered levels of cholesterol, phospholipids, triglycerides, and free fatty acids in plasma, liver, and kidney were reverted back to normal with KA and SA treatment. Likewise, angiogenic factors, lipid peroxidation, and antioxidant levels in tumor-bearing rats were also decreased with SA and KA treatment. Further, compared to SA, KA was more effective [215–217]. Bio-assay-guided fractionation of *Semecarpus lehyam* hexane extract (hSL), effectively sensitized both ER(+) MCF7 and ER(−) MDA-MB-231 cancer cells. The three compounds (7;*Z*,10;*Z*)-3-pentadeca-7,10-dienyl-benzene-1,2-diol, (8;*Z*)-3-pentadec-10-enyl-benzene-1,2-diol, and 3-pentadecyl-benzene-1,2-diol were responsible for the antitumor effect of SL [218].

8.11 *Punica granatum* L.

Punica granatum, commonly known as pomegranate, has a long history of pharmacological properties attributed to a broad spectrum of phytochemicals. Pomegranate emulsion reduced DMBA-induced mammary tumorigenesis by suppressing the expression of cyclooxygenase 2 (COX-2), HSP 90, and intratumor ER (α and β) and lowered the ratio of ERα to ERβ. Further, it prevented degradation of IkBα, translocation of NF-kB from cytosol to nucleus, cytoplasmic accumulation and nuclear translocation of β-catenin (a transcription cofactor for Wnt signaling), suppressed cyclin D1 expression (downstream target for both ER and Wnt/β-catenin signaling pathways), and increased Nrf2 expression. These results display the involvement of anti-inflammatory, antiproliferative, and pro-apoptotic mechanisms involved in PE-mediated prevention of rat mammary tumorigenesis [219, 220].

Pomegranate peel extract (PPE) enhanced intracellular adhesion molecules (ICAM1) and E-cadherin proteins, whereas it suppressed cell migration-associated proteins such as MMP9, fibronectin, VEGF, vimentin, ZEB1, and β-catenin. This represents the antimetastatic properties of PPE treatment for TNBC [221, 222]. Punic acid isolated from pomegranate seed induced MCF7 and MDA-MB-231 cell cycle arrest at the G0/G1 phase. Further, it also suppressed the levels of VEGF and pro-inflammatory cytokines such as IL-2, IL-6, IL-12, IL-17, IP-10, MIP-1α, MIP-1β, MCP-1, and TNFα [223].

9. Conclusion

Breast cancer is one of the leading causes of death among women. Therapeutic strategies have been associated with adverse side effects and therefore there is a need for effective alternative therapy. Medicinal plants form the basis of modern medicine and contribute largely to commercial drug preparations manufactured today. About 25 percent of drugs prescribed worldwide are derived from plants. Currently, there is a growing interest in developing potential drugs for the treatment of breast cancer. Several attempts have been made to evaluate the therapeutic efficacy of medicinal plants and their bioactive components. However, it is very important to critically evaluate the potential of biologically active natural products for their toxicity, efficacy, and safety, ensuring the absence of any adverse effects.

Conflicts of Interest

Authors declare that there is no conflict of interest.

References

[1] Bray F, et al. Global cancer statistics 2018: GLOBOCAN estimates of incidence and mortality worldwide for 36 cancers in 185 countries. CA Cancer J Clin 2018;68(6):394–424.

[2] Chen JG, et al. Trends in the mortality of liver cancer in Qidong, China: an analysis of fifty years. Zhonghua Zhong Liu Za Zhi 2012;34(7):532–7.

[3] Torre LA, et al. Global cancer incidence and mortality rates and trends—an update. Cancer Epidemiol Biomarkers Prev 2016;25(1):16–27.

[4] Kamangar F, Dores GM, Anderson WF. Patterns of cancer incidence, mortality, and prevalence across five continents: defining priorities to reduce cancer disparities in different geographic regions of the world. J Clin Oncol 2006;24(14):2137–50.

[5] Parkin DM, et al. Global cancer statistics, 2002. CA Cancer J Clin 2005;55(2):74–108.

[6] Parkin DM, et al. Part I: cancer in indigenous Africans—burden, distribution, and trends. Lancet Oncol 2008;9(7):683–92.

[7] Singh M, et al. Distinct breast cancer subtypes in women with early-onset disease across races. Am J Cancer Res 2014;4(4):337–52.

[8] Dogra A, et al. Clinicopathological characteristics of triple negative breast cancer at a tertiary care hospital in India. Asian Pac J Cancer Prev 2014;15(24):10577–83.

[9] Shetty P. India faces growing breast cancer epidemic. Lancet 2012;379(9820):992–3.

[10] Sen S, et al. A clinical and pathological study of triple negative breast carcinoma: experience of a tertiary care centre in eastern India. J Indian Med Assoc 2012;110(10):686–9 705.

[11] Kumar P, Aggarwal R. An overview of triple-negative breast cancer. Arch Gynecol Obstet 2016;293 (2):247–69.

[12] Thummuri D, et al. Epigenetic regulation of protein tyrosine phosphatase PTPN12 in triple-negative breast cancer. Life Sci 2015;130:73–80.

[13] Lakshmaiah KC, et al. A study of triple negative breast cancer at a tertiary cancer care center in southern India. Ann Med Health Sci Res 2014;4(6):933–7.

[14] Zubeda S, et al. Her-2/neu status: a neglected marker of prognostication and management of breast cancer patients in India. Asian Pac J Cancer Prev 2013;14(4):2231–5.

[15] Nabi MG, et al. Clinicopathological comparison of triple negative breast cancers with non-triple negative breast cancers in a hospital in North India. Niger J Clin Pract 2015;18(3):381–6.

[16] Cherbal F, et al. Distribution of molecular breast cancer subtypes among Algerian women and correlation with clinical and tumor characteristics: a population-based study. Breast Dis 2015;35 (2):95–102.

[17] Nigam JS, Yadav P, Sood N. A retrospective study of clinico-pathological spectrum of carcinoma breast in a West Delhi, India. South Asian J Cancer 2014;3(3):179–81.

[18] Singh R, et al. Evaluation of ER, PR and HER-2 receptor expression in breast cancer patients presenting to a semi urban cancer centre in Western India. J Cancer Res Ther 2014;10(1):26–8.

[19] Sood N, Nigam JS. Correlation of CK5 and EGFR with clinicopathological profile of triple-negative breast cancer. Patholog Res Int 2014;2014:141864.

[20] Lehmann BD, et al. Identification of human triple-negative breast cancer subtypes and preclinical models for selection of targeted therapies. J Clin Invest 2011;121(7):2750–67.

[21] Thomas DB. Factors that promote the development of human breast cancer. Environ Health Perspect 1983;50:209–18.

[22] Wu S, et al. Evaluating intrinsic and non-intrinsic cancer risk factors. Nat Commun 2018;9(1):3490.

[23] Kate RJ, Nadig R. Stage-specific predictive models for breast cancer survivability. Int J Med Inform 2017;97:304–11.

[24] Teshome M. Role of operative management in stage IV breast cancer. Surg Clin North Am 2018;98 (4):859–68.

[25] Vuong D, et al. Molecular classification of breast cancer. Virchows Arch 2014;465(1):1–14.

[26] Maughan KL, Lutterbie MA, Ham PS. Treatment of breast cancer. Am Fam Physician 2010;81 (11):1339–46.

[27] Saikolappan S, et al. Reactive oxygen species and cancer: a complex interaction. Cancer Lett 2019;452:132–43.

[28] Valko M, et al. Free radicals and antioxidants in normal physiological functions and human disease. Int J Biochem Cell Biol 2007;39(1):44–84.

[29] Liao Z, Chua D, Tan NS. Reactive oxygen species: a volatile driver of field cancerization and metastasis. Mol Cancer 2019;18(1):65.

[30] Karar J, Maity A. PI3K/AKT/mTOR pathway in angiogenesis. Front Mol Neurosci 2011;4:51.

[31] Woo CC, et al. Thymoquinone inhibits tumor growth and induces apoptosis in a breast cancer xenograft mouse model: the role of p38 MAPK and ROS. PLoS One 2013;8(10)e75356.

[32] Hussain AR, et al. Thymoquinone suppresses growth and induces apoptosis via generation of reactive oxygen species in primary effusion lymphoma. Free Radic Biol Med 2011;50(8):978–87.

[33] Sosa V, et al. Oxidative stress and cancer: an overview. Ageing Res Rev 2013;12(1):376–90.

[34] Del Principe MI, et al. Apoptosis and immaturity in acute myeloid leukemia. Hematology 2005;10 (1):25–34.

[35] Wong RSY. Apoptosis in cancer: from pathogenesis to treatment. J Exp Clin Cancer Res 2011;30 (1):87.

[36] Kadam CY, Abhang SA. Chapter five: apoptosis markers in breast cancer therapy. In: Makowski GS, editor. Advances in clinical chemistry. Elsevier; 2016. p. 143–93.

[37] Schaeffer HJ, Weber MJ. Mitogen-activated protein kinases: specific messages from ubiquitous messengers. Mol Cell Biol 1999;19(4):2435–44.

[38] Hanahan D, Weinberg RA. The hallmarks of cancer. Cell 2000;100(1):57–70.

[39] Dhillon AS, et al. MAP kinase signalling pathways in cancer. Oncogene 2007;26(22):3279–90.

[40] Yue W, et al. Activation of the MAPK pathway enhances sensitivity of MCF-7 breast cancer cells to the mitogenic effect of estradiol. Endocrinology 2002;143(9):3221–9.

[41] Fresno Vara JA, et al. PI3K/Akt signalling pathway and cancer. Cancer Treat Rev 2004;30 (2):193–204.

[42] Paplomata E, O'Regan R. The PI3K/AKT/mTOR pathway in breast cancer: targets, trials and biomarkers. Ther Adv Med Oncol 2014;6(4):154–66.

[43] Huang S, Houghton PJ. Targeting mTOR signaling for cancer therapy. Curr Opin Pharmacol 2003;3 (4):371–7.

[44] Faivre S, Kroemer G, Raymond E. Current development of mTOR inhibitors as anticancer agents. Nat Rev Drug Discov 2006;5(8):671–88.

[45] Zhou BP, et al. HER-2/neu blocks tumor necrosis factor-induced apoptosis via the Akt/NF-kappaB pathway. J Biol Chem 2000;275(11):8027–31.

[46] Pópulo H, Lopes JM, Soares P. The mTOR signalling pathway in human cancer. Int J Mol Sci 2012;13(2):1886–918.

[47] Sharma VR, et al. PI3K/Akt/mTOR intracellular pathway and breast cancer: factors, mechanism and regulation. Curr Pharm Des 2017;23(11):1633–8.

[48] Malumbres M, Barbacid M. Cell cycle, CDKs and cancer: a changing paradigm. Nat Rev Cancer 2009;9(3):153–66.

[49] Abdullah C, Wang X, Becker D. Expression analysis and molecular targeting of cyclin-dependent kinases in advanced melanoma. Cell Cycle 2011;10(6):977–88.

[50] Georgieva J, Sinha P, Schadendorf D. Expression of cyclins and cyclin dependent kinases in human benign and malignant melanocytic lesions. J Clin Pathol 2001;54(3):229–35.

[51] Peyressatre M, et al. Targeting cyclin-dependent kinases in human cancers: from small molecules to peptide inhibitors. Cancer 2015;7(1):179–237.

[52] Husdal A, Bukholm G, Bukholm IRK. The prognostic value and overexpression of cyclin A is correlated with gene amplification of both cyclin A and cyclin E in breast cancer patient. Cell Oncol 2006;28(3):107–16.

[53] Ekberg J, et al. Expression of cyclin A1 and cell cycle proteins in hematopoietic cells and acute myeloid leukemia and links to patient outcome. Eur J Haematol 2005;75(2):106–15.

[54] Gillett C, et al. Amplification and overexpression of cyclin D1 in breast cancer detected by immunohistochemical staining. Cancer Res 1994;54(7):1812–7.

[55] Shaye A, et al. Cyclin E deregulation is an early event in the development of breast cancer. Breast Cancer Res Treat 2009;115(3):651–9.

[56] Wang M, et al. 8-Chloro-adenosine sensitizes a human hepatoma cell line to TRAIL-induced apoptosis by caspase-dependent and -independent pathways. Oncol Rep 2004;12(1):193–9.

[57] Morales J, et al. Review of poly (ADP-ribose) polymerase (PARP) mechanisms of action and rationale for targeting in cancer and other diseases. Crit Rev Eukaryot Gene Expr 2014;24(1):15–28.

[58] DiGiovanni J, Juchau MR. Biotransformation and bioactivation of 7,12-dimethylbenz[a]anthracene (7,12-DMBA). Drug Metab Rev 1980;11(1):61–101.

[59] Torre LA, et al. Global cancer statistics, 2012. CA Cancer J Clin 2015;65(2):87–108.

[60] Jazayeri SB, et al. Incidence of primary breast cancer in Iran: ten-year national cancer registry data report. Cancer Epidemiol 2015;39(4):519–27.

[61] Hamidinekoo A, et al. Deep learning in mammography and breast histology, an overview and future trends. Med Image Anal 2018;47:45–67.

[62] Gao J, et al. p53 and ATM/ATR regulate 7,12-dimethylbenz[a]anthracene-induced immunosuppression. Mol Pharmacol 2008;73(1):137–46.

[63] Nakatsugi S, et al. Chemoprevention by nimesulide, a selective cyclooxygenase-2 inhibitor, of 2-amino-1-methyl-6-phenylimidazo[4,5-b]pyridine (PhIP)-induced mammary gland carcinogenesis in rats. Jpn J Cancer Res 2000;91(9):886–92.

[64] Alfred LJ, et al. A chemical carcinogen, 3-methylcholanthrene, alters T-cell function and induces T-suppressor cells in a mouse model system. Immunology 1983;50(2):207–13.

[65] Lasfargues EY, Ozzello L. Cultivation of human breast carcinomas. J Natl Cancer Inst 1958;21(6):1131–47.

[66] Dai X, et al. Breast cancer cell line classification and its relevance with breast tumor subtyping. J Cancer 2017;8(16):3131–41.

[67] Gazdar AF, et al. Characterization of paired tumor and non-tumor cell lines established from patients with breast cancer. Int J Cancer 1998;78(6):766–74.

[68] Cailleau R, Olivé M, Cruciger QV. Long-term human breast carcinoma cell lines of metastatic origin: preliminary characterization. In Vitro 1978;14(11):911–5.

[69] Ghosh D. Chapter 64: Seed to patient in clinically proven natural medicines**Partly adapted from Zangara and Ghosh (2014), with permission from CCR Press. In: Gupta RC, editor. Nutraceuticals. Boston: Academic Press; 2016. p. 925–31.

[70] Boadu AA, Asase A. Documentation of herbal medicines used for the treatment and management of human diseases by some Communities in Southern Ghana. Evid Based Complement Alternat Med 2017;2017:3043061.

[71] Sundarraj S, et al. γ-Sitosterol from *Acacia nilotica* L. induces G2/M cell cycle arrest and apoptosis through c-Myc suppression in MCF-7 and A549 cells. J Ethnopharmacol 2012;141(3):803–9.

[72] Rabi T, et al. Novel drug amooranin induces apoptosis through caspase activity in human breast carcinoma cell lines. Breast Cancer Res Treat 2003;80(3):321–30.

[73] Kumar S, et al. Andrographolide inhibits osteopontin expression and breast tumor growth through down regulation of PI3 kinase/Akt signaling pathway. Curr Mol Med 2012;12(8):952–66.

[74] Banerjee M, et al. Cytotoxicity and cell cycle arrest induced by andrographolide lead to programmed cell death of MDA-MB-231 breast cancer cell line. J Biomed Sci 2016;23:40.

[75] Mishra SK, et al. Andrographolide and analogues in cancer prevention. Front Biosci 2015;7:255–66.

[76] Menon V, Bhat S. Anticancer activity of andrographolide semisynthetic derivatives. Nat Prod Commun 2010;5(5):717–20.

[77] Roham PH, et al. Induction of mitochondria mediated apoptosis in human breast cancer cells (T-47D) by *Annona reticulata* L. Leaves methanolic extracts. Nutr Cancer 2016;68(2):305–11.

[78] Shafi G, et al. Artemisia absinthium (AA): a novel potential complementary and alternative medicine for breast cancer. Mol Biol Rep 2012;39(7):7373–9.

[79] Rautray S, et al. Anticancer activity of *Adiantum capillus* veneris and *Pteris quadriureta* L. in human breast cancer cell lines. Mol Biol Rep 2018;45(6):1897–911.

[80] Kamalanathan D, Natarajan D. Anticancer potential of leaf and leaf-derived callus extracts of *Aerva javanica* against MCF-7 breast cancer cell line. J Cancer Res Ther 2018;14(2):321–7.

[81] Diab KAE, et al. In vitro anticancer activities of *Anogeissus latifolia, Terminalia bellerica, Acacia catechu* and *Moringa oleifera* indian plants. Asian Pac J Cancer Prev 2015;16(15):6423–8.

[82] George VC, et al. Quantitative assessment of the relative antineoplastic potential of the n-butanolic leaf extract of *Annona muricata* Linn. in normal and immortalized human cell lines. Asian Pac J Cancer Prev 2012;13(2):699–704.

[83] Prabhakaran K, et al. Polyketide natural products, acetogenins from graviola (*Annona muricata* L.), its biochemical, cytotoxic activity and various analyses through computational and bio-programming methods. Curr Pharm Des 2016;22(34):5204–10.

[84] Bhagat J, et al. Cytotoxic potential of *Anisochilus carnosus* (L.f.) wall and estimation of luteolin content by HPLC. BMC Complement Altern Med 2014;14:421.

[85] Sahu N, et al. Extraction, fractionation and re-fractionation of *Artemisia nilagirica* for anticancer activity and HPLC-ESI-QTOF-MS/MS determination. J Ethnopharmacol 2018;213:72–80.

[86] Kant K, Lal UR, Ghosh M. In silico prediction and wet lab validation of *Arisaema tortuosum* (Wall.) schott extracts as antioxidant and anti-breast cancer source: a comparative study. Pharmacogn Mag 2018;13(Suppl 4):S786–90.

[87] Khazaei S, et al. Cytotoxicity and proapoptotic effects of *Allium atroviolaceum* flower extract by modulating cell cycle arrest and caspase-dependent and p53-independent pathway in breast cancer cell lines. Evid Based Complement Alternat Med 2017;2017:1468957.

[88] Kaur V, et al. Inhibitory activities of butanol fraction from *Butea monosperma* (Lam.) Taub. Bark against free radicals, genotoxins and cancer cells. Chem Biodivers 2017;14(6). https://doi.org/10.1002/cbdv.201600484.

[89] Kumar S, et al. Antiproliferative and apoptotic effects of black turtle bean extracts on human breast cancer cell line through extrinsic and intrinsic pathway. Chem Cent J 2017;11(1):56.

[90] Johnson Inbaraj J, et al. Cytotoxicity, redox cycling and photodynamic action of two naturally occurring quinones. Biochim Biophys Acta 1999;1472(3):462–70.

[91] Bassan P, et al. Extraction, profiling and bioactivity analysis of volatile glucosinolates present in oil extract of *Brassica juncea* var. raya. Physiol Mol Biol Plants 2018;24(3):399–409.

[92] Sreeja S, Sreeja S. An in vitro study on antiproliferative and antiestrogenic effects of *Boerhaavia diffusa* L. extracts. J Ethnopharmacol 2009;126(2):221–5.

[93] Thummuri D, et al. Boswellia ovalifoliolata abrogates ROS mediated NF-κB activation, causes apoptosis and chemosensitization in triple negative breast cancer cells. Environ Toxicol Pharmacol 2014;38(1):58–70.

[94] Mishra T, et al. Isolation, characterization and anticancer potential of cytotoxic triterpenes from betula utilis bark. PLoS One 2016;11(7):e0159430.

[95] Park SE, et al. Induction of apoptosis in MDA-MB-231 human breast carcinoma cells with an ethanol extract of *Cyperus rotundus* L. by activating caspases. Oncol Rep 2014;32(6):2461–70.

[96] Mannareddy P, et al. Cytotoxic effect of *Cyperus rotundus* rhizome extract on human cancer cell lines. Biomed Pharmacother 2017;95:1375–87.

[97] Wang F, et al. The treatment role of *Cyperus rotundus* L. to triple-negative breast cancer cells. Biosci Rep 2019;39(6)BSR20190502.

[98] Khoobchandani M, et al. Chenopodium album prevents progression of cell growth and enhances cell toxicity in human breast cancer cell lines. Oxid Med Cell Longev 2009;2(3):160–5.

[99] Neethu PV, Suthindhiran K, Jayasri MA. Methanolic extract of costus pictus D. DON induces cytotoxicity in liver hepatocellular carcinoma cells mediated by histone deacetylase inhibition. Pharmacogn Mag 2017;13(Suppl 3):S533–8.

[100] Pitchai D, Roy A, Banu S. In vitro and in silico evaluation of NF-κB targeted costunolide action on estrogen receptor-negative breast cancer cells—a comparison with normal breast cells. Phytother Res 2014;28(10):1499–505.

[101] Prasanna R, et al. Anti-cancer effect of *Cassia auriculata* leaf extract in vitro through cell cycle arrest and induction of apoptosis in human breast and larynx cancer cell lines. Cell Biol Int 2009;33(2):127–34.

[102] Lambertini E, et al. Effects of extracts from Bangladeshi medicinal plants on in vitro proliferation of human breast cancer cell lines and expression of estrogen receptor alpha gene. Int J Oncol 2004;24(2):419–23.

[103] Looi CY, et al. Induction of apoptosis in human breast cancer cells via caspase pathway by vernodalin isolated from *Centratherum anthelminticum* (L.) seeds. PLoS One 2013;8(2)e56643.

[104] Ananda Sadagopan SK, et al. Forkhead Box Transcription Factor (FOXO3a) mediates the cytotoxic effect of vernodalin in vitro and inhibits the breast tumor growth in vivo. J Exp Clin Canc Res 2015;34:147.

[105] Arya A, et al. Chloroform fraction of *Centratherum anthelminticum* (L.) seed inhibits tumor necrosis factor alpha and exhibits pleotropic bioactivities: inhibitory role in human tumor cells. Evid Based Complement Alternat Med 2012;2012:627256.

[106] Moirangthem DS, et al. *Cephalotaxus griffithii* Hook.f. needle extract induces cell cycle arrest, apoptosis and suppression of hTERT and hTR expression on human breast cancer cells. BMC Complement Altern Med 2014;14:305.

[107] Chowdhury K, et al. Colocynth extracts prevent epithelial to mesenchymal transition and stemness of breast cancer cells. Front Pharmacol 2017;8:593.

[108] Naik Bukke A, et al. In vitro studies data on anticancer activity of *Caesalpinia sappan* L. heartwood and leaf extracts on MCF7 and A549 cell lines. Data Brief 2018;19:868–77.

[109] Masood Ur R, et al. Synthesis and biological evaluation of novel 3-O-tethered triazoles of diosgenin as potent antiproliferative agents. Steroids 2017;118:1–8.

[110] Ghate NB, et al. Sundew plant, a potential source of anti-inflammatory agents, selectively induces G2/M arrest and apoptosis in MCF-7 cells through upregulation of p53 and Bax/Bcl-2 ratio. Cell Death Dis 2016;2:15062.

[111] Mohanakumara P, et al. *Dysoxylum binectariferum* Hook.f (Meliaceae), a rich source of rohitukine. Fitoterapia 2010;81(2):145–8.

[112] Shriram V, et al. Cytotoxic activity of 9,10-dihydro-2,5-dimethoxyphenanthrene-1,7-diol from *Eulophia nuda* against human cancer cells. J Ethnopharmacol 2010;128(1):251–3.

[113] Kumar S, Kashyap P. Antiproliferative activity and nitric oxide production of a methanolic extract of *Fraxinus micrantha* on Michigan cancer foundation-7 mammalian breast carcinoma cell line. J Intercult Ethnopharmacol 2015;4(2):109–13.

[114] Haneef J, et al. Bax translocation mediated mitochondrial apoptosis and caspase dependent photosensitizing effect of *Ficus religiosa* on cancer cells. PLoS One 2012;7(7):e40055.

[115] Sharma R, et al. Glycyrrhiza glabra extract and quercetin reverses cisplatin resistance in triple-negative MDA-MB-468 breast cancer cells via inhibition of cytochrome P450 1B1 enzyme. Bioorg Med Chem Lett 2017;27(24):5400–3.

[116] Sebastian KS, Thampan RV. Differential effects of soybean and fenugreek extracts on the growth of MCF-7 cells. Chem Biol Interact 2007;170(2):135–43.

[117] Jamila N, et al. In vivo carbon tetrachloride-induced hepatoprotective and in vitro cytotoxic activities of garcinia hombroniana (seashore mangosteen). Afr J Tradit Complement Altern Med 2017;14(2):374–82.

[118] Shoja MH, et al. *Glycosmis pentaphylla* (Retz.) DC arrests cell cycle and induces apoptosis via caspase-3/7 activation in breast cancer cells. J Ethnopharmacol 2015;168:50–60.

[119] Hasan TN, et al. Anti-proliferative effects of organic extracts from root bark of *Juglans Regia* L. (RBJR) on MDA-MB-231 human breast cancer cells: role of Bcl-2/Bax, caspases and Tp53. Asian Pac J Cancer Prev 2011;12(2):525–30.

[120] Mohanty SK, et al. Evaluation of antioxidant, in vitro cytotoxicity of micropropagated and naturally grown plants of *Leptadenia reticulata* (Retz.) Wight & Arn.-an endangered medicinal plant. Asian Pac J Trop Med 2014;7S1:S267–71.

[121] Rawat P, et al. Phytochemicals and cytotoxicity of launaea procumbens on human cancer cell lines. Pharmacogn Mag 2016;12(Suppl 4):S431–5.

[122] Ismail A, et al. Cytotoxicity and proteasome inhibition by alkaloid extract from *Murraya koenigii* leaves in breast cancer cells-molecular docking studies. J Med Food 2016;19(12):1155–65.

[123] Noolu B, et al. *Murraya koenigii* leaf extract inhibits proteasome activity and induces cell death in breast cancer cells. BMC Complement Altern Med 2013;13:7.

[124] Yeap SK, et al. Chemopreventive and immunomodulatory effects of *Murraya koenigii* aqueous extract on 4T1 breast cancer cell-challenged mice. BMC Complement Altern Med 2015;15:306.

[125] Sharma K, et al. Anticancer effects of extracts from the fruit of *Morinda citrifolia* (noni) in breast cancer cell lines. Drug Res 2016;66(3):141–7.

[126] Sinha S, et al. *Mucuna pruriens* (L.) DC chemo sensitize human breast cancer cells via downregulation of prolactin-mediated JAK2/STAT5A signaling. J Ethnopharmacol 2018;217:23–35.

[127] Deepa M, Priya S. Purification and characterization of a novel anti-proliferative lectin from *Morus alba* L. leaves. Protein Pept Lett 2012;19(8):839–45.

[128] Deepa M, et al. Antioxidant rich Morus alba leaf extract induces apoptosis in human colon and breast cancer cells by the downregulation of nitric oxide produced by inducible nitric oxide synthase. Nutr Cancer 2013;65(2):305–10.

[129] Sodde VK, et al. Cytotoxic activity of *Macrosolen parasiticus* (L.) Danser on the growth of breast cancer cell line (MCF-7). Pharmacogn Mag 2015;11(Suppl 1):S156–60.

[130] Tiwary BK, et al. The in vitro cytotoxic activity of ethno-pharmacological important plants of Darjeeling district of West Bengal against different human cancer cell lines. BMC Complement Altern Med 2015;15:22.

[131] Ghosh P, et al. *Madhuca indica* inhibits breast cancer cell proliferation by modulating COX-2 expression. Curr Mol Med 2018;18(7):459–74.

[132] Chaudhary S, et al. Evaluation of antioxidant and anticancer activity of extract and fractions of *Nardostachys jatamansi* DC in breast carcinoma. BMC Complement Altern Med 2015;15:50.

[133] Pandeti S, et al. Synthesis of novel anticancer iridoid derivatives and their cell cycle arrest and caspase dependent apoptosis. Phytomedicine 2014;21(3):333–9.

[134] Singh S, et al. Antiproliferative and antimicrobial efficacy of the compounds isolated from the roots of *Oenothera biennis* L. J Pharm Pharmacol 2017;69(9):1230–43.

[135] Singh MK, Dhongade H, Tripathi DK. *Orthosiphon pallidus*, a potential treatment for patients with breast cancer. J Pharm 2017;20(4):265–73.

[136] Naveen Kumar DR, et al. Cytotoxicity, apoptosis induction and anti-metastatic potential of *Oroxylum indicum* in human breast cancer cells. Asian Pac J Cancer Prev 2012;13(6):2729–34.

[137] Ghate NB, et al. An antioxidant extract of tropical lichen, *Parmotrema reticulatum*, induces cell cycle arrest and apoptosis in breast carcinoma cell line MCF-7. PLoS One 2013;8(12):e82293.

[138] Rajan V, et al. Mechanism of cytotoxicity by *Psoralea corylifolia* extract in human breast carcinoma cells. J Environ Pathol Toxicol Oncol 2014;33(3):265–77.

[139] Sharma M. Selective cytotoxicity and modulation of apoptotic signature of breast cancer cells by *Pithecellobium dulce* leaf extracts. Biotechnol Prog 2016;32(3):756–66.

[140] Vini R, Juberiya AM, Sreeja S. Evidence of pomegranate methanolic extract in antagonizing the endogenous SERM, 27-hydroxycholesterol. IUBMB Life 2016;68(2):116–21.

[141] Joseph MM, et al. Evaluation of antioxidant, antitumor and immunomodulatory properties of polysaccharide isolated from fruit rind of *Punica granatum*. Mol Med Rep 2012;5(2):489–96.

[142] Rajkumar V, Guha G, Kumar RA. Antioxidant and anti-neoplastic activities of *Picrorhiza kurroa* extracts. Food Chem Toxicol 2011;49(2):363–9.

[143] Ganie SA, et al. Long dose exposure of hydrogen peroxide (H_2O_2) in albino rats and effect of *Podophyllum hexandrum* on oxidative stress. Eur Rev Med Pharmacol Sci 2011;15(8):906–15.

[144] Kaur P, et al. Suppression of SOS response in *E. coli* PQ 37, antioxidant potential and antiproliferative action of methanolic extract of *Pteris vittata* L. on human MCF-7 breast cancer cells. Food Chem Toxicol 2014;74:326–33.

[145] Sharma R, et al. Furanoflavones pongapin and lanceolatin B blocks the cell cycle and induce senescence in CYP1A1-overexpressing breast cancer cells. Bioorg Med Chem 2018;26(23–24):6076–86.

[146] Kumar Singh S, Patra A. Evaluation of phenolic composition, antioxidant, anti-inflammatory and anticancer activities of *Polygonatum verticillatum* (L.). J Integr Med 2018;16(4):273–82.

[147] Suma HK, et al. *Pyrenacantha volubilis* Wight, (Icacinaceae) a rich source of camptothecine and its derivatives, from the Coromandel Coast forests of India. Fitoterapia 2014;97:105–10.

[148] Rajkumar V, Guha G, Kumar RA. Apoptosis induction in MDA-MB-435S, Hep3B and PC-3 cell lines by Rheum emodi rhizome extracts. Asian Pac J Cancer Prev 2011;12(5):1197–200.

[149] Dutta S, Khanna A. Aglycone rich extracts of phytoestrogens cause ROS-mediated DNA damage in breast carcinoma cells. Biomed Pharmacother 2016;84:1513–23.

[150] Mathivadhani P, Shanthi P, Sachdanandam P. Apoptotic effect of *Semecarpus anacardium* nut extract on T47D breast cancer cell line. Cell Biol Int 2007;31(10):1198–206.

[151] Chakraborty S, et al. Cytotoxic effect of root extract of *Tiliacora racemosa* and oil of *Semecarpus anacardium* nut in human tumour cells. Phytother Res 2004;18(8):595–600.

[152] Sowmyalakshmi S, et al. Investigation on *Semecarpus lehyam*—a Siddha medicine for breast cancer. Planta 2005;220(6):910–8.

[153] Dwivedi V, et al. Comparative anticancer potential of clove (*Syzygium aromaticum*)—an Indian spice—against cancer cell lines of various anatomical origin. Asian Pac J Cancer Prev 2011;12(8):1989–93.

[154] Yadav NK, et al. Saraca indica bark extract shows in vitro antioxidant, antibreast cancer activity and does not exhibit toxicological effects. Oxid Med Cell Longev 2015;2015:205360.

[155] Musini A, Rao JP, Giri A. Phytochemicals of *Salacia oblonga* responsible for free radical scavenging and antiproliferative activity against breast cancer cell lines (MDA-MB-231). Physiol Mol Biol Plants 2015;21(4):583–90.

[156] Ansari JA, et al. ROS mediated pro-apoptotic effects of *Tinospora cordifolia* on breast cancer cells. Front Biosci 2017;9:89–100.

[157] Johnson Inbaraj J, Gandhidasan R, Murugesan R. Cytotoxicity and superoxide anion generation by some naturally occurring quinones. Free Radic Biol Med 1999;26(9–10):1072–8.

[158] Chattopadhyay SK, et al. Absolute configuration and anticancer activity of taxiresinol and related lignans of *Taxus wallichiana*. Bioorg Med Chem 2003;11(23):4945–8.

[159] Appadath Beeran A, et al. The enriched fraction of *Vernonia cinerea* L. induces apoptosis and inhibits multi-drug resistance transporters in human epithelial cancer cells. J Ethnopharmacol 2014;158(Pt A):33–42.

[160] Chakravarti B, et al. In vitro anti-breast cancer activity of ethanolic extract of *Wrightia tomentosa*: role of pro-apoptotic effects of oleanolic acid and urosolic acid. J Ethnopharmacol 2012;142(1):72–9.

[161] Dar PA, et al. An anti-cancerous protein fraction from *Withania somnifera* induces ROS-dependent mitochondria-mediated apoptosis in human MDA-MB-231 breast cancer cells. Int J Biol Macromol 2019;135:77–87.

[162] Vyry Wouatsa NA, et al. Zantholic acid, a new monoterpenoid from *Zanthoxylum zanthoxyloides*. Nat Prod Res 2013;27(21):1994–8.

[163] Mallick MN, et al. Exploring the cytotoxic potential of triterpenoids-enriched fraction of bacopa monnieri by implementing in vitro, in vivo, and in silico approaches. Pharmacogn Mag 2017;13 (Suppl 3):S595–606.

[164] Karia P, Patel KV, Rathod SSP. Breast cancer amelioration by *Butea monosperma* in-vitro and in-vivo. J Ethnopharmacol 2018;217:54–62.

[165] Singh SK, et al. Chemically standardized isolates from *Cedrus deodara* stem wood having anticancer activity. Planta Med 2007;73(6):519–26.

[166] Choudhury B, et al. *Garcinia morella* fruit, a promising source of antioxidant and anti-inflammatory agents induces breast cancer cell death via triggering apoptotic pathway. Biomed Pharmacother 2018;103:562–73.

[167] Noolu B, et al. In vivo inhibition of proteasome activity and tumour growth by *Murraya koenigii* leaf extract in breast cancer xenografts and by its active flavonoids in breast cancer cells. Anticancer Agents Med Chem 2016;16(12):1605–14.

[168] Yuan C, et al. Inhibition on the growth of human MDA-MB-231 breast cancer cells in vitro and tumor growth in a mouse xenograft model by Se-containing polysaccharides from *Pyracantha fortuneana*. Nutr Res 2016;36(11):1243–54.

[169] Majumder M, et al. *Ricinus communis* L. fruit extract inhibits migration/invasion, induces apoptosis in breast cancer cells and arrests tumor progression in vivo. Sci Rep 2019;9(1):14493.

[170] Choi BY, Kim B-W. Withaferin-A inhibits colon cancer cell growth by blocking STAT3 transcriptional activity. J Cancer Prev 2015;20(3):185–92.

[171] Ray SD, Dewanjee S. Isolation of a new triterpene derivative and in vitro and in vivo anticancer activity of ethanolic extract from root bark of *Zizyphus nummularia* Aubrev. Nat Prod Res 2015;29(16):1529–36.

[172] Sareer O, Ahmad S, Umar S. *Andrographis paniculata*: a critical appraisal of extraction, isolation and quantification of andrographolide and other active constituents. Nat Prod Res 2014;28(23):2081–101.

[173] Akbar S. *Andrographis paniculata*: a review of pharmacological activities and clinical effects. Altern Med Rev 2011;16(1):66–77.

[174] Okhuarobo A, et al. Harnessing the medicinal properties of *Andrographis paniculata* for diseases and beyond: a review of its phytochemistry and pharmacology. Asian Pac J Trop Dis 2014;4(3):213–22.

[175] Islam MT, et al. Andrographolide, a diterpene lactone from *Andrographis paniculata* and its therapeutic promises in cancer. Cancer Lett 2018;420:129–45.

[176] Dar NJ, Hamid A, Ahmad M. Pharmacologic overview of *Withania somnifera*, the Indian Ginseng. Cell Mol Life Sci 2015;72(23):4445–60.

[177] Rai M, et al. Anticancer activities of *Withania somnifera*: current research, formulations, and future perspectives. Pharm Biol 2016;54(2):189–97.

[178] Dutta R, et al. *Withania Somnifera* (Ashwagandha) and Withaferin A: potential in integrative oncology. Int J Mol Sci 2019;20(21):5310.

[179] Dar PA, et al. Unique medicinal properties of *Withania somnifera*: phytochemical constituents and protein component. Curr Pharm Des 2016;22(5):535–40.

[180] Palliyaguru DL, Singh SV, Kensler TW. *Withania somnifera*: from prevention to treatment of cancer. Mol Nutr Food Res 2016;60(6):1342–53.

[181] Khazal KF, Hill DL. *Withania somnifera* extract reduces the invasiveness of MDA-MB-231 breast cancer and inhibits cytokines associated with metastasis. J Cancer Metastasis Treatment 2015;1(2):94–100.

[182] Ghosh K, et al. Withaferin A induced impaired autophagy and unfolded protein response in human breast cancer cell-lines MCF-7 and MDA-MB-231. Toxicol In Vitro 2017;44:330–8.

[183] Singh D, Chaudhuri PK. Chemistry and pharmacology of *Tinospora cordifolia*. Nat Prod Commun 2017;12(2):299–308.

[184] Sharma P, et al. The chemical constituents and diverse pharmacological importance of *Tinospora cordifolia*. Heliyon 2019;5(9)e02437.

[185] Rashmi KC, et al. A new pyrrole based small molecule from *Tinospora cordifolia* induces apoptosis in MDA-MB-231 breast cancer cells via ROS mediated mitochondrial damage and restoration of p53 activity. Chem Biol Interact 2019;299:120–30.

[186] Samanta SK, et al. Phytochemical portfolio and anticancer activity of *Murraya koenigii* and its primary active component, mahanine. Pharmacol Res 2018;129:227–36.

[187] Utaipan T, et al. Carbazole alkaloids from *Murraya koenigii* trigger apoptosis and autophagic flux inhibition in human oral squamous cell carcinoma cells. J Nat Med 2017;71(1):158–69.

[188] Iman V, et al. Anticancer and anti-inflammatory activities of girinimbine isolated from *Murraya koenigii*. Drug Des Devel Ther 2016;11:103–21.

[189] Kamala A, Middha SK, Karigar CS. Plants in traditional medicine with special reference to *Cyperus rotundus* L.: a review. 3 Biotech 2018;8(7):309.

[190] Pirzada AM, et al. *Cyperus rotundus* L.: traditional uses, phytochemistry, and pharmacological activities. J Ethnopharmacol 2015;174:540–60.

[191] Suryavanshi S, et al. A polyherbal formulation, HC9 regulated cell growth and expression of cell cycle and chromatin modulatory proteins in breast cancer cell lines. J Ethnopharmacol 2019;242:112022.

[192] Arya A, et al. Anti-diabetic effects of *Centratherum anthelminticum* seeds methanolic fraction on pancreatic cells, β-TC6 and its alleviating role in type 2 diabetic rats. J Ethnopharmacol 2012;144 (1):22–32.

[193] Delaviz H, et al. A review study on phytochemistry and pharmacology applications of *Juglans regia* plant. Pharmacogn Rev 2017;11(22):145–52.

[194] Cosmulescu S. Seasonal variation of total phenols in leaves of walnut (*Juglans regia* L.). J Med Plant Res 2011;52:4938.

[195] Li Q, et al. Azacyclo-indoles and phenolics from the flowers of *Juglans regia*. J Nat Prod 2017;80 (8):2189–98.

[196] Luo J-J, et al. Chemical constituents from the flower of *Juglans regia*. Zhong Yao Cai 2012;35 (10):1614–6.

[197] Sun Z-L, Dong J-L, Wu J. Juglanin induces apoptosis and autophagy in human breast cancer progression via ROS/JNK promotion. Biomed Pharmacother 2017;85:303–12.

[198] Salimi M, et al. Anti-proliferative and apoptotic activities of constituents of chloroform extract of *Juglans regia* leaves. Cell Prolif 2014;47(2):172–9.

[199] Salimi M, et al. Cytotoxicity effects of various *Juglans regia* (walnut) leaf extracts in human cancer cell lines. Pharm Biol 2012;50(11):1416–22.

[200] Vanden Heuvel JP, et al. Mechanistic examination of walnuts in prevention of breast cancer. Nutr Cancer 2012;64(7):1078–86.

[201] Flower G, et al. Flax and breast cancer: a systematic review. Integr Cancer Ther 2014;13(3):181–92.

[202] Hu T, et al. Flaxseed extract induces apoptosis in human breast cancer MCF-7 cells. Food Chem Toxicol 2019;127:188–96.

[203] Mason JK, Thompson LU. Flaxseed and its lignan and oil components: can they play a role in reducing the risk of and improving the treatment of breast cancer? Appl Physiol Nutr Metab 2014;39 (6):663–78.

[204] Di Y, et al. Flaxseed lignans enhance the cytotoxicity of chemotherapeutic agents against breast cancer cell lines MDA-MB-231 and SKBR3. Nutr Cancer 2018;70(2):306–15.

[205] Delman DM, et al. Effects of flaxseed lignan secoisolariciresinol diglucosideon preneoplastic biomarkers of cancer progression in a model of simultaneous breast and ovarian cancer development. Nutr Cancer 2015;67(5):857–64.

[206] Czemplik M, et al. Flavonoid C-glucosides derived from flax straw extracts reduce human breast cancer cell growth in vitro and induce apoptosis. Front Pharmacol 2016;7:282.

[207] Lee J, Cho K. Flaxseed sprouts induce apoptosis and inhibit growth in MCF-7 and MDA-MB-231 human breast cancer cells. In Vitro Cell Dev Biol Anim 2012;48(4):244–50.

[208] Kubatka P, et al. Antineoplastic effects of clove buds (*Syzygium aromaticum* L.) in the model of breast carcinoma. J Cell Mol Med 2017;21(11):2837–51.

[209] Yan X, et al. Eugenol inhibits oxidative phosphorylation and fatty acid oxidation via downregulation of c-Myc/PGC-1β/ERRα signaling pathway in MCF10A-ras cells. Sci Rep 2017;7(1):12920.

[210] Anita Y, et al. Structure-based design of eugenol analogs as potential estrogen receptor antagonists. Bioinformation 2012;8(19):901–6.

[211] Kumar PS, et al. Anticancer potential of *Syzygium aromaticum* L. in MCF-7 human breast cancer cell lines. Pharm Res 2014;6(4):350–4.

[212] Khan FA, et al. Extracts of clove (*Syzygium aromaticum*) potentiate FMSP-nanoparticles induced cell death in MCF-7 cells. Int J Biomater 2018;2018:8479439.

[213] Mathivadhani P, Shanthi P, Sachdanandam P. Effect of *Semecarpus anacardium* Linn. nut extract on mammary and hepatic expression of xenobiotic enzymes in DMBA-induced mammary carcinoma. Environ Toxicol Pharmacol 2007;23(3):328–34.

[214] Mathivadhani P, Shanthi P, Sachdanandam P. Hypoxia and its downstream targets in DMBA induced mammary carcinoma: protective role of *Semecarpus anacardium* nut extract. Chem Biol Interact 2007;167(1):31–40.

[215] Veena K, Shanthi P, Sachdanandam P. The biochemical alterations following administration of Kalpaamruthaa and *Semecarpus anacardium* in mammary carcinoma. Chem Biol Interact 2006;161 (1):69–78.

[216] Sathish S, Shanthi P, Sachdanandam P. Mitigation of DMBA-induced mammary carcinoma in experimental rats by antiangiogenic property of Kalpaamruthaa. J Diet Suppl 2011;8(2):144–57.

[217] Veena K, Shanthi P, Sachdanandam P. Therapeutic efficacy of Kalpaamruthaa on reactive oxygen/nitrogen species levels and antioxidative system in mammary carcinoma bearing rats. Mol Cell Biochem 2007;294(1–2):127–35.

[218] Zhao W, et al. Identification of urushiols as the major active principle of the Siddha herbal medicine *Semecarpus lehyam*: anti-tumor agents for the treatment of breast cancer. Pharm Biol 2009;47 (9):886–93.

[219] Mandal A, Bhatia D, Bishayee A. Anti-inflammatory mechanism involved in pomegranate-mediated prevention of breast cancer: the role of NF-κB and Nrf2 signaling pathways. Nutrients 2017;9(5):436.

[220] Mandal A, Bishayee A. Mechanism of breast cancer preventive action of pomegranate: disruption of estrogen receptor and Wnt/β-catenin signaling pathways. Molecules 2015;20(12):22315–28.

[221] Ahmadiankia N, Bagheri M, Fazli M. Gene expression changes in pomegranate peel extract-treated triple-negative breast cancer cells. Rep Biochem Mol Biol 2018;7(1):102–9.

[222] Bagheri M, et al. Pomegranate peel extract inhibits expression of β-catenin, epithelial mesenchymal transition, and metastasis in triple negative breast cancer cells. Cell Mol Biol 2018;64(7):86–91.

[223] Costantini S, et al. Potential anti-inflammatory effects of the hydrophilic fraction of pomegranate (*Punica granatum* L.) seed oil on breast cancer cell lines. Molecules 2014;19(6):8644–60.

CHAPTER 8

Carboplatin and paclitaxel: Role in the treatment of endometrial cancer

Sreedevi Muttathuveliyil Sivadasan, Pavan Kumar Kancharla, and Neelakantan Arumugam
Department of Biotechnology, School of Life Sciences, Pondicherry University, Puducherry, India

Abstract

Endometrial cancer is a common type of cancer affecting women. It is the most common form of uterine cancer and originates from the endometrium often after menopause. Over the last 10 years, on an average 1.9% increase in mortality rate has been reported. Based on extent of severity endometrial cancer is classified into eight stages: IA, IB, II, IIIA, IIIB, IIIC, IVA, and IVB. The major risk factors leading to endometrial cancer are obesity, type 2 diabetes, not giving birth, breast cancer treatment with tamoxifen, estrogen hormone replacement therapy, polycystic ovarian syndrome (PCOS), abnormal metabolic conditions, and family history of the disease. The treatments available to cure endometrial cancer include surgery, radiation therapy, hormone therapy, chemotherapy, and targeted drug therapy. Carboplatin in combination with paclitaxel has been the most efficient and least toxic for chemotherapy. These drugs are also highly administered in several other cancer types because of their unique properties. Carboplatin, a derivative of cisplatin, is a less toxic and more stable compound. This drug is capable of causing alkylation leading to production of reactive platinum complexes that damage DNA. Paclitaxel, on the other hand, interacts with the microtubules of the cell and hyperstabilizes them. This disrupts the normal cell cycle allowing cells to undergo apoptosis. Although solo drug therapy induced a lower level of overall response rate, it increased significantly when administered in combination. Favorable results were also obtained when these two drugs were administered in conjunction with other forms of cancer therapy such as radiation, surgery, and hormone therapy.

Keywords: Cancer, Endometrial cancer, Carboplatin, Paclitaxel, Chemotherapy.

Abbreviations

AIs	aromatase inhibitors
AUC	area under the curve
DNA	deoxyribonucleic acid
EGCG	epigallocatechin gallate
LHRHA	luteinizing hormone-releasing hormone agonists
mTOR	mechanistic target of rapamycin
NIH	National Institutes of Health
PLGA	poly-L-lactide-*co*-glycolic acid
SEER	surveillance epidemiology and end result program

G.P. Nagaraju, R.R. Malla (eds.)
A Theranostic and Precision Medicine Approach for Female-Specific Cancers
ISBN 978-0-12-822009-2
https://doi.org/10.1016/B978-0-12-822009-2.00008-X

1. Introduction

Endometrial cancer (EC) is the fourth most common type of cancer in women worldwide. It originates from the inner lining of the uterus and therefore is termed endometrial carcinoma. In fact, 95% of all uterine cancers originate from the endometrium [1]. Uterine sarcoma is different from EC and it occurs in the uterine muscles. Therefore, the term uterine cancer can't be used interchangeably with endometrial cancer. Based on histology, there are six types of endometrial cancers recognized. They are adenocarcinoma, uterine carcinosarcoma, squamous cell carcinoma, small cell carcinoma, transitional carcinoma, and serous carcinoma.

According to the 2019 Surveillance Epidemiology and End Result Program (SEER) report, uterine cancers account for 3.5% of all new cancers. The report indicated that there were 61,880 new cases of uterine cancers and 12,160 people died in 2019. Over the last 10 years, new cases have increased by 1% and the death rate has increased 1.9% per year.

The common symptoms of EC are abnormal menstrual bleeding, bleeding after menopause, bleeding unrelated to menstruation, pain in the pelvic area, and pain during intercourse. Most of the time, EC is diagnosed postmenopause. However there are also cases where it is diagnosed before menopause. In extremely rare cases it appears to coexist during pregnancy with the pregnancy observed to have a protective role of containing tumor growth. Treatment of pregnant women with EC is more difficult due to the lack of availability of clinical trial data [2].

2. Major risk factors for developing endometrial cancer

The risk factors of EC include obesity, type 2 diabetes, polycystic ovarian syndrome (PCOS), metabolic syndrome, family history of endometrial cancer, genetic conditions like Lynch syndrome, and endometrial hyperplasia. Studies have earlier confirmed that patients undergoing estrogen therapy have 4.5 times greater risk of EC [3]. Breast cancer patients under treatment with tamoxifen are also at risk of EC. This drug affects the endometrial cells through estrogen pathways [4]. But there are studies that show treatment with estrogen in combination with progesterone reduced the risk of getting EC [5].

The chance of getting EC is greater for women who have never given birth to a child. In addition, start of menstruation at an early age or a late menopause also contribute to chances of increasing the exposure of the endometrium to estrogen, leading to EC [5]. Advanced age is another important risk factor. The chances of getting cancer increases with advancing age. This holds true not only for EC but other cancers too [5].

3. Regimens for treatment of endometrial cancer

The choice of treatment for EC depends on various factors such as age of the patient, stage of cancer, type of cancer, location of cancer, overall health condition, and other personal considerations. Surgery is the main treatment for endometrial cancer. However, other forms of treatment, especially the combination of different types of treatments, are performed to achieve better efficiency. There are five standard treatment measures that are followed for treatment of endometrial cancer. Treatment after surgery follows radiation therapy, hormone therapy, targeted therapy, and chemotherapy.

3.1 Surgery

Surgical removal of the tumor is the most commonly used treatment method in EC. The choice of surgical procedures differs depending on the location, size, and nature of EC [6]. Total hysterectomy is performed to remove the uterus and the cervix. Both the ovaries and fallopian tubes are removed by bilateral salpingo-oophorectomy. Radical hysterectomy is performed to remove the uterus, cervix, vagina, ovaries, fallopian tube, and lymph nodes. The later in the pelvic area are often removed by lymph node dissection.

3.2 Radiation therapy

Tumor regions are exposed to high-energy radiation to cause death of cancer cells. Radiation in general is emitted from an external device and focused on the cancer. Alternatively, the radioactive material is placed near the cancer region. Although treatment using radiation is an effective way of curing cancer, the chances of reoccurrence of the tumor are highest as compared to other treatment methods [7, 8].

3.3 Hormone therapy

For hormone therapy of EC, one uses compounds like progestin that slow down the growth of EC. Tamoxifens, luteinizing hormone-releasing hormone agonists (LHRHA), and aromatase inhibitors (AIs) that reduce estrogen levels in the body are also used. Hormone therapy is the choice of treatment for patients with hormone-positive receptors, low-grade endometrioid tumors, and long disease-free interval. Hormone therapy is ideal for those the patients who have no life-threatening conditions. The first line of treatment is usually progestogens followed by tamoxifen as the second line of treatment [9].

3.4 Targeted therapy

Targeted therapy is used for specifically killing cancer cells without harming normal cells. Monoclonal antibodies, mTOR inhibitors, and signal transduction inhibitors could be the drugs of choice for targeted therapy of EC. Monoclonal antibodies against epitopes specific to cancer cells can be raised and administered. They can be used alone or in combination with chemotherapeutics or radioactivity to mediate the death of the cancer cells. mTOR inhibitors such as everolimus and ridaforolimus prevent cell division and formation of new blood vessels in cancers and thus may prevent metastasis [6].

3.5 Chemotherapy

In chemotherapy, drugs that are capable of killing cancers are administered through the bloodstream to reach the targeted cancer cells. This is a systemic approach. Drugs can also be administered directly to the organ or body cavity, in which case the method is termed regional chemotherapy [6]. Chemotherapy is the choice of treatment for patients with high-grade tumors, short treatment-free intervals, histopathologic types such as serous or clear cell tumors, and negative hormone receptor levels. Patients with serious life-threatening conditions and those who cannot be treated with hormone therapy can benefit the most from chemotherapy [10].

Chemotherapy-naive patients exhibit a greater response rate to treatment than patients who had received prior chemotherapy. Hence it is important to take into account prior treatments before opting for chemotherapy in recurrent cases. There are various chemotherapy drugs available for the treatment of EC. They can be administered either as a single drug or in a combination to be a more effective therapy. The increased efficiency of combination therapy is attributed to the difference in the modes of action of the individual drugs used. Depending on age, overall health status, previous medical history, and present diseases, the treatment parameters such as the choice of drugs, method of treatment, dosage, and so on vary from patient to patient. The dosage of chemotherapeutic drugs is often calculated based on a parameter called Area under the Curve (AUC). It is a plot of drug concentration in blood plasma versus time. AUC reflects actual body exposure to the drug after the administration of a specified dose of the drug. AUC, in fact, predicts the rate of elimination of the drug from the body after administration of a certain dose of the drug. Dose reduction is advised for patients who exhibit adverse side effects.

Paclitaxel is one of the most active single agent drugs used in chemotherapy [11, 12]. In a phase II trial paclitaxel showed a response rate of 77% in chemotherapy-naive patients and a response rate of 37% was observed when it is used as second-line treatment in the clinical trial [13]. Combination therapy involves the use of a combination of more than one drug. The common drugs used are paclitaxel, doxorubicin, and cisplatin. The combination drugs often have a greater level of toxicity, making them unideal for treatment [14]. A combination of carboplatin and paclitaxel was considered to be least toxic [13, 15, 16].

Various studies have been conducted to assess the efficiency and toxicity of carboplatin–paclitaxel combination for chemotherapy at different stages of EC. In patients with surgical stages I–II serous EC, Intravaginal Radiation Therapy (IVRT) followed by carboplatin–paclitaxel treatment yielded a disease-free condition with an overall survival rate of 88% after 5 years [17, 18]. Patients suffering from stage II uterine papillary serous carcinoma (UPSC), an uncommon form of EC occurring postmenopause, are at a greater risk of extra pelvic recurrence. Paclitaxel–carboplatin treatment was found to reduce this risk and also improve progression-free survival outcomes. The different studies on EC using carboplatin and paclitaxel, their response rates, and side effects are listed in Table 1.

Sovak et al. [21] studied the overall survival as well as progression-free survival among patients with stage III and stage IV EC upon treatment with carboplatin and paclitaxel after complete surgical tumor resection. Paclitaxel was administered at a dosage of 175 mg/m^2 and carboplatin at a dosage of AUC 5–6. Six cycles of the drug combinations were administered every 3–4 weeks. At the end of the study, an overall survival rate of 56% was observed over 3 years. It was concluded that paclitaxel-carboplatin combination is a well-tolerated active regimen for the treatment of stage III and stage IV EC. In a couple of similar studies, patients with recurrent endometrial adenocarcinoma were treated with paclitaxel at a dose of 135 mg/m^2 and carboplatin at an AUC of 5. At an interval of 21 days the treatment was repeated for six cycles. An overall response rate of about 63% was observed in patients with stage IV cancer [19, 22].

Carboplatin and paclitaxel yielded a response rate of 60% in the case of advanced and recurrent uterine malignant mixed Mullerian tumors. Here carboplatin was administered at an AUC of 5–6 and paclitaxel at a dose of 175 mg/m^2. This had similar efficacy to that of phase III clinical trial with ifosfamide combinations in malignant mixed Müllerian tumors. Also, carboplatin-paclitaxel is cheaper and easier to deliver, making it a more suitable option [20]. Michener et al. [10] reported an 87% overall response rate in patients with advanced, metastatic, and recurrent EC when carboplatin and paclitaxel were administered at a 21-day interval (dosage of 4–6 AUC and 135–175 mg/m^2, respectively). Although hematotoxicity was observed, it was considered as the most efficient and the safest chemotherapy regimen. The study suggested that this treatment can be used as first-line therapy in patients with advanced or recurrent EC.

Surgery followed by chemotherapy, hormone therapy, or radiation therapy is the most commonly suggested effective combination method. Even after the first line of treatment with any of these combinations, there are reoccurrences of cancer in later stages of life. For EC, the chances of reoccurrence are approximately 13% [27] and more than half of the reoccurrence has been observed to develop within 2 years. But the chances of getting good results in second-line therapy after reoccurrence is less than that of first-line therapy.

4. Mechanism and chemical characteristics of carboplatin

Carboplatin is a small molecule possessing broad-spectrum antineoplastic activity. It is a cisplatin-derived organometallic compound. Carboplatin activity is similar to that of

Table 1 Studies on endometrial cancer using carboplatin and paclitaxel drugs.

Type of cancer	Compounds	Response rate	Side effects observed	References
Papillary serous endometrial cancer	Carboplatin and paclitaxel	90%	Nonhematologic toxicities	[18]
Serous endometrial cancer	Carboplatin and paclitaxel	88%	Neurotoxicity, peripheral Neuropathy, tinnitus, smell alteration, fatigue, constipation, dehydration, myalgia, bladder toxicity	[17]
Endometrial carcinoma	Carboplatin and paclitaxel	62%	Myelotoxicity, neutropenia, thrombocytopenia, anemia, neuropathy	[19]
Endometrial cancer	Carboplatin and paclitaxel	60%—newly diagnosed, 55%—recurrent	None	[20]
Endometrial cancer	Paclitaxel and carboplatin	56%	Neurotoxicity, nonlife threatening hypersensitivity	[21]
Endometrial adenocarcinoma	Carboplatin and paclitaxel	87%	Granulocytopenia, emesis, neuropathy, neocortical toxicity, arthralgia	[10]
Endometrial adenocarcinoma	Paclitaxel, carboplatin, and amifostine	40%	Neutropenia	[16]
Endometrial cancer	Carboplatin and paclitaxel	63%	None	[22]
Endometrial carcinoma	Paclitaxel	27.3%	Neutropenia, neurotoxicity, thrombocytopenia, weakness, gastrointestinal disorders	[23]
Endometrial adenocarcinoma	Carboplatin	13%	Nausea, vomiting, white blood cell (WBC) toxicity (grade 3)	[24]
Endometrial carcinoma	Paclitaxel and cisplatin	67%	Granulocytopenia, neurotoxicity, anemia, renal toxicity	[15]
Uterine papillary serous carcinoma	Paclitaxel	23%	Gastrointestinal toxicity, ototoxicity, neurotoxicity, granulocytopenia	[12]

Cancer type	Drug(s)	Response	Side effects	Reference
Uterine adenocarcinoma	Cisplatin, epirubicin, and paclitaxel	73%	Neutropenia, peripheral neuropathy, stomatitis, acute congestive heart failure	[25]
Endometrial adenocarcinoma	Paclitaxel	35.7%	Leukopenia, thrombocytopenia, gastrointestinal toxicity, neurotoxicity, anemia, cardiac toxicity, alopecia	[11]
Adenocarcinoma, adenosquamous carcinoma	Paclitaxel	35%	Neutropenia, thrombocytopenia, anemia, alopecia Nausea/vomiting, peripheral neuropathy, myalgia/arthralgia, itching	[13]
Endometrial adenocarcinoma	Carboplatin	28%	Nausea, emesis, anorexia, diarrhea, alopecia, renal toxicity, paresthesia	[26]

cisplatin, but it is less toxic and more stable. It was discovered by scientists at the Institute for Creation Research (ICR), who tested around 300 derivatives of cisplatin throughout the 1970s to see if they could produce a milder form. This led to the discovery of carboplatin, which is an extremely effective anticancer drug but with significantly reduced side effects. It has a molecular weight of 371.25 g/mol and its molecular formula is $C_6H_{12}N_2O_4$ Pt. It consists of a platinum atom complexed with two ammonia groups and a 1,1-cyclobutanedicarboxylic acid residues. In addition to EC, carboplatin is used for the treatment of various other cancers like cervical cancer, esophageal cancer, head and neck cancer, advanced melanoma, lung cancer, sarcoma, retinoblastoma, ovarian cancer, bladder cancer, testicular cancer, and others [28] (Fig. 1).

Carboplatin is an alkylating agent and can add an alkyl group to electronegative groups of various biomolecules in the cell. The addition of the alkyl group can increase the mismatches of base pairs in DNA. Also, once they are intracellular, they get activated by hydrolysis forming 1,1-cyclobutanedicarboxylate and reactive positively charged equated platinum complexes. These reactive platinum complexes can stably bind to G-C rich regions of the DNA. In this way, they directly attack DNA and form intrastrand or interstrand cross-links of the guanine bases of DNA and DNA-protein cross-links (Fig. 2). This prevents the unwinding of the DNA double helix during DNA replication and disrupts cell division, leading to apoptosis of the cells [29, 30].

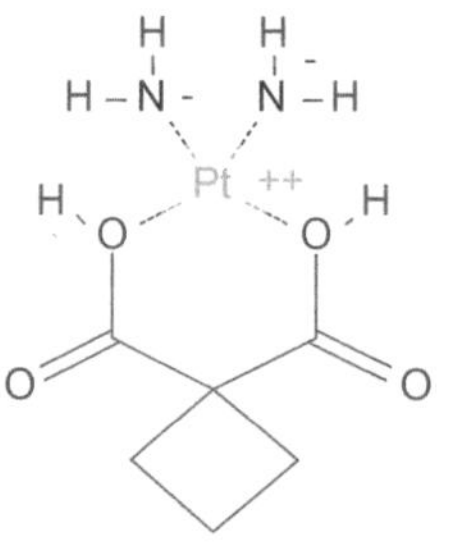

Fig. 1 Chemical structure of Carboplatin. *Source: From PubChem.*

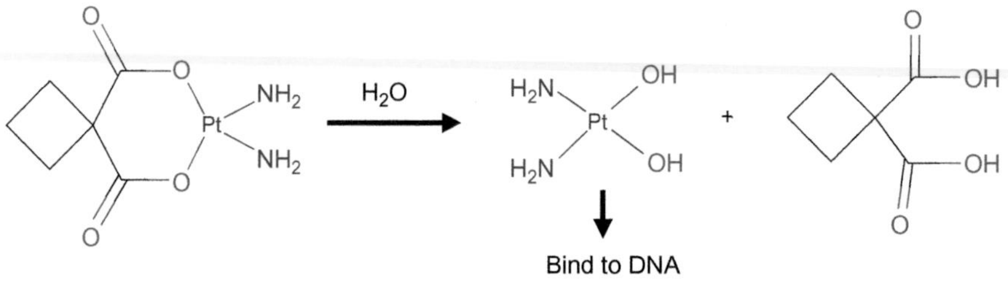

Fig. 2 Hydrolysis of carboplatin inside cell.

The co-existence of EC and pregnancy occurs only in extremely rare cases. There are no human trial data available with respect to the effects of carboplatin in pregnant women. However, there were animal trials that showed indication of mutagenesis, embryotoxicity, and teratogenicity. Effect of this drug on the reproductive capacity of humans is also unknown. Due to the positive evidence of risks to the fetus, the use of carboplatin is not recommended during pregnancy. In breastfeeding women, carboplatin is capable of reaching the breast milk. This can cause potential side effects like myelosuppression, hypersensitivity reactions, nephrotoxicity, and neurotoxicity in the feeding infant [31].

4.1 Nanoencapsulation of carboplatin

Attempts were made to encapsulate the molecule for targeted drug delivery. Phospholipid-based nanocapsules of carboplatin rendered very high cytotoxicity to cancer cells in vitro. The improvement in cytotoxicity was because of preferential intake of the nanoparticles by the cancer cells via phagocytosis, which in turn resulted in increased accumulation of platinum in the cells [32]. In a study by Karanam et al. [33], carboplatin loaded into a nanoparticle made from a biodegradable polymer poly (epsilon-caprolactone) showed a biphasic pattern of drug release in vitro when treated with U–87 MG cell lines. It was found that 30%–40% of the drug get released in the first hour and the remaining 70%–80% get released in 8–10 h of incorporation. The efficiency of killing cancer cells was three times greater as compared to the free drug. One of the most important side effects of carboplatin is hemolysis. The study found that nanoencapsulated carboplatin caused no hemolysis in vitro, suggesting it could be a better alternative to free drugs.

Carboplatin encapsulated in carbon nanotubes has also been tested for therapeutic efficiency against cancer cells. Carbon nanotubes are made by rolling a thin sheet of graphene. They have a distinctive hexagonal latticework that interacts with cells and causes acute inflammation, the formation of reactive oxygen species, and cell death by autophagy. When carboplatin-filled carbon nanotubes were tested for their effect on cell proliferation and cytotoxicity, positive results were obtained [34]. Another study on using carbon nanotubes conjugated with hyaluronic acid and carboplatin on murine lung cancer cells revealed that the hyaluronic acid was bound on the outer side of the carbon nanotubes. Tumor cells usually overexpress hyaluronic acid receptors, which leads to more specific targeting of the nanoencapsulated carboplatin to the tumor cells. The carboplatin–hyaluronic acid conjugated carbon nanotubes were endocytosed to a greater extent by the tumor cells than the normal cells and thus the cytotoxicity remains higher in tumor cells. The tumor cells also showed a decrease in metabolic activity in a dose-dependent manner. The cytotoxic effect of the drug, when delivered through carbon nanotubes, was twice as efficient as carboplatin delivered alone [35].

5. Mechanism and chemical characteristics of paclitaxel

Paclitaxel is an antineoplastic drug commonly known by its brand name Taxol. It was first isolated in 1971 from the bark of the Pacific yew tree. The drug is intravenously administrated; a tablet or pill form of paclitaxel is not available [36]. The molecular formula of paclitaxel is $C_{47}H_{51}NO_{14}$ and it has a molecular mass of 853.9 g/mol. Its structure is composed of a diterpene taxane ring with a four-membered oxetane ring and an ester side chain at position C-13. This complex structure contributes to its unusual properties [37, 38].

A newer formulation of paclitaxel called Abraxane is popular now and it contains paclitaxel bound to albumin [39]. In addition to antineoplastic properties, paclitaxel is also an immunosuppressive and myelosupressive agent. It is used for the treatment of cancers in the cervix, head and neck, esophagus, fallopian tubes, lungs, bladder, breasts, ovaries, and testicles in addition to EC [38] (Fig. 3).

The cytotoxic activity of paclitaxel lies in its interference with the cell microtubules. In contrast to many other drugs that cause depolymerization of microtubules, like colchicine, paclitaxel acts by hyperstabilization of the microtubules. In vitro studies have shown that paclitaxel enhances the polymerization of tubulin to form stable microtubules. Along with this, it also interacts with microtubules and prevents them from depolymerization. This interaction with microtubules is due to the presence of a high-affinity specific binding site for paclitaxel in the microtubules. The normal mitotic apparatus cannot be formed in the cell as the ability of the microtubules to shorten and lengthen are disrupted, leading to cell cycle arrest at G2/M phase. The cells, therefore, cannot divide but rather undergo apoptosis [38].

Animal studies have shown that paclitaxel causes adverse side effects like embryotoxicity and fetotoxicity. The drug may cause human fetal malformations or other irreversible

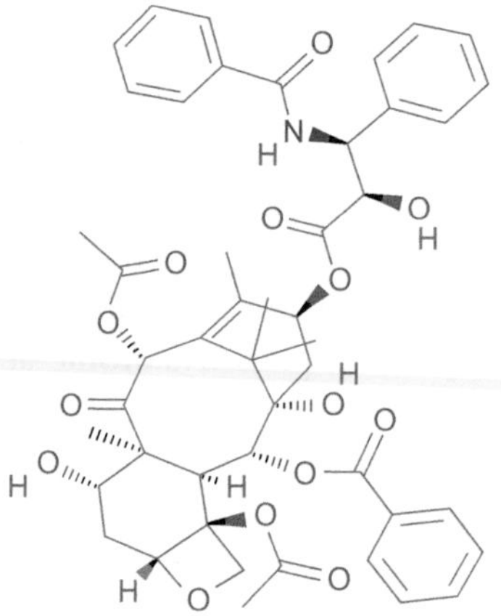

Fig. 3 Chemical structure of Paclitaxel. *Source: From PubChem.*

damages. Similar to carboplatin, paclitaxel is also found excreted into breast milk in relatively large amounts. This phenomenon adversely affects the normal microbiome and chemical makeup of breast milk and causes side effects of myelosuppression, hypersensitivity reactions, nephrotoxicity, and neurotoxicity to the infants [40].

5.1 Nanoencapsulation of paclitaxel

Paclitaxel is a poorly water-soluble drug. So to improve its solubility and improve its specificity and efficacy, nanoencapsulation in poly-ethyl oxazoline (PEOX) is suggested. The nanocapsules are capable of targeting tumor tissues, avoiding normal healthy tissue. This specificity is because of the enhanced permeation and retention (EPR) effect. As the PEOX nanoparticle loaded with paclitaxel reaches closer to the tumor site, it penetratea through the leaky walls of the blood vessels and releases the drug at the tumor site. Use of this method has been found to improve tumor penetration and efficacy of paclitaxel at the same time minimizing the toxicity [41]. A combination of paclitaxel and epigallocatechin gallate (EGCG), which is a multiple signaling inhibitor drug, encapsulated in poly-L-lactide-*co*-glycolic acid (PLGA)–casein nanoparticle has been used to treat breast cancer. This way of delivering the drugs in nanoparticles aided in the sequential release of EGCG followed by paclitaxel. This early release of EGCG substantially increased the sensitivity of paclitaxel to cancer and sensitized paclitaxel-resistant cells to paclitaxel, leading to induction of their apoptosis. Paclitaxel-induced expression of P-glycoprotein was repressed by the nanocombination both at the transcription and translation levels. This combination offered a significant cytotoxic response to breast cancer primary cells [42].

6. Conclusion

Carboplatin and paclitaxel in combination is one of the most effective chemotherapeutic drug therapies for the treatment of EC. Though these drugs show a lower response rate in solo drug therapy, the rate increases drastically when they are administered in combination. In addition to a high overall survival rate, as the combination is less toxic when compared with other chemotherapy drugs, favorable results have been obtained when carboplatin and paclitaxel were given with other forms of cancer therapy like radiation, surgery, and hormone therapy. In light of the low survival advantage associated with this regimen, the use of less toxic combinations of paclitaxel with carboplatin may be tested but require further study. Nanoencapsulation of carboplatin and paclitaxel for therapy also shows promising effects, although this technique is in the preliminary stage. Encapsulating these drugs in specialized polymers enhances their specificity to tumor cells, and the release of the drug from the nanoparticle occurs at the tumor site only. This reduces the toxic effects of these drugs to normal healthy tissues and at the same time increases the cytotoxicity of tumor cells. This method is very promising for better and efficient cancer treatment and therefore more research and studies should be conducted on it.

References

[1] Endometrial Cancer Facts: Seattle Cancer Care Alliance. Retrieved 27 December 2019, from: https://www.seattlecca.org/diseases/endometrial-cancer/endometrial-cancer-facts; 2019.

[2] Skrzypczyk-Ostaszewicz A, Rubach M. Gynaecological cancers coexisting with pregnancy—a literature review. Contemp Oncol 2016;20(3):193.

[3] Smith DC, Prentice R, Thompson DJ, Herrmann WL. Association of exogenous estrogen and endometrial carcinoma. N Engl J Med 1975;293(23):1164–7.

[4] Hu R, Hilakivi-Clarke L, Clarke R. Molecular mechanisms of tamoxifen-associated endometrial cancer. Oncol Lett 2015;9(4):1495–501.

[5] Endometrial Cancer Risk Factors. Retrieved 27 December 2019, from: https://www.cancer.org/cancer/endometrial-cancer/causes-risks-prevention/risk-factors.html; 2019.

[6] Endometrial Cancer Treatment (PDQ®)–Patient Version. Retrieved 27 December 2019, from: https://www.cancer.gov/types/uterine/patient/endometrial-treatment-pdq; 2019.

[7] Lin LL, Grigsby PW, Powell MA, Mutch DG. Definitive radiotherapy in the management of isolated vaginal recurrences of endometrial cancer. Int J Radiat Oncol Biol Phys 2005;63(2):500–4.

[8] Tewari K, Cappuccini F, Brewster WR, DiSaia PJ, Berman ML, Manetta A, Puthawala A, Syed AN, Kohler MF. Interstitial brachytherapy for vaginal recurrences of endometrial carcinoma. Gynecol Oncol 1999;74(3):416–22.

[9] Van Wijk FH, Van der Burg MEL, Burger CW, Vergote I, van Doorn HC. Management of recurrent endometrioid endometrial carcinoma: an overview. Int J Gynecol Cancer 2009;19(3):314–20.

[10] Michener CM, Peterson G, Kulp B, Webster KD, Markman M. Carboplatin plus paclitaxel in the treatment of advanced or recurrent endometrial carcinoma. J Cancer Res Clin Oncol 2005;131(9):581–4.

[11] Ball HG, Blessing JA, Lentz SS, Mutch DG. A phase II trial of paclitaxel in patients with advanced or recurrent adenocarcinoma of the endometrium: a gynecologic oncology group study. Gynecol Oncol 1996;62(2):278–81.

[12] Ramondetta L, Burke TW, Levenback C, Bevers M, Bodurka-Bevers D, Gershenson DM. Treatment of uterine papillary serous carcinoma with paclitaxel. Gynecol Oncol 2001;82(1):156–61.

[13] Lissoni A, Zanetta G, Losa G, Gabriele A, Parma G, Mangioni C. Phase II study of paclitaxel as salvage treatment in advanced endometrial cancer. Ann Oncol 1996;7(8):861–3.

[14] Fleming GF, Brunetto VL, Cella D, Look KY, Reid GC, Munkarah AR, Kline R, Burger RA, Goodman A, Burks RT. Phase III trial of doxorubicin plus cisplatin with or without paclitaxel plus filgrastim in advanced endometrial carcinoma: a Gynecologic Oncology Group Study. J Clin Oncol 2004;22(11):2159–66.

[15] Dimopoulos MA, Papadimitriou CA, Georgoulias V, Mouloupoulos LA, Aravantinos G, Gika D, Karpathios S, Stamatelopoulos S. Paclitaxel and cisplatin in advanced or recurrent carcinoma of the endometrium: long-term results of a phase II multicenter study. Gynecol Oncol 2000;78(1):52–7.

[16] Scudder SA, Liu PY, Wilczynski SP, Smith HO, Jiang C, Hallum III AV, Smith GB, Hannigan EV, Markman M, Alberts DS. Paclitaxel and carboplatin with amifostine in advanced, recurrent, or refractory endometrial adenocarcinoma: a phase II study of the Southwest Oncology Group. Gynecol Oncol 2005;96(3):610–5.

[17] Alektiar KM, Makker V, Abu-Rustum NR, Soslow RA, Chi DS, Barakat RR, Aghajanian CA. Concurrent carboplatin/paclitaxel and intravaginal radiation in surgical stage I–II serous endometrial cancer. Gynecol Oncol 2009;112(1):142–5.

[18] Kiess AP, Damast S, Makker V, Kollmeier MA, Gardner GJ, Aghajanian C, Abu-Rustum NR, Barakat RR, Alektiar KM. Five-year outcomes of adjuvant carboplatin/paclitaxel chemotherapy and intravaginal radiation for stage I–II papillary serous endometrial cancer. Gynecol Oncol 2012;127(2):321–5.

[19] Pectasides D, Xiros N, Papaxoinis G, Pectasides E, Sykiotis C, Koumarianou A, Psyrri A, Gaglia A, Kassanos D, Gouveris P, Panayiotidis J. Carboplatin and paclitaxel in advanced or metastatic endometrial cancer. Gynecol Oncol 2008;109(2):250–4.

[20] Hoskins PJ, Le N, Ellard S, Lee U, Martin LA, Swenerton KD, Tinker AV. Carboplatin plus paclitaxel for advanced or recurrent uterine malignant mixed mullerian tumors. The British Columbia Cancer Agency experience. Gynecol Oncol 2008;108(1):58–62.

[21] Sovak MA, Hensley ML, Dupont J, Ishill N, Alektiar KM, Abu-Rustum N, Barakat R, Chi DS, Sabbatini P, Spriggs DR, Aghajanian C. Paclitaxel and carboplatin in the adjuvant treatment of patients

with high-risk stage III and IV endometrial cancer: a retrospective study. Gynecol Oncol 2006;103 (2):451–7.

[22] Akram T, Maseelall P, Fanning J. Carboplatin and paclitaxel for the treatment of advanced or recurrent endometrial cancer. Am J Obstet Gynecol 2005;192(5):1365–7.

[23] Lincoln S, Blessing JA, Lee RB, Rocereto TF. Activity of paclitaxel as second-line chemotherapy in endometrial carcinoma: a Gynecologic Oncology Group Study. Gynecol Oncol 2003;88(3):277–81.

[24] Van Wijk FH, Lhomme C, Bolis G, Di Palumbo VS, Tumolo S, Nooij M, De Oliveira CF, Vermorken JB. Phase II study of carboplatin in patients with advanced or recurrent endometrial carcinoma. A trial of the EORTC Gynaecological Cancer Group. Eur J Cancer 2003;39(1):78–85.

[25] Lissoni A, Gabriele A, Gorga G, Tumolo S, Landoni F, Mangioni C, Sessa C. Cisplatin-, epirubicin- and paclitaxel-containing chemotherapy in uterine adenocarcinoma. Ann Oncol 1997;8(10):969–72.

[26] Long HJ, Pfeifle DM, Wieand HS, Krook JE, Edmonson JH, Buckner JC. Phase II evaluation of carboplatin in advanced endometrial carcinoma. J Natl Cancer Inst 1988;80(4):276–8.

[27] Fung-Kee-Fung M, Dodge J, Elit L, Lukka H, Chambers A, Oliver T, Cancer Care Ontario Program in Evidence-based Care Gynecology Cancer Disease Site Group. Follow-up after primary therapy for endometrial cancer: a systematic review. Gynecol Oncol 2006;101(3):520–9.

[28] Carboplatin—DrugBank. Retrieved 27 December 2019, from: https://www.drugbank.ca/drugs/DB00958; 2019.

[29] Gerson SL, Caimi PF, William BM, Creger RJ. Pharmacology and molecular mechanisms of antineoplastic agents for hematologic malignancies. In: Hematology. Elsevier; 2018. p. 849–912.

[30] Sousa GFD, Wlodarczyk SR, Monteiro G. Carboplatin: molecular mechanisms of action associated with chemoresistance. Braz J Pharm Sci 2014;50(4):693–701.

[31] Griffin SJ, Milla M, Baker TE, Liu T, Wang H, Hale TW. Transfer of carboplatin and paclitaxel into breast milk. J Hum Lact 2012;28(4):457–9.

[32] Hamelers IH, van Loenen E, Staffhorst RW, de Kruijff B, de Kroon AI. Carboplatin nanocapsules: a highly cytotoxic, phospholipid-based formulation of carboplatin. Mol Cancer Ther 2006;5 (8):2007–12.

[33] Karanam V, Marslin G, Krishnamoorthy B, Chellan V, Siram K, Natarajan T, Bhaskar B, Franklin G. Poly (ε-caprolactone) nanoparticles of carboplatin: preparation, characterization and in vitro cytotoxicity evaluation in U-87 MG cell lines. Colloids Surf B Biointerfaces 2015;130:48–52.

[34] Kumari A, Singla R, Guliani A, Yadav SK. Nanoencapsulation for drug delivery. EXCLI J 2014;13:265.

[35] Salas-Treviño D, Saucedo-Cárdenas O, Loera-Arias MDJ, Rodríguez-Rocha H, García-García A, Montes-de-Oca-Luna R, Piña-Mendoza EI, Contreras-Torres FF, García-Rivas G, Soto-Domínguez A. Hyaluronate functionalized Multi-Wall carbon nanotubes filled with carboplatin as a novel drug nanocarrier against murine lung cancer cells. Nanomaterials 2019;9(11):1572.

[36] Success Story: Taxol. Retrieved 27 December 2019, from: https://dtp.cancer.gov/timeline/flash/success_stories/S2_taxol.htm; 2019.

[37] Paclitaxel, 99+%. Retrieved 27 December 2019, from https://pubchem.ncbi.nlm.nih.gov/compound/133640187; 2019.

[38] Paclitaxel—DrugBank. Retrieved 27 December 2019, from: https://www.drugbank.ca/drugs/DB01229; 2019.

[39] Miele E, Spinelli GP, Miele E, Tomao F, Tomao S. Albumin-bound formulation of paclitaxel (Abraxane® ABI-007) in the treatment of breast cancer. Int J Nanomedicine 2009;4:99.

[40] Paclitaxel Use During Pregnancy. (n.d.). Retrieved 30 January 2020, from: https://www.drugs.com/pregnancy/paclitaxel.html

[41] Yin R, Pan J, Zhou B, Zhang Y, Dougherty J, Liu J, Riesenberger TA, Qin D, Vallejo YR. Evaluation of the toxicity and efficacy of paclitaxel nanoencapsulated with polyethyloxazoline polymers. J Clin Oncol 2013;31:e13538.

[42] Narayanan S, Mony U, Vijaykumar DK, Koyakutty M, Paul-Prasanth B, Menon D. Sequential release of epigallocatechin gallate and paclitaxel from PLGA-casein core/shell nanoparticles sensitizes drug-resistant breast cancer cells. Nanomedicine 2015;11(6):1399–406.

CHAPTER 9

Female-specific cancer

P.S. Pradeep[a], D. Sivaraman[a], Jayshree Nellore[b], and Sujatha Peela[c]

[a]Centre for Laboratory Animal Technology and Research, Sathyabama Institute of Science and Technology, Chennai, Tamil Nadu, India
[b]Department of Biotechnology, School of Bio and Chemical Engineering, Sathyabama Institute of Science and Technology, Chennai, Tamil Nadu, India
[c]Department of Biotechnology, Dr. BR Ambedkar University, Srikakulam, Andhra Pradesh, India

Abstract

Female-specific cancers are perhaps the most prevalent cancers in women worldwide. For most of these cancers, early diagnosis strategies remain inaccessible, ensuring that many are identified at later stages. Current treatment methods for female-specific cancers include surgery, radiation, and chemotherapy. In this context, this chapter focuses on female-specific breast cancer, peritoneal cancer, uterine cancer, and vaginal cancer.

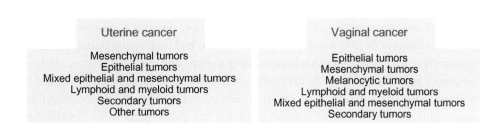

Keywords: Breast cancer, Peritoneal cancer, Uterine cancer, Vaginal cancer.

G.P. Nagaraju, R.R. Malla (eds.)
A Theranostic and Precision Medicine Approach for Female-Specific Cancers
ISBN 978-0-12-822009-2
https://doi.org/10.1016/B978-0-12-822009-2.00009-1

Abbreviations

AMFB	angiomyofibroblastoma
AML	acute myeloid leukemia
CNS	central nervous system
DCIS	ductal carcinoma in situ
IARC	International Agency for Research on Cancer
ILC	invasive lobular carcinoma
MMMT	malignant mixed Mullerian tumor
MS	myeloid sarcoma
US	ultrasound

1. Introduction

Female-specific cancers, such as breast or gynecological (ovary, vaginal, uterine) cancers, occur in sensitive places of the female body, making them difficult to treat due to the risk of destroying healthy cells. Breast cancer, for example, constitutes the most common female malignancy, which accounts for up to 30% of all cancer diagnoses in women [1] in both developed and developing countries worldwide [2]. Early diagnosis in women seeks to limit morbidity involved with advanced stages of cancer. Indeed, the advantages and disadvantages of mammography screening have also been widely debated throughout the last decades [3]. After cautious assessment of the consistency between benefits and adverse events of mammography screening, the International Agency for Research on Cancer (IARC)'s latest analysis indicated that there is a net benefit from recruiting women aged 50 to 74 years to obtain a screening test. In 2018, breast cancer was diagnosed in 2 million women around the world, with 627,000 deaths attributed to the disease [4]. Pharmaceutical safety and pharmaceutical effectiveness are both issues in the treatment of elderly people with chemotherapy drugs, and much more targeted clinical trials must be performed. For instance, in a comparative analysis of tamoxifen aromatase inhibitors, aromatase inhibitors decreased the risk of chronic disease in postmenopausal women with breast cancer [5]. The Early Breast Cancer Trialists Collaborative Group reviewed findings for 46,000 women in 91 studies who had undergone hormone replacement therapy over 5 years following breast cancer therapy and reported that recurrence of breast cancer would develop up to 20 years post hormonal therapy. Scientists found that even those with the smallest, least invasive tumors had a 14% incidence threat after 20 years, and the likelihood of cancer recurring for those with more severe illness increased as much as 47%, indicating that 5 years of cancer therapy is not sufficient [6]. In fact, the current data lead to the conclusion that perhaps the majority of peritoneum, uterine, and vaginal cancers may have a common source and that the precursor cyst is located in the fallopian tube [7]. The most common gynecologic malignancy is uterine cancer, while breast cancer is highly confined to female malignancy. It has also been seen in several epidemiological studies that women diagnosed with breast cancer are always at greater possibility of acquiring secondary uterine cancer. Enhanced vascularization arises in peritoneum-related tumor deposit biopsies, a phenomenon that is based on the assumption that angiogenesis is triggered by cancer cells and tumor-related mesothelial cells [8]. This chapter attempts to inform readers about female-specific cancer, including breast cancer, peritoneal cancer, uterine cancer, and vaginal cancer, along with discussion of the interplay between cancer treatment and cancer care outcomes.

1.1 Breast cancer

Symptoms of breast cancer might include a lump in the breast, a variation in the shape of the breast, skin dimpling, a newly twisted nipple, a red or scaly skin patch, or discharge from the nipple. Breast cancer tends to occur when an abnormal growth starts in some breast cells. These cells divide faster than normal cells, and keep accumulating, forming a lump or mass [9]. Breast cancer commonly starts with cells in the ducts that produce milk (invasive ductal carcinoma). Signs of breast cancer include a new lump in either the breast or underarm, inflammation of a portion of the breast, discomfort of the breast skin, flaky skin in the nipple or breast, and nipple discharge apart from breast milk, including blood. Categories of breast cancer include invasive ductal carcinoma, in situ ductal carcinoma, metastatic breast cancer, and inflammatory breast cancer [10].

1.1.1 Ductal tumor in breast

Ductal carcinoma in situ (DCIS) is a noninvasive carcinoma. DCIS is the most common form of breast cancer, accounting for 80% among all breast cancer illnesses, where the cancer cells are observed in the breast milk duct liner. DCIS is a very early cancer, which is curable, but if left undiagnosed, it may spread to the neighboring breast tissue with no obvious symptoms. DCIS can often cause indications like breast lumps and blood discharge from the nipple. Characteristics that could increase the chance of DCIS include age, history of breast disease, having never been pregnant, delivering first baby after age 30, menopause after age 55, and gene variations (BRCA1 and BRCA2) that increase breast cancer risk. Management of DCIS has a significant rate of success, in several cases by eliminating the tumor and minimizing any recurrence [11]. In many other patients, DCIS therapies involve breast-preservation surgery (lumpectomy), radiation therapy, and breast-removal surgery (mastectomy).

1.1.2 Lobular tumor in breast

Invasive lobular carcinoma (ILC) is a form of breast cancer that occurs in the breast milk-producing glands (lobules). Invasive cancer are cells that have spread out from the lobule where they originated and have the potential to invade the lymph nodes as well as other soft tissues. Patients with earlier stage breast cancers have a greater chance of survival than those with mature breast cancers. Overall, with advanced medications, the diagnosis of lobular carcinoma is closely related to that of ductal carcinoma. ILC typically will not develop a lump, but propagates through the fat tissue of the breast in single-file lines [12]. As per the American Cancer Society, each of the following abnormal breast transitions could be the first indication of breast cancer: inflammation in the whole or in a portion of the breast, skin irritation, and breast pain. ILC arises when cells of breast milk-producing glands acquire mutations in the DNA, which controls the cell growth, resulting in cell division and rapid growth.

1.1.3 Tubular tumor in breast

Tubular breast carcinoma is a subclass of invasive ductal carcinoma (a cancer that commences within the breast milk duct and propagates beyond it to normal tissues). Although tubular carcinoma is an invasive breast cancer, it manages to be a less virulent type that reacts well to treatment. Tubular adenoma is a category of breast adenoma distinguished by tightly packed tubular and acinar (bud-like) cellular configurations. Adenoma is an unorganized neoplasm composed of a variety of glandular, fibrous, and fat tissues that may lead to an increase in density or pseudo-injury. Tubular breast cancer is indeed a form of invasive ductal breast cancer that accounts for less than 2% of all breast cancers [13]. As with other forms of invasive ductal cancer, tubular breast cancer develops in the breast milk duct before it spreads to the tissues across the gland. The lump can multiply rapidly to be sensed during breast self-examination or a doctor's examination. Tubular carcinomas are generally small, 1 cm or less in diameter, and are hard to feel.

1.1.4 Mucinous tumor in breast

Mucinous breast carcinoma, sometimes called colloid carcinoma, is a rare condition of invasive ductal carcinoma (a cancer that occurs in the milk duct and grows outside of it to surrounding normal tissue). Mucinous carcinoma is an invasive cancer that could also spread to the rest of the body. Nonetheless, it is less destructive than many other invasive cancers and usually reacts well to medication [14]. The rate of survival for mucinous breast carcinoma is greater than that of most other forms of invasive breast cancer with a 5-year survival rate of 87%. Mucinous tumors are a component of the epithelial-stromal ovary neoplasm cancer community, which accounts for nearly 36% of all ovary tumors. Total mucinous carcinomas have a lower level of genetic uncertainty. Genome patterns of the cancer constituents of mixed mucinous tumors are strikingly identical to those of total mucinous carcinomas.

1.1.5 Medullary tumor in breast

Medullary breast carcinoma is a subtype of invasive ductal carcinoma. This is also a form of breast cancer that starts in the milk duct. Medullar carcinoma is typically less capable of spreading to the lymph nodes and is more receptive to therapy than the more common types of invasive breast cancer. In short, they resemble very aggressive, abnormal tumor cells, but do not behave like them [15]. The overall survival rate of breast cancer for lobular carcinoma penetration, when compared by phase, is substantially greater than that of ductal carcinoma during the first 5 years. Although the health outcomes vary between 77% and 93%, the 5-year survival rate was reported at about 90% on average. Medullar breast carcinomas may be unique for diagnosis. Occasionally, a woman might be more inclined to feel a cancerous lesion than can be diagnosed by an imaging system. For this purpose, it is necessary for women to perform monthly breast self-examinations, whereby they sense their breast tissue and nipples. Many women may encounter other symptoms similar to medullar carcinoma, including breast tenderness, discomfort, swelling, bleeding, and so on.

1.1.6 Papillary tumor in breast

A very rare type of invasive ductal breast cancer is papillary breast cancer. An intraductal papilloma is a small, benign tumor developed within a breast milk duct. Such cancers are composed of fibrous tissue and prostate as well as blood vessels. These tumors appear as long, thin nodules, much like "fingers." Such tumors develop out of a tissue covering the inner wall of an organ. Papillary tumors can be either benign or cancerous and most often occur in the bladder, thyroid, and breast, but could also arise in several other parts of the human body [16]. Generally, an invasive papillary carcinoma has a very well-defined boundary and is comprised of thin, finger-like projections. As with other types of invasive ductal cancer, papillary breast cancer starts in the breast milk-secreting duct. Papillary breast cancers are typically small and active for receptors of estrogen and/or progesterone (ER/PR +) and inactive for receptors (HER2). Primary papillomas are more prone to be single and to cause bleeding nipple discharge [17]. In younger women, multiple lesions occur more frequently than solitary papillomas and are therefore more determined to be bilateral, asymptomatic, and persist after ablation.

1.2 Peritoneum cancer

Cancer of the peritoneum is a rare cancer. It grows in a thin tissue matrix surrounding the uterus including the rectum and bladder and its layer is comprised of epithelial cells called the peritoneum. It creates a fluid that enables cells to efficiently travel within the abdominal matrix [18]. If left undiagnosed, the mean life span for patients with peritoneal cancer may be as limited as 6 weeks. Where appropriate care is given, life expectancy could be significantly longer, but most patients are diagnosed with peritoneal cancer at advanced stages, which makes them difficult to treat. The 5-year rate of survival is 47% for women with all forms of reproductive cancers. Advanced-stage peritoneal cancer signs may result in complete urinary or bowel blockage. As the tumor develops, the abdominal cavity (ascites) produces a watery fluid that is suppressed during treatments like chemotherapy.

1.2.1 Smooth muscle tumor in peritoneum cancer

Uterine smooth muscle tumors are smooth muscle neoplasms ranging from benign leiomyomas to low-risk leiomyosarcomas. There are several histological subtypes, including common (spindled), epithelioid, and tumors of the myxoids. Leiomyosarcomas are soft-muscle malignant cancers that are cutaneous, and probably originate from the pilorum muscle. Cutaneous leiomyosarcomas often reoccur after excision, but are rarely metastasized. Benign soft tissue tumors are more prevalent than benign bone tumors. They can exist at all locations, both between and within tissues, muscles, nerves, and blood vessels [19]. These cancers vary greatly in behavior patterns and appearance. The majority of cancers throughout this group are benign fat tumors termed as lipomas.

1.2.2 Mesothelial tumor in peritoneum cancer

Mesothelial cell tumors may be benign (noncancerous) or malignant (cancerous). Mesothelioma is a malignant tumor that is induced by inhalation of asbestos fibers and arises in the inner lining of the abdomen, heart, or lungs [20]. Clinical signs of this type of tumor may include breathing problems and chest pain. Life span for most patients with mesothelioma is about 12 months after detection. Mesothelial cells are a monolayer of specialized pavement-like cells that often line the body's serous cavities and vital organs. The primary role of this layer, called the mesothelium, is to make a nonadhesive, slippery, and protective surface. The major risk element for mesothelioma is the use of asbestos.

1.2.3 Epithelial tumors in peritoneum cancer

Ovarian epithelial cancer, fallopian tube cancer, and primary peritoneal cancer are diseases in which cancerous cells develop in the ovarian or fallopian tube or peritoneal tissue [21]. Peritoneum is a tissue that forms the lining of the abdomen and surrounds the organs in the abdomen. Malignant histologies typically involve undifferentiated carcinomas, tubular adenocarcinomas, and papillary adenocarcinomas. Surface epithelial-stromal tumors are indeed a type of benign or malignant ovarian neoplasm. Neoplasms in this category are believed to originate from the epithelium of the ovarian surface (modified peritoneum) or from the tissue of the ectopic endometrial or fallopian tube. One of several subtypes of ovarian carcinoma is clear cell ovarian carcinoma. Epithelial and nonepithelial are both types of ovarian carcinoma. Many forms of benign epithelial cancers exist, including serous adenomas, mucinous adenomas, and tumors of Brenner. Cancerous epithelial tumors are carcinomas, that is, they originate in the tissue lining the ovaries.

1.2.4 Tumor-like lesions in peritoneum cancer

Gliomatosis peritonei may include most of the serosal surface. Various types of reactive peritoneum alterations may cause minor tumor-like lesions. Mainly, these contain different types of granulomas, metaplasia, deciduosis, and endometriosis. Peritoneal cancer is a rare type of cancer that grows within a thin tissue layer in the abdomen including the uterus, rectum, and bladder. It emits a fluid that enables the organs to smoothly relocate within the abdomen. Tumor and tumor-like lesions, by nature, are lesions that look alike in reports of ultrasound (US), computed tomography, or magnetic resonance imaging. Peritoneal metastasis includes cancers that have progressed or metastasized to the peritoneal cavity. Often cancer cells may split off from the primary tumor and transplant directly into other organs and tissues in the peritoneum [22]. The patient may experience pain and painful symptoms once the tumor starts to grow. Typically, tumor-like lesions are recorded as observations consistent with a tumor-like lesion, but wherein a neoplasm could not be excluded. Lesions could be classified by whether the cancer causes them or not. A coin lesion is a round spot on a chest X-ray that resembles a coin. Therefore, one must include the standard epiphysis entities like chondroblastoma, giant cell tumors, and aneurysmal bone cysts for lucent lesions in these regions.

1.2.5 Secondary tumor in peritoneum cancer

Secondary peritoneal tumors tend to start in other abdominal organs and propagate into the peritoneum. Such cancers may be initially gynecologic, genitourinary, or gastrointestinal (stomach, small intestine, colorectal, appendix). Both men and women can be diagnosed with secondary peritoneal cancers. As of 2019, women with all forms of ovarian, fallopian tube, and peritoneal cancers have a 5-year survival rate of 47%. Among women younger than 65 years (60%), the survival rate is greater, and for women older than 65 years (29%), 5-year survival is less. The most prevalent terminal feature of abdominal cancers is peritoneal carcinomatosis. This is a troubling situation for gastrointestinal surgeons and medical oncologists because, even though the condition is confined to the peritoneal surface, total surgical excision is extremely difficult and systemic chemotherapy is functionless. Peritoneal metastasis includes tumors that have progressed to the peritoneal cavity. Occasionally cancerous cells may split off from the primary tumor and transplant themselves into other organs and tissues in the peritoneum [23]. Patients experience pain and crippling complications as the tumor begins to grow. The indications of peritoneal cancer may include abdominal swelling or gas pain, indigestion, pressure, inflammation, bloating or cramping, feeling full even after a small meal, vomiting or diarrhea, constipation, frequent urination, loss of appetite, abnormal weight gain or loss, and irregular vaginal bleeding.

1.2.6 Other primary tumor in peritoneum cancer

A primary tumor is a cyst that develops at the anatomical site where tumor progression began, producing a cancerous mass. Many tumors grow at their primary site and spread over to metastasize or spread to other parts of the body [24]. Such cancers are termed as secondary tumors. Metastases are also known as secondary cancers. Unknown primary cancer is a disease that may have metastasized (spread) from any other parts of the body. The site where it originated, also described as the primary site, is unidentified. Such cases account for about 2%–5%% of diagnosed cancers in the United States.

1.3 Vaginal cancer

Vaginal cancer is a disorder in the vagina where malignant (cancerous) cells form. The vagina is the canal that leads from the cervix (the uterine opening) to the outside of the body. At birth, a baby crosses through the vagina (also called the birth canal) out of the body. Vaginal cancer is uncommon. Together, both vulvar and vaginal cancers account for less than 7% of all reproductive organ cancers treated annually in women. Cancer can form in the vagina in some very rare cases, without spreading from another location. Factors for vaginal cancer include being 60 years of age or older, being exposed to diethyl-stilbestrol while in the womb of the mother, being diagnosed with human papilloma virus, and having a history of abnormal or cancerous cervical cells [25].

1.3.1 Epithelial tumor in vagina

Epithelial ovarian cancers grow out of the cells that make up the outer layer of the ovary. Most of the ovarian epithelial tumors are benign (noncancerous). Numerous categories of benign epithelial tumors exist, such as serous adenomas, mucinous adenomas, and tumors of Brenner. Surface epithelial-stromal cancers are indeed a class of benign or malignant ovarian neoplasms. Neoplasms among this community are assumed to derive from the epithelium of the ovarian surface (modified peritoneum) or from the tissue of the ectopic endometrial or fallopian tube [26]. Carcinoma refers to a malignant epithelial neoplasm, or cancer of the body's inner or outer lining. Epithelial tissue malignancies account for 80%–90% of all cancer cases. Epithelial ovarian cancer, fallopian tube cancer, and primary peritoneal cancer are diseases in which malignant (cancerous) cells develop in the tissue that covers the ovary or lines the fallopian tube or peritoneum. Fallopian tube cancer and primary peritoneal cancer are identical to ovarian epithelial cancer and are given the same treatment.

1.3.2 Mesenchymal tumor in vagina

The superficial myofibroblastoma in the lower female genital tract is a rare mesenchymal tumor. Neoplasms of the mesenchymal tissue are soft tissue tumors, often known as connective tissue tumors, that are extremely common in domestic animals and have a greater prevalence in some species [27]. These cancers could be in all organs, with a greater or lesser occurrence in certain tissues. The precise mechanism of action remains unclear for superficial myofibroblastoma. In some particular types of soft tissue tumors the correlation of viral infection and mesenchymal tumors has been well known. Mesenchymal tumors involve entities that arise from mesodermal-derived precursor cells that grow into bone, cartilage, or other connective tissues, such as blood vessels, adipose tissue, smooth muscle, or fibroblasts; in the central nervous system (CNS), they more commonly occur from meninges rather than CNS parenchyma [28]. Angiomyofibroblastoma (AMFB) is an uncommon, indolent mesenchymal tumor that occurs most frequently in the genital tracts of premenopausal women, most often in the vulva and the vagina. Usually AMFB measures less than 5 cm, but scientific studies have identified tumors up to 23 cm in size. A sarcoma is a tumor of mesenchymal (connective tissue) origin that develops from transformed cells. It is comparable to a secondary (or "metastatic") connective tissue tumor that arises when a cancer spreads to the connective tissue from elsewhere in the body (such as the lungs, breast tissue, or prostate).

1.3.3 Lymphoid and myeloid tumor in vagina

Hematopoietic and lymphoid tissue tumors are abnormalities that damage the blood, bone marrow, lymph nodes, and lymphatic system [29]. Chromosomal translocations are a common cause of these diseases, though rare in solid tumors. The myeloid and lymphoid lineages are both involved in the production of dendritic cells. Myeloid cells may include platelet monocytes, neutrophils, macrophages, erythrocytes, basophils, eosinophils, and megakaryocytes. Lymphoid cells involve B cells, T cells, and natural killer cells. Myeloid malignancies are clonal disorders of the progenitor cells or hematopoietic stem cells, which are comprised of chronic phases including myeloproliferative neoplasms, myelodysplastic disorders, chronic myelomonocytic leukemia, and acute stages, i.e., acute myeloid leukemia (AML). Lymphoma (also known as lymphatic or lymphocytic cancer) is a form of cancer that includes immune cells called lymphocytes. Even though AML is a serious condition, chemotherapy with or without a bone marrow/stem cell transplant is curable.

1.3.4 Secondary tumor in vagina

Secondary tumors are more frequent in the vagina than primary vaginal tumors, which means that the cancer has spread from any other organ, such as the cervix, uterus, vulva, liver, intestine, or other organs nearby [30]. Secondary cancer of the vagina is treated differently than primary cancer where treatment may take a long time to control and ease the side effects. This would be termed palliative therapy. Vaginal cancer signs and symptoms include pain or unusual vaginal bleeding that is not associated with menstrual periods, pain in the pelvic area, a lump in the vagina, and pain during urination. Due to the rare nature of vaginal cancer, its survival rates are given in ranges. If cancer is detected early before spreading (stage I), the 5-year rate of survival ranges from 75% to 95%. If the cancer has not progressed outside of the vagina (stage II), the survival rate for 5 years is 50%–80%.

1.3.5 Mixed epithelial and mesenchymal tumor in vagina

A biphasic malignant tumor with high-grade epithelial and mesenchymal components is carcinosarcoma, also known as malignant mixed Mullerian tumor (MMMT). Carcinosarcoma is highly identified as a large polypoid mass clogging the uterine cavity, with necrosis and hemorrhage [31]. The tumors are associated with vaginal bleeding that arises in menopause. Overall, knowledge about the origin of MMMTs is very limited at present. MMMT is a type of cancer that includes two types of cancer cells: carcinoma and sarcoma. Such tumors typically form in female genital tissues and are associated with poor outcome.

1.4 Uterine cancer

Endometrial cancer is a malignancy that arises in the uterus. The uterus is the hollow pelvic organ in pear-shaped form where fetal development takes place. Endometrial cancer starts in the cell layer that forms the uterine lining (endometrium). Endometrial cancer is often referred to as uterine cancer [32]. Early warning indications of uterine cancer are unexpected vaginal discharge that has no indications of blood, hard or difficult urination, pelvic pain and/or mass, and unexpected weight loss. For women with uterine cancer, the 5–year survival rate is 81%. The irregular vaginal bleeding is the most common symptom of endometrial cancer, varying from a watery to a blood-streaked discharge that produces more blood. Vaginal bleeding both during and following menopause is often a sign of an issue. Uterine cancer will usually metastasize to the rectum or bladder. Certain areas where it could propagate include the vagina, ovaries, and fallopian tubes. Typically this form of cancer grows slowly and is often identified before it has disseminated to more distant parts of the body. Transvaginal US, uterine lining examination (hysteroscopy), and tissue sampling (biopsy) are the main tests for diagnosing uterine cancer. A Pap test for diagnosing uterine cancer is not used. Total hysterectomy with bilateral salpingo-oophorectomy is the most successful treatment for early cancer, where the uterus, cervix, ovaries, and fallopian tubes are removed [33]. Sufferers with widespread endometrial cancer are generally treated with hormone therapy, usually progesterone, to slow progress of the cancer.

1.4.1 Mesenchymal tumor in uterine cancer

Mesenchymal uterine cancers are identified in women from a broad range of age groups. The most common neoplasms among the mesenchymal uterine corpus tumors are uterine leiomyomas. They emerge after menarche, usually develop during reproductive years, and after menopause they stabilize or decline. Fibroids are anomalous tumors found in or on the uterus, which sometimes become very large, causing severe abdominal pain with irregular periods. Neoplasms arising from or distinguishing against mesodermally related tissues are uterine mesenchymal cancers. Generally, differentiation is toward the endometrial stromal and myometrial smooth muscle cells in the uterine corpus. Uterine mesenchymal tumors are often a heterogeneous group of neoplasms that can pose a therapeutic challenge. A large number of these cancers (smooth uterine muscle and stromal endometrial) exhibit homologous separation in mesodermal tissues. At times, however, distinction between heterologous elements such as cartilage, skeletal muscle, bone, and so on is not unusual. Differentiation among benign and malignant mesenchymal tumor counterparts is significant due to changes in diagnostic outcome, and it is not possible to underrate the task of the surgical pathologist in making these statements (particularly in complex cases). Epithelial and mesenchymal tumors are primarily categorized depending on their epithelial and mesenchymal components after being morphologically evaluated [34]. Even though immunohistochemical stains support the establishment of a final diagnosis, the morphological characteristics of this category of neoplasms trump other ancillary strategies. Therefore, this analysis emphasizes the main morphological characteristics needed to diagnose and differentiate uterine mesenchymal tumors against their imitations, and provides a brief summary of the related immunohistochemical characteristics.

1.4.2 Epithelial tumor in uterine cancer

Endometrial carcinoma is characterized as an epithelial tumor that occurs in the endometrium, typically with glandular division, that has the ability to penetrate the myometrium and propagate to distant locations [35]. Benign uterine leiomyomas (fibroids) are by far the most prevalent pelvic neoplasm in females (70%–80% lifetime risk). Once healthy cells in the uterus modify and expand out of range, uterine cancer emerges as a lump (tumor). Uterine cancer may generally metastasize to the rectum or bladder. Certain areas where it could spread also include vagina, ovaries, and fallopian tubes. Frequently, this type of cancer tends to grow and is often identified before it propagates to more distant parts of the body. However, the prevention of all cancer in patients with stage IV disease cannot usually be accomplished. Women with stage IV uterine cancer are treated with the goal of reducing symptoms and prolonging survival. Few cases receiving existing standard therapies are recovered.

1.4.3 Mixed epithelial and mesenchymal tumors in uterine cancer

Mixed uteral epithelial and mesenchymal tumors represent a heterogeneous community of neoplasms with adenomyoma, carcinosarcoma, atypical polypoid adenomyoma, and adenofibroma categories. Uterine carcinosarcoma or MMMT is a rare uterine tumor that accounts for less than 5% of uterine tumors and usually occurs in older women [36]. Deep invasion of the myometrium and cervical involvement can occur. Combined uterine epithelial and mesenchymal tumors comprise adenofibroma, carcinosarcoma, adenosarcoma, adenomyoma, and atypical polypoid adenomyoma, the last two lesions being a mixture of benign epithelial and mesenchymal elements with a predominant smooth muscular portion.

1.4.4 Lymphoid and myeloid tumors in uterine cancer

Myeloid sarcoma (MS) is a rare neoplasm in which the information is largely based on clinical studies and/or observations that are dated. In one study, 92 MSs with accessible clinical data were tested morphologically as well as immunohistochemically. MS has been noted to pertain more frequently to the skin, bone, or lymph nodes, though it can affect almost every body site [37]. Lymphocytic leukemia (also referred to as lymphoid or lymphoblastic leukemia) evolves in the white blood cells (lymphocytes) of the bone marrow. In white blood cells other than lymphocytes, along with red blood cells and platelets, myeloid (also defined as myelogenous) leukemia can begin. The myeloid and lymphoid lineages have both been prominent in the production of dendritic cells. Throughout adults, lymphoid and myeloid cells as well as tumor cells develop in a specific organ and ultimately disperse and reside in another organ; these processes are referred to as homing or metastasis.

1.4.5 Secondary tumors in uterine cancer

Secondary tumors are cancers of the cervix that arise beyond the cervix. Cryosurgery, cauterization, or laser surgery can treat precancerous changes in the cervix. Cervical cancer in the early stages and precancerous cervical disorders are curable by almost 100%. However, after all the medication that a patient receives, the cancer can keep arising in the pelvis or abdomen; this is known as recurrent cancer [38]. Less frequently, womb cancer can grow to other body parts from where it started. This is termed as metastasis (secondary cancer). Metastatic uterine tumor (endometrial) is often a type of cancer that emerged in the uterus lining (endometrium) and has distributed to distant parts of the body. Uterine cancer will usually metastasize to the rectum or bladder. Certain places where it might grow include the vagina, ovaries, and fallopian tubes. According to a new Canadian report, based on age, women suffering with uterine cancer may also have an increased risk of developing colon cancer later.

1.4.6 Other tumors in uterine cancer

Endometrial cancer usually starts in the cell membrane that forms the uterine lining (endometrium). Endometrial cancer is occasionally referred to as uterine cancer. Certain kinds of cancer, like uterine sarcoma, can arise in the uterus, but they are far less prevalent than endometrial cancer. Endometrial cancer develops in the uterine lining. This type of uterine cancer is the most widespread, accounting for more than 90% of patients. For women with uterine cancer, the 5-year survival rate is 81%. The 5-year rate of survival is about 69% if the cancer has spread regionally. If diagnosed after the cancer has spread to the rest of the body, the rate of survival is 16%.

2. Conclusion

Female-specific cancer continues to be a major health concern globally. This chapter outlined the broad range of breast cancer, peritoneum cancer, uterine cancer, and cervical cancer, taking into account their indications and diagnosis to provide a realistic perspective on the potential of effective methods of transmission and care. Although cancer prevention measures are needed to lower mortality rates, the availability of safer and more effective anticancer drugs is urgently necessary.

References

[1] Siegel RL, Miller KD, Jemal A. Cancer statistics, 2019. CA Cancer J Clin 2019;69(1):7–34.

[2] Bray F, Ferlay J, Soerjomataram I, Siegel RL, Torre LA, Jemal A. Global cancer statistics 2018: GLO-BOCAN estimates of incidence and mortality worldwide for 36 cancers in 185 countries. CA Cancer J Clin 2018;68:394e424.

[3] Lauby-Secretan B, Scoccianti C, Loomis D, BenbrahimTallaa L, Bouvard V, Bianchini F, et al. Breast-cancer screening–viewpoint of the IARC working group. N Engl J Med 2015;372:2353e8.

[4] Office for National Statistics. Cancer statistics regulations, England (Series MB1), http://www.ons.gov.uk/ons/rel/vsob1/cancerstatistics-registrationseenglandeseries-mb1-/index.html.

[5] The Breast International Group (BIG). A comparison on letrozole and tamoxifen in postmenopausal women with early breast cancer. N Engl J Med 2005;353:2747–57.

[6] Pan H, Gray R, Braybrooke J, Davies C, Taylor C, McGale P, Peto R, Pritchard KI, Bergh J, Dowsett M, Hayes DF, EBCTCG. 20-year risks of breast-cancer recurrence after stopping endocrine therapy at 5 years. N Engl J Med 2017;377(19):1836–46.

[7] Dahm-Kähler P, Borgfeldt C, Holmberg E, Staf C, Falconer H, Bjurberg M, Kjölhede P, Rosenberg P, Stålberg K, Högberg T, Åvall-Lundqvist E. Population-based study of survival for women with serous cancer of the ovary, fallopian tube, peritoneum or undesignated origin-on behalf of the Swedish gynecological cancer group (SweGCG). Gynecol Oncol 2017;144(1):167–73.

[8] Sandoval P, et al. Carcinoma-associated fibroblasts derive from mesothelial cells via mesothelial-to-mesenchymal transition in peritoneal metastasis. J Pathol 2013;231(4):517–31.

[9] Barlow WE, Lehman CD, Zheng Y, Ballard-Barbash R, Yankaskas BC, Cutter GR, Carney PA, Geller BM, Rosenberg R, Kerlikowske K, Weaver DL. Performance of diagnostic mammography for women with signs or symptoms of breast cancer. J Natl Cancer Inst 2002;94(15):1151–9.

[10] Weigelt B, Reis-Filho JS. Histological and molecular types of breast cancer: is there a unifying taxonomy? Nat Rev Clin Oncol 2009;6(12):718.

[11] Fisher ER, Leeming R, Anderson S, Redmond C, Fisher B. Conservative management of intraductal carcinoma (DCIS) of the breast. J Surg Oncol 1991;47(3):139–47.

[12] Boelens MC, Nethe M, Klarenbeek S, de Ruiter JR, Schut E, Bonzanni N, Zeeman AL, Wientjens E, van der Burg E, Wessels L, van Amerongen R. PTEN loss in E-cadherin-deficient mouse mammary epithelial cells rescues apoptosis and results in development of classical invasive lobular carcinoma. Cell Rep 2016;16(8):2087–101.

[13] Steponavičienė L, Gudavičienė D, Meškauskas R. Rare types of breast carcinoma. Acta Med Lituanica 2012;19(2).

[14] Acs G. Serous and mucinous borderline (low malignant potential) tumors of the ovary. Pathol Patterns Rev 2005;123(Suppl_1):S13–57.

[15] Stelmach A, Ryś J, Mituś JW, Patla A, Skotnicki P, Reinfuss M, Pluta E, Walasek T, Sas-Korczyńska B. Typical medullary breast carcinoma: clinical outcomes and treatment results. Nowotwory J Oncol 2017;67(1):7–13.

[16] Rakha EA, Ellis IO. Diagnostic challenges in papillary lesions of the breast. Pathology 2018;50(1):100–10.

[17] Han Y, Li J, Han S, Jia S, Zhang Y, Zhang W. Diagnostic value of endoscopic appearance during ductoscopy in patients with pathological nipple discharge. BMC Cancer 2017;17(1):300.

[18] Salo SA, Ilonen I, Laaksonen S, Myllärniemi M, Salo JA, Rantanen T. Epidemiology of malignant peritoneal mesothelioma: a population-based study. Cancer Epidemiol 2017;51:81–6.

[19] Gadducci A, Zannoni GF. Uterine smooth muscle tumors of unknown malignant potential: a challenging question. Gynecol Oncol 2019;154(3):631–7.

[20] Røe OD, Stella GM. Malignant pleural mesothelioma: history, controversy, and future of a man-made epidemic. In: Asbestos and mesothelioma. Cham: Springer; 2017. p. 73–101.

[21] Berek JS, Friedlander ML, Bast Jr RC. Epithelial ovarian, fallopian tube, and peritoneal cancer. In: Holland-Frei cancer medicine; Wiley Online Library; 2016. p. 1–27.

[22] Ly T, Chan RC, Lau C. Peritoneal mesothelioma. Pathology 2018;50:S79–80.

[23] Mikuła-Pietrasik J, Uruski P, Tykarski A, Książek K. The peritoneal "soil" for a cancerous "seed": a comprehensive review of the pathogenesis of intraperitoneal cancer metastases. Cell Mol Life Sci 2018;75(3):509–25.

[24] Kotha NV, Baumgartner JM, Veerapong J, Cloyd JM, Ahmed A, Grotz TE, Leiting JL, Fournier K, Lee AJ, Dineen SP, Dessureault S. Primary tumor sidedness is predictive of survival in colon cancer patients treated with cytoreductive surgery with or without hyperthermic intraperitoneal chemotherapy: a US HIPEC collaborative study. Ann Surg Oncol 2019;26(7):2234–40.

[25] Hayes SC, Janda M, Ward LC, Reul-Hirche H, Steele ML, Carter J, Quinn M, Cornish B, Obermair A. Lymphedema following gynecological cancer: results from a prospective, longitudinal cohort study on prevalence, incidence and risk factors. Gynecol Oncol 2017;146(3):623–9.

[26] Deng S, Young B, Vilain R. Primary vaginal melanoma—report of 2 cases. Pathology 2018;50:S68–9.

[27] Fritchie KJ. Genital mesenchymal tumors. In: Soft tissue tumors of the skin. New York: Springer; 2019. p. 383–403.

[28] Prindull G, Zipori D. Environmental guidance of normal and tumor cell plasticity: epithelial mesenchymal transitions as a paradigm. Blood 2004;103(8):2892–9.

[29] Hernández JA, Navarro JT, Rozman M, Ribera JM, Rovira M, Bosch MA, Fantova MJ, Mate JL, Millá F. Primary myeloid sarcoma of the gynecologic tract: a report of two cases progressing to acute myeloid leukemia. Leuk Lymphoma 2002;43(11):2151–3.

[30] Adams T, Denny L. Abnormal vaginal bleeding in women with gynaecological malignancies. Best Pract Res Clin Obstet Gynaecol 2017;40:134–47.

[31] McCluggage WG. A practical approach to the diagnosis of mixed epithelial and mesenchymal tumours of the uterus. Mod Pathol 2016;29(1):S78–91.

[32] Felix AS, Brinton LA. Cancer progress and priorities: uterine cancer. Cancer Epidemiol Biomarkers Prev 2018;27(9):985–94.

[33] Kruse AJ, ter Brugge HG, de Haan HH, Van Eyndhoven HW, Nijman HW. Vaginal hysterectomy with or without bilateral salpingo-oophorectomy may be an alternative treatment for endometrial cancer patients with medical co-morbidities precluding standard surgical procedures: a systematic review. Int J Gynecol Cancer 2019;29(2):299–304.

[34] Howitt BE, Nucci MR, Quade BJ. Uterine mesenchymal tumors. In: Diagnostic gynecologic and obstetric pathology. Elsevier; 2018. p. 652–715.

[35] Koh WJ, Greer BE, Abu-Rustum NR, Apte SM, Campos SM, Chan J, Cho KR, Cohn D, Crispens MA, DuPont N, Eifel PJ. Uterine neoplasms, version 1.2014. J Natl Compr Canc Netw 2014;12(2):248–80.

[36] McCluggage WG. Mesenchymal and mixed epithelial-mesenchymal neoplasms of the cervix. In: Pathology of the cervix. Cham: Springer; 2017. p. 201–11.

[37] Cohen PR, Kurzrock R. Sarcoidosis and malignancy. Clin Dermatol 2007;25(3):326–33.

[38] Lee NK, Cheung MK, Shin JY, Husain A, Teng NN, Berek JS, Kapp DS, Osann K, Chan JK. Prognostic factors for uterine cancer in reproductive-aged women. Obstet Gynecol 2007;109(3):655–62.

CHAPTER 10

Therapeutic options for the management of cervical cancer

S. Shinde[a], N.K. Vishvakarma[a], A.K. Tiwari[b], V. Dixit[c], S. Saxena[d], and D. Shukla[a]
[a]Department of Biotechnology, Guru Ghasidas Vishwavidyalaya, Bilaspur, Chhattisgarh, India
[b]Department of Zoology, Bhanwar Singh Porte Government Science College, Pendra, India
[c]Department of Botany, Guru Ghasidas Vishwavidyalaya, Bilaspur, Chhattisgarh, India
[d]Department of Medical Laboratory Sciences, Lovely Professional University, Phagwara, India

Abstract

Cervical cancer is one of leading causes of cancer-related deaths worldwide. Incidence and mortality rates of cervical cancer are greater in developing countries. The main risk factor for cervical cancer is human papillomavirus (HPV) infection and poor hygiene practices. Delayed diagnosis in developing countries is the main cause of cervical cancer-related death. Other risk factors are smoking, obesity, lack of exercise, and immunomodulation by infection or drug. These factors indirectly help HPV to establish infection or modulate immune functions to increase susceptibility to cervical cancer. The treatment strategies largely depend on the stage of diagnosis and include surgery, chemotherapy, radiotherapy, or combination therapy. With the advancement of research new therapeutic modalities, such as immunotherapy, have shown effective response in controlling cervical cancer. Preventive measures such as HPV vaccination have also been successfully used for reducing cervical cancer incidence in the past few years. Other preventive measures such as awareness, healthy diet, exercise, and early diagnosis may help in reducing HPV infection and cervical cancer in developing countries.

Keywords: Cervical cancer, HPV, Chemotherapy, Immunotherapy.

Abbreviations

AC	adenocarcinoma
AP-1	activator protein-1
CIN	cervical intraepithelial neoplasia
CLR	C-type lectin receptor
CRT	chemoradiotherapy
DFS	disease-free survival
FDA	Food and Drug Administration
FIGO	International Federation of Gynecology and Obstetrics
HC2	Hybrid Capture2
HIV	human immunodeficiency virus
HPV	human papillomavirus
IFN	interferon
IL	interleukin
OS	overall survival
PAMP	pathogen-associated molecular pattern
PRR	pattern recognition receptor

G.P. Nagaraju, R.R. Malla (eds.)
A Theranostic and Precision Medicine Approach for Female-Specific Cancers
ISBN 978-0-12-822009-2
https://doi.org/10.1016/B978-0-12-822009-2.00010-8

RH	radical hysterectomy
RLR	RIG-I-like receptor
RT	radiotherapy
STAT	signal transducer and activator of transcription
STD	sexually transmitted disease
TLR	toll like receptor

1. Introduction

Cervical cancer is one of the leading causes of death in women worldwide. Approximately 570,000 cases were reported in 2018 with 311,000 deaths. Moreover, more than 85% of deaths due to cervical cancer were reported from low- and middle-income countries [1, 2]. Cervical cancer is the most frequently diagnosed cancer in women with 270,000 deaths and 500,000 new cases diagnosed every year [3]. With an annual incidence of 122,844 and 67,477 mortality [4], cervical cancer ranks second after breast cancer in women. In most of the developing countries like India and others, cervical cancer is the most common neoplasm among women aged between 15 and 44 years [4]. About 17% of the world's population is contributed by India, but it shares more than one fourth of the global burden of cervical cancer. Every fifth women suffering from cervical cancer is in India. Most of the women affected are from rural backgrounds, which reflects the proviso of socioeconomic status of India. The soaring incidence of in rural India is mainly due to lack of awareness about symptoms as well as severity of disease, poor hygiene, postponed diagnosis, and test availability [5]. It can also be hypothesized that social stigma attached with such diseases of the reproductive system especially in women also affects well-timed diagnostic measures. Wide disparity in the occurrence of cervical cancer across the globe is reported in its frequency, disease spread, and severity. In the United States, the death toll is reduced to three quarters after implementation of Pap smear test, through which early diagnosis is achieved with relative ease. This suggests a role of early detections in management of diseases as the progression of cervical cancer in women without any treatment leads to poor prognosis. Contrary to other malignant disorders, cancer of the cervix is potentially prophylactic. Moreover, treatment of this neoplastic disorder is achievable based on its early detection through effective screening. Explicit variants of oncogenic human papillomaviruses (HPVs) are now well established as a major cause of cervical cancer [6]. In addition, various epidemiological risk factors contribute to cervical cancer, such as sexual intercourse with multiple partners, number of pregnancies, poor genital hygiene, smoking, undernourishment, use of oral contraceptives, inadequate or no awareness, early age at first sexual intercourse, and being immunocompromised by other diseases such as human immunodeficiency virus (HIV) infection, immunosuppressive drugs, and organ transplantation [7]. The Word Health Organization (WHO) has estimated the prevalence of HPV infection in 630 million people worldwide, which accounts for 9%–13% of the world population [8, 9]. Recently, HPV infection in men has shown an impact in men, specifically in their genitalia in the form of genital warts and carcinoma of the penis [10]. HPV is one of the frequent sexually transmitted infections without a specific treatment. Moreover, HPV is asymptomatic at the early stage with symptoms manifesting later. Many studies have indicated that the majority of cervical cancers are sourced to HPV infection of genitalia [11]. The presence of HPV 16 or 18 in about 82.7% of invasive cervical cancer cases among women is reported by various studies of Indian origin [12].

HPV infection offers the opportunity for detection of infection at early and late stages along with cellular changes associated with it, which can be detected through simple cytological tests such as the Pap smear, or through visual inspection with acetic acid. The Pap test detects normal, benign, and cancerous cell lesions. Detections of HPV DNA or RNA through molecular techniques are frequently used measures for biopsy of tissue and exfoliated cell samples. Accurate HPV test can also be done by signal amplified hybridization assays [13]. Furthermore, detection of E6/E7 mRNA of HPV and other oncogenes can be successfully detected by reverse transcriptase polymerase chain reaction (RT-PCR) [14, 15]. Other nucleic acid amplification–based tests include nucleic acid sequence–based amplification. Apart from cost-effective screening of HPV infection, primary prevention can also be achieved by prophylactic measures through vaccination against HPV. Three vaccines has been approved by the Food and Drug Administration (FDA) that thwart the infection with disease-causing HPV types 16 and 18. Gardasil, Gardasil 9, and Cervarix have been approved for clinical use in several countries including India (NIH: National cancer institute). However, HPV vaccines are expensive and cannot be afforded by all groups/individuals at potential risk. Moreover, these vaccines cannot protect from all types of HPV with potential pathogenicity. This makes routine screening and follow-up of woman for cervical cancer necessary, even if vaccines are available. However, uses of vaccines curtails the probability of infection with recognized variants of HPV.

2. Pathophysiology

In approximately 99.7% of adenocarcinomas (AC) and squamous cell carcinomas, HPV infection is detected [16]. Cervical carcinoma usually begins within the squamocolumnar junction in the cervix. The HPV virion utilizes the squamocolumnar junction as an active transformation zone, which bears most cases of dysplastic lesions due to high-risk HPV infection. Various hormonal and physiological as well as physical changes in women limit the transformation zone in the canal of the cervix, which results in failure to spot infection through visual vaginal examination. Further, the acidic environment in the vagina induces destruction of columnar cells of the transformation zone, leading to promotion of squamous metaplasia [17]. Over the last few decades, occurrence of cervical AC has increased in various countries. About 25% of cervical cancers arise in the glandular cells of the endocervix, which secrets mucus and therefore makes early detection by Pap test difficult. A number of investigations demonstrate worse prognosis for AC of the cervix compared to other cancers. Moreover, cumulative data cannot be overlooked that indicate that incidence rate has also been declining owing to early screening of cervical lesions including both premalignant and malignant ones [18]. Mostly, adults with frequent sexual activity are likely to be exposed to HPV; about half of infected adults are 20–24 years old [19].

3. Risk factors

A number of factors influence the development of cervical cancer. One of the major risk factors is persistent prolonged infection with high-risk oncogenic HPV types. Other risk factors either influence the acquisition of the virus or cause immune dysfunction, which supports virus growth. Host genetic makeup, exposure to mutagens, and hormonal factors are also shown to influence the etiology of cervical cancer. One of the notable risk factors includes sexual activity at an early age and having multiple sexual partners. Passive or active smoking is also a critical risk factor for HPV infection-driven cervical carcinoma. Exposure to other sexually transmitted diseases (STDs), HIV infection, and use of immunosuppressive drugs can also increase risk of cervical cancer in women. Use of oral contraceptives is also correlated with frequent occurrence of HPV infection and cervical carcinoma. This may be due to coercion alteration of hormone levels. In countries with lower socioeconomic status, lack of education, poor hygiene, and limited availability/access to screening services are also important risk factors for cervical cancer.

The incidence rate of HPV infection is significantly higher in women younger than 25 years. Worldwide about 50%–80% of sexually active women face at least one HPV infection in their lifetime. Interestingly, reports also suggest that cervical cancer is usually found/detected after age 45 years [20]. In the majority of cervical cancer cases (more than 90%), HPV infection is found, making it the foremost responsible factor. However, other factors such as long-term use of hormonal contraceptives, use of tobacco, smoking, multiple pregnancies, HIV and/or STD infections (such as *Chlamydia trachomatis* or herpes simplex virus type 2) as well as certain nutritional deficiencies have also been acknowledged as important cofactors [21].

3.1 Human papillomavirus

Adequate evidences have been generated through epidemiological studies on the causative association of human HPV infections in the onset and development of cancer of the cervix. The worldwide load of HPV-related diseases is great, and as per estimates more than 5% of total cancer burden arises from it. In different organs, cellular tropism for squamous epithelial cells is highly exhibited by papillomaviruses, which can range from dysplastic, benign, and hyperplastic to invasive carcinomas. Persistent infection of HPV could lead to cervical intraepithelial neoplasia (CIN3) or cervical cancer [22]. HPV is classified into 30 types, of which 15 are high risk (HR) and 15 are low risk (LR). High-risk HPV types (HPV 16, 18, 31, 35, 39, 45, 51, 52, 56, 66, 68, 69, and 73) are frequently associated with lesions of high-grade and invasive cancer. Low-risk HPV types include HPV 6, 11, 40, 42, 43, 44, 54, 61, 70, 72, 81, CP6108. These low-risk HPV types are mostly detected in skin warts, low-grade lesions, and sometimes in condyloma accuminata [23]. Among all these types of HPV, the most carcinogenic types are HPV 16 and HPV 18, which are found in almost 70% of cancer lesions of the cervix.

The genome of HPV consists of ~8000 bp of single-stranded circular DNA and codes different genes expressed at different stages of the virus life cycle. These are categorized into six early genes (E1, E2, E4, E5, E6, and E7) and two late genes (L1 and L2) [24]. Among these, E6 and E7 have a number of cellular targets and are principal HPV oncoproteins. E6 of HPV binds with tumor-suppressor protein p53 and blocks apoptosis [25], whereas E7 binds with pRb and makes it relatively inaccessible to E2F transcription factor, leading to abrogation of cell cycle arrest [26]. Further downregulation of p53 by E6 stabilizes hypoxia-inducing factor-1 (HIF-1) and favors the expression of its downstream genes such as vascular endothelial growth factor (VEGF), erythropoietin, and glucose transporters. VEGF is responsible for increased angiogenesis, which ensures the nutrient supply to cells undergoing transformation and later serves as a roadway for invasive cancer cells. Erythropoietin, known for its role in increased RBC formation, upregulates the glucose transporters, resulting in amplified glucose uptake. Both angiogenesis and metabolic alteration are recognized as hallmarks of cancer [27]. These together and with other factors help in transforming infected cells into cancer cells, leading to tumor formation and survival via metabolic adaptation. Keeping in mind their important role in tumor formation and survival, E6 and E7 proteins are being therapeutically targeted.

3.1.1 Life cycle of HPV

HPV-induced cervical cancer involves a series of events that can be divided into four steps, namely, transmission, viral persistence, progression of persistently infected cells to the precancerous stage, and invasion. Viral persistence occurs in cervical epithelial cells, which undergo transformation. Precancerous cells arise from mild lesions that are cytologically normal, while HPV-infected cells/tissues are not discriminated as almost all of the initial lesions preserve viral components as an episome and support viral replication under regulation of products of the viral genome [28]. Cell-mediated immunity eliminates most HPV infections of the cervix, causing cytological alterations within 1–2 years of primary exposure [29]. The transmission of HPV and viral persistence in intraepithelial locales establishes the viral genome into the infected host-cell. This stage serves as a therapeutic window to provide prophylaxis or vaccination to prevent further propagation of the viral load in cervical epithelium. In the precancerous stage, early lesions of low grade have a homogeneous cell population with genetically stability. These cells replace full thickness of cervical epithelium. The next most important step in the life cycle of HPV is the integration of viral genome into the genome of human cervical cells. This creates high-grade lesions with heterogeneous cells attaining genetic instability along with expression of oncogenic proteins of viral origin. It is frequently observed in CIN3 or higher-grade lesions. Low-grade lesions or CIN1 are not precancerous and histological studies have confirmed that such lesions are at lesser risk of progression to cervical cancer. The last step of the viral life cycle is upregulated expression of viral oncogenes in order to maintain the cancer cells and differentiate the cervical epithelium. Cervical epithelial differentiation and other associated cellular factors are found to strengthen the viral gene expression of HPV. Several transcription factors of host cell origin, their interaction, and keratinocyte-specific enhancers (AP-1, AP-2, KRF-1, NF-1, Oct-1, Sp1, STAT3, TEF-1, TEF-2, YY1, and glucocorticoid-responsive elements) have been well known to be present in the upstream regulatory region of HPV genes. This ensures the expression of genes of viral as well as host cell origin as per the stage of infected cell differentiation, specificity, and tissue tropism of HPV [30].

3.1.2 Immunomodulation by HPV

The role of the immune system in modulation of cellular transformation is well established in eliminating as well as stabilization transforming cells. It is customary that evasion of immune tumor surveillance is a feature of cancer [27]; initial transforming cells are mostly cleared by cells of the immune system. In HPV–infected cervical epithelial cells and transformation, the initial stage of cervical intraepithelial carcinoma is cleared by cells innate in the immune system. Like cancers of other origins, HPV–induced cervical cancer cells also undergo immune escape by targeting initial steps of the innate immune system, including keratinocytes, which take up toll-like receptors (TLRs) on the surface and release cytokines, thus compromising the immune response against invasive agents [31]. Pattern recognition receptors (PRRs) of immune cells, such as TLR, CLR, RLR, and XLR, have the capability to identify microbial pathogens or damage signals. These PRRs recognize the pathogen through their ability to interact with pathogen-associated molecular patterns (PAMPs) or damage-associated molecular patterns (DAMPs) present on the pathogen or in damage-associated state of affairs. When PRRs are engaged with PAMPs or DAMPs, they promote immune response by synthesizing and secreting cytokines [32]. PRR receptors critical in antiviral immunity include TLRs and nucleotide-binding oligomerization domain-like receptors that detect nucleic acids during viral replication [33]. Viral nucleic acid can be recognized by TLR3, TLR7, TLR8, and TLR9 and increased expression of these TLRs is associated with HPV elimination and can be used as predictive marker for HPV 16 clearance in infected women [34]. However, some high-risk HPV components are capable of decreasing TLR expression, nuclear factor-kappaB (NF-κB), and interferon regulatory factor 3 (IRF3) activation/nuclear translocation, which contribute to viral immune evasion and persistence [35, 36]. Polymorphism of genes responsible for immune response against viral infection is often associated with outcome or progression of infection. Gene polymorphisms related to innate immunity genes such as interleukin (IL)-1β, IL-18, NLR1, and NLR3 are associated with HPV infection and persistence. However, polymorphism of TLR9, recognized as a DNA sensor, is not related with viral clearance or persistence [37]. There are several mechanisms through which HPV evades immune response. One such mechanism is modulation of cytokines and chemokines and downregulation of interferon (IFN) pathways. Defacement of antigen presentation and decrease in expression of various cell adhesion molecules, especially intercellular adhesion molecule 1, also affect immune evasion. These modulations for achieving immune escape are directed mainly by the E6 and E7 oncoproteins of HPV origin [38]. IFN-mediated evasion of the immune system is another mechanism used by viral oncoproteins that reduces the secretion of IFN by keratinocytes by phosphorylating intermediates of IFN through interferon regulatory factors (IRFs). IRF1 expression is reduced in cervical tissues. High risk of HPV infection has been associated with reduced phosphorylation of IRF3 as well [39]. Downregulated expression of inflammasome components and PRRs in IL-1β network correlates with high-risk HPV. This affects response of innate immunity as well as adaptive immunity, which are mediated by IL-1β [40]. The function of antigen-presenting cells (APCs), macrophages, and natural killer (NK) cells are also compromised by HPV infection. Dendritic cells (DCs) are found to be reduced in CINs [41]. In addition, programmed death-ligand 1 (PD-L1) is overexpressed by tumor cells, which carry out anergy in cells of the immune system through interaction with its PD-1 receptor expressed on T cells. High-risk HPV-positive cervical cancer-associated DCs were also reported to overexpress PD-L1 as compared to high-risk HPV-negative cervical cancer DCs [42].

3.2 Smoking

Smoking has been associated with a number of health ailments, especially those related to bronchopulmonary dysfunction. However, its role in pathophysiological manifestation in other organs cannot be overruled. In HPV-induced cervical carcinoma, persistent infection, especially that of high-risk HPV, is one of the most important threats. However, other factors are also required for various triggers of disease development and progression. Smoking has been listed as causally related to cervical cancer in 2004 by the International Agency for Research on Cancer (IARC) [43]. Smoking is not directly related with cervical cancer, but rather it interferes with HPV infection and progression and is therefore associated with cervical cancer incidences. Further, tobacco affects cervical cancer by its direct local and to some extent systemic carcinogenic as well as immunosuppressive consequence. Findings also indicate that smoking can influence systemic immunity, contraception, and nutrient absorption/digestion. These smoking-driven alterations can impel cervical carcinogenesis, especially in individuals who have infection of HPV.

3.3 Weight

A strong relation is found between obesity (BMI > 30) and incidence of cancer. Researchers have shown that obesity increases the risk of various types of cancer, including leukemia, lymphoma, multiple myeloma, and liver and gall bladder cancers by influencing immunosuppression [44]. Obesity also affects diagnosis sensitivity. Investigations have shown that obese women have elevated risk of cervical cancer [45].

3.4 Exercise

A number of health benefits are demonstrated for regular exercise and one of them reduction of cancer risk. Exercise improves the overall health status by affecting the biochemical, metabolic, inflammatory, and epigenetic changes in the body that are responsible for induction of different anticancer pathways [46]. A few important pathways are expression of growth factor promoting apoptosis, activation of antioxidant pathway, and maintenance of telomere length, inflammation, and heat shock proteins. Various studies have demonstrated that exercise may affect the degree of risk for ovarian and breast cancer in women. In a case-controlled study, Szender et al. [47] demonstrated that not engaging in a regular exercise regimen increases the risk of cervical cancer in women.

3.5 Diet

Healthy diet is one of the important factors for lowering cancer incidence and progression. Overnutrition causes obesity, which increases the risk of various types of cancer. Undernutrition has not been directly associated with cancer occurrence, but reduced immune health due to undernutrition may contribute to cancer progression [48]. However, undernutrition is generally observed in patients with low socioeconomic statuses who also have compromised hygiene conditions and are therefore prone to HPV infection and cervical cancer [49]. Studies has also shown that a Western diet is associated with greater risk of HPV infection [48].

3.6 Infection

High-risk HPV infection is a foremost cause of cervical cancer among women. Other infections of the reproductive tract or genitalia such as HIV, chlamydia, and so on indirectly affect the risk of cervical cancer. Non-HPV infections render hosts immunocompromised due to infection-induced immunosuppression, as in the case of HIV or due to immunosuppressive action of medication [50]. Moreover, infections also pave the way for invasion and establishment of secondary infections. It has been reported that coinfection with HIV or other STDs with HPV increases the risk of cervical cancer in women. It is well established that a decrease in rate of infection reduces the chances of HPV infection, which may also help in lowering the incidence of cervical cancer. Nevertheless, sexual transmission is a mode of distribution and spread for HPV. Preventive measures such as using prophylactics during sex reduce HPV transmission and limit cervical cancer incidence.

4. Therapeutic strategies

A broad range of clinical outcomes are associated with cervical cancer, including small untraceable microinvasive lesions to large tumors. Improvements have been made for early detection and screening through molecular biomarker-based assays. Stage I tumors are stratified in two categories in the cervical cancer staging system by the International Federation of Gynecology and Obstetrics (FIGO). These include Stage IA (microinvasive) and Stage IB (gross tumor). The treatment choice for cervical cancer largely depends on the stage of diagnosis. Early detection of cervical cancer offers several options for treatment with real probability of curing. Radical hysterectomy (RH) with pelvic lymphadenectomy is the standard choice of treatment for women with early-stage cervical cancer. However, this procedure results in infertility. Moreover, radical trachelectomy with pelvic lymphadenectomy could be a viable option for patients who want to remain fertile [51]. Depending upon the severity, late-stage cervical cancer patients may require surgical intervention, radio and/or chemotherapy or a combination of treatments. With advancement of diagnostics and treatment measures, more targeted strategies are being developed, including immunotherapy as well as preventive vaccination. Table 1 summarizes stage-wise treatment options and their therapeutic outcomes.

4.1 Conventional therapies

Most of cervical cancer arises from HPV infection-driven cellular transformation, both anticancer and antiviral therapeutic regimens are implemented for treatment. Treatment strategies preferably aim to remove lesions rather than eliminate HPV infection. However, microlesions cannot be detected and removed through conventional strategies and elimination of HPV infection cannot be achieved. For precancerous cells, the entire transformation zone of CIN is treated as per requirements. Hysterectomy requires general anesthesia, which was recommended for most CIN3 lesions until the 1970s. Cold-knife conization is also utilized for removal of extended tissue from affected areas [59]. Most CIN lesions do not progress to the invasive stage of cancer. Therefore, surgical hysterectomy is a preventable procedure at the stage of CIN. In many countries, treatment options for precancers of the cervix do not include hysterectomy. Cold-knife conization and cryotherapy under local anesthesia is the preferred choice of treatment for cervical intraepithelial lesions.

Table 1 Treatment strategies in different stages of cervical carcinoma.

Stage	Carcinoma characteristics	Treatment strategies	Survival rate/ treatment efficacy	References
I	Carcinoma is strictly confined to cervix uteri (microinvasive)			
IA	Invasive carcinoma, invasion is limited with maximum depth of <5 mm			
IA_1	Measured stromal invasion <3 mm in depth	Hysterectomy and patients who desire fertility (cone biopsy with negative surgical margins)	Ovarian preservation in early AC is safe and not associated with an increased risk of overall mortality, 100% survival	[52]
IA_2	Measured stromal invasion >3 mm and <5 mm in depth	Lymphovascular space, lymph node metastasize–radical hysterectomy and pelvic lymphadenectomy		[51]
IB	Invasive carcinoma with deepest invasion >5 mm, lesion limited to cervix uteri	Radical hysterectomy including pelvic lymphadenectomy	5-year OS = 88.2% and radiation-free survival = 83.8%, 90.7% (FIGO)	[53]
IB_1	>5 mm depth of stromal invasion and <2 cm in greatest dimension	Radical hysterectomy or radiotherapy		
IB_2	>2 cm and <4 cm in greatest dimension	Radiation therapy, chemotherapy		
IB_3	>4 cm in greatest dimension			
II	Carcinoma extends beyond uterus, but not to lower third of the vagina or pelvic wall			

Table 1 Treatment strategies in different stages of cervical carcinoma—cont'd

Stage	Carcinoma characteristics	Treatment strategies	Survival rate/ treatment efficacy	References
IIA	Involvement limited to upper two-thirds of vagina without parametrial invasion	Radical surgery, radiotherapy	5-year OS = 87%–92%, FIGO = 82.8%	[54]
IIA_1	Invasive carcinoma <4 cm in greatest dimension			
IIA_2	Invasive carcinoma >4 cm in size in greatest dimension			
IIB	Parametrial invasion with no spread in pelvic side wall		FIGO = 55.6%	
III	Carcinoma extends to pelvic wall, kidney, pelvic, or paraaortic lymph nodes			
IIIA	Carcinoma involves lower third of vagina, no extension to pelvic wall	Chemoradiation alone or chemoradiation first than radical hysterectomy	3-year (CRT) OS = 63.2% or OS = 67.7%. RH after (CRT) is an effective approach	[55]
IIIB	Extends to pelvic wall and causes hydronephrosis or nonfunctioning kidney	Cisplatin in combination (CRT) or alone (RT) in squamous cell carcinoma of uterine cervix	Cisplatin (CRT) 5-year DFS = 52.3%, OS = 54%, RT 5-year DFS = 43.8%, OS = 46%; CRT has better DFS and OS	[56]
IIIC	Involvement of pelvic and paraaortic lymph nodes			
$IIIC_1$	Pelvic lymph node metastasis	Adjuvant radiation and chemotherapy	Adequate OS in patients with only one metastatic node, OS outcomes are poor in patients with two or more metastatic nodes	[57]

Continued

Table 1 Treatment strategies in different stages of cervical carcinoma—cont'd

Stage	Carcinoma characteristics	Treatment strategies	Survival rate/ treatment efficacy	References
$IIIC_1r$	Identified/ characterized through diagnostic imaging			
$IIIC_1p$	Identified/ characterized through histo/ cytopathology			
$IIIC_2$	Paraaortic lymph node metastasis			
IV	Extends beyond the true pelvis and involved in mucosa of bladder or rectum			
IVA	Spread to adjacent organs	Brachytherapy, chemoradiation		
IVB	Spread to distant organs	Radiation, combinational therapy (paclitaxel + cisplatin), mitomycin, irinotecan		[58]

OS, overall survival; *CRT*, chemoradiotherapy; *RT*, radiotherapy; *DFS*, disease-free survival.

Large loop excision or loop electrosurgical excision of the transformation zone of the virus was used to execute cone excision of the affected cervical component in outpatient clinics [60]. It has now been established that a fewer number of women with precancerous lesions of the cervix require surgical intervention. The WHO has also recommended ablation as the first treatment option for cervical intraepithelial lesions (CIN1/CIN2). In the 1990s, a cryotherapeutic ablative technique using CO_2 and N_2O was introduced. A cooled-down probe was applied to cervical tissue inducing necrosis of the target cells and destroying them. It is a low-cost treatment used to treat precancerous lesions in approachable benign hyperplasia. However, a long recovery time over weeks is associated with cryo-ablation. Thermal coagulation is also used, which employs a heated probe in place of a cold probe. However, the thermal probe has not been given as much recognition as the cold probe.

An important advancement in treatment of invasive cervical cancer is RH for Stage I. Over the past few years, it has been the preferred choice of treatment for women with small tumors. Moreover, fertility remains unaffected for women undergoing treatment with radical vaginal trachelectomy with laparoscopic pelvic lymphadenectomy. Nevertheless, this procedure is relatively safe. Plante et al. [61] reported the occurrence of pregnancy in 31 (43%) of 72 patients treated in this way.

4.1.1 Radiotherapy

Radiotherapy (RT) follows surgical interventions in women with cervical cancer who were diagnosed at the early stages of cancer. A study has shown radiation decreases progression of cervical cancer compared with no further treatment. Although decreased cervical cancer progression was achieved, improvement in overall survival (OS) in Stage IB cervical cancer after surgery is not evident [62]. Moreover, mortality was found to be reduced in women who received chemoradiation after surgery, according to a Cochrane study of 401 women affected with IA2-IIA early stage cervical cancer [63]. This suggests the efficacy of combination treatment with radiation. Chemoradiation is usually the choice of treatment for locally advanced cervical cancer patients. Organometallic compounds used in cancer therapies are also used in HPV-induced cervical cancer. Platinum-based regimens in combination treatment have achieved striking success. An external beam is used for radiation, and for treatment of cervical cancer, a sealed radiation source is positioned in close proximity to the affected tissue [64].

4.1.2 Chemotherapy

Chemotherapeutic intervention has been considered for advanced stages of metastatic cancer, including cervical cancer. For the treatment of recurrent/advanced cervical cancer, cisplatin is one of the chemotherapeutic drugs of choice that has shown improved OS in patients [65]. However, findings related to the development of chemoresistance hindered the use of cisplatin in clinical practice. More targeted drugs like bevacizumab (Avastin), an antibody-targeting VEGF, inhibit tumor angiogenesis and culminate in shrinkage of cervical tumors with low toxicity. A clinical trial with oral pazopanib (VEGF inhibitor) and lapatinib (Her2/neu inhibitor) has been done [66]. Pazopanib was found to impose less toxicity along with prolonged progression free survival as compared to lapatinib. Another drug, cediranib, a potent tyrosine kinase inhibitor of VEGF 1–3, was combined with standard chemotherapeutic agents carboplatin and paclitaxel in patients with metastatic/recurrent cervical cancer in a randomized double-blind placebo controlled phase II trial. The study showed efficacy of cediranib with carboplatin, but an increase in toxic effects [67]. These collectively suggest that combination treatment has an advantage over single treatment. A better therapeutic outcome can be achieved if toxicity of combination treatments can be controlled, which may necessitate application of in silico predictive tools of bioinformatics. Drug details, their mechanism of action, and patient outcomes are listed in Table 2.

Table 2 Drugs and their targets for cervical cancer.

Drug	Mechanism of action	Patients/outcomes	References
Bevacizumab (Avastin)	Recognizes VEGF and prevents interaction to its receptors (Flt-1 and KDR) on the surface of endothelial cells	Metastatic, persistent, recurrent cervical carcinoma; shrinkage of cervical tumors and low toxicity	[68]
Pazopanib	Inhibits tyrosine kinase activity of VEGFR, PDGFR, and c-kit	Stage IVB, adenocarcinoma of the cervix	[66]
Cediranib + carboplatin	Inhibits tyrosine kinase activity of VEGF1, 2, and 3	Metastatic/recurrent cervical cancer; increase in toxicity among patients	[67]

4.1.3 Combinational therapy

With advancement of our understanding of the molecular alterations in cervical cancer, new therapeutic modalities are being developed after clinical trials, such as immune modulators, therapeutic vaccines, and monoclonal antibody therapy. Current research is now focused on restricting the growth of cancer cells by targeting multiple molecules, known as combination therapy. Various clinical trials have shown that combination therapy is more effective in patients with metastatic, persistent/recurrent cervical cancer and increases the OS rate compared to monotherapy. For example, the best platinum-based cytotoxic regimen for advanced cervical cancer was identified by comparing cisplatin with a combination of cisplatin/paclitaxel, cisplatin/topotecan, and gemcitabine/vinorelbine. The most effective response was observed in cisplatin/paclitaxel [66]. However, further studies have identified cisplatin resistance in cervical cancer and therefore new strategies were developed to identify and target molecular pathways that are critical for the growth of cervical cancer. A combination of carboplatin and paclitaxel has been approved as first-line therapy in metastatic cervical cancer. Bevacizumab is the anti-VEGF monoclonal antibody that targets angiogenesis and therefore restricts the nutrient and oxygen supply to growing tumors. The combination therapy of bevacizumab with carboplatin-paclitaxel was also found to be effective in treating advanced metastatic cervical cancer. Another combinational drug, topotecan-paclitaxel-bevacizumab, has showed no impact on OS, although it is beneficial in patients with a history of platinum hypersensitivity or renal insufficiency [69].

5. Advance therapeutics

Early stage cervical cancer patients are generally treated with surgery and the only chemotherapeutic drug approved that has shown effectiveness for metastatic, persistent/

recurrent cervical cancer is bevcizumab combined with chemotherapy. However, due to a lack of effective second-line treatment, patients who progress after first-line treatment have high mortality rates [70]. For this reason, new and advanced therapeutic options are being considered, such as immunotherapy. Regulation of cellular immunity occurs by both activatory signals like costimulatory molecules and inhibitory signals such as immune checkpoints [71]. In normal conditions, the immune checkpoint maintains self-tolerance, prevents autoimmunity, and protects normal cells from immune attack during infections. Cancer cells escape this immune surveillance by changing the tumor microenvironment. Therefore, immune checkpoint inhibition could be one of the potential methods to enhance antitumor immunity in cervical cancers. One of the important targets for immunotherapy in cervical cancer is cytotoxic T lymphocyte–associated antigen 4 (CTLA-4). After activation of T cells in the lymph nodes, expression of CTLA4 begins and downregulates the activated T cells [72]. Due to this, T-cell activation within secondary lymphoid organs is also inhibited [73]. Another potential target is PD-1 expressed on effector T cells in peripheral tissues. PD-1 binds with PD-L1 expressed on DCs, TAFs, or tumor cells and prevents T cells from attacking the cancer [74]. Normally IFN-γ-induced PD-L1 protects DCs from T cell-mediated cytotoxicity. In tumor tissue a higher percentage of PD-1 expressing CD8+ T cells is observed. Currently, the FDA has approved only CTLA-4 and PD-1/PD-L1 inhibitors [75]. B7-1 and B7-2 costimulatory molecules expressed on the surface of APCs are also integrated with the help of CTLA-4 [76]. Further, it is clear that by blocking the CTLA-4 expression on the surface of naïve T cells in lymph nodes and PD-1 on effector T cells could be potential therapeutic targets. The details of immunotherapy drugs are listed in Table 3.

Table 3 Immunotherapy drugs for cervical cancer.

Monoclonal antibody	Study	Significance	References
Anti-CTLA-4			
Ipilimumab (FDA approved)	Women with metastatic or recurrent HPV-related cervical carcinoma	Drug was tolerable, but didn't show significant activity	[77, 78]
Anti-PD-1			
Cemiplimab	REGN2810	Response rate is greater when combined with radiation, suggesting abscopal effects	[79]
Pembrolizumab	Keynote-028 Trial	No deaths were observed, drug showed antitumor activity and exhibited safety profile	[80]
Nivolumab	Checkmate358	Durable responses in cervical cancer patients	[81]

6. Therapeutic HPV vaccines

One of the important therapeutic measures to prevent cervical cancer is HPV vaccination, which has been clinically shown to prevent the formation of precursor lesions and high-grade CIN. It has been validated that vaccinating girls (around age 11 or 12 years) effectively protects them against major types of HPV that can cause cervical cancer later in life [82]. The jeopardy of inclusive transmission rate and HPV-associated cancer is probably decreased by immunizing boys as well [83]. The hepatitis B vaccine reduces liver cancer risk, and the growing use of HPV vaccines worldwide is expected to result in similar benefits [84]. Furthermore, successful screening protocols for cervical cancer and treatment have led to a decline in mortality rates. In response to the clear, causative relation between high-risk HPV and cervical cancer, HPV vaccines utilize virus-like particles to generate antibody response. The FDA approved three HPV vaccines. Gardasil, Gardasil 9, and Cervarix are currently available in the market. Gardasil provides protection against HPV 6 and 11, which cause genital warts. Gardasil 9 covers additional strains of HPV and is expected to increase protection against cervical cancer by up to 90% [85]. Ceravix uses aluminum and an LPS derivative as an adjuvant that stimulates the innate immune system, and hence activates TLR4, which promotes death of HPV-infected cells through activation of DCs and NK cells [85]. It also provides protection against HPV 16 and 18, which are responsible for 70% of cervical cancers.

6.1 Prevention measures for cervical cancer

Research from the last 200 years have identified that exposure to mutagens present in the environment or diet causes cancer and are responsible for increasing cancer incidence and mortality in both developed and underdeveloped countries. However, research over the past 50 years has shown that different preventive measures can be taken to decrease the incidence of cancer. One such measure is lifestyle change, which plays an important role in cancer progression by modulating the immune system. Other measures include avoidance of exposure to carcinogens from the environment, healthy diet, and maintaining good hygiene, all of which may help prevent cancer. Cervical cancer is a preventable disease and therefore increasing awareness of the public to possible causes may reduce the burden of cervical cancer.

References

[1] Balasubramaniam SD, et al. Key molecular events in cervical cancer development. Medicina (Kaunas) 2019;55(7):384.
[2] World Health Organization. WHO guidelines for screening and treatment of precancerous lesions for cervical cancer prevention: supplemental material: GRADE evidence-to-recommendation tables and evidence profiles for each recommendation. World Health Organization; 2013.
[3] Bray F, et al. Global cancer statistics 2018: GLOBOCAN estimates of incidence and mortality worldwide for 36 cancers in 185 countries. CA Cancer J Clin 2018;68(6):394–424.

[4] Sreedevi A, Javed R, Dinesh A. Epidemiology of cervical cancer with special focus on India. Int J Womens Health 2015;7:405–14.

[5] Tripathi N, et al. Barriers for early detection of cancer amongst Indian rural women. South Asian J Cancer 2014;3(2):122.

[6] Faridi R, et al. Oncogenic potential of human papillomavirus (HPV) and its relation with cervical cancer. Virol J 2011;8(1):269.

[7] Opoku CA, et al. Perception and risk factors for cervical cancer among women in northern Ghana. Ghana Med J 2016;50(2):84–9.

[8] Pandhi D, Sonthalia S. Human papilloma virus vaccines: current scenario. Indian J Sex Transm Dis 2011;32(2):75.

[9] World Health Organization. WHO guidelines for screening and treatment of precancerous lesions for cervical cancer prevention. World Health Organization; 2013.

[10] Sichero L, Giuliano AR, Villa LL. Human papillomavirus and genital disease in men: what we have learned from the HIM study. Acta Cytol 2019;63(2):109–17.

[11] Wardak S. Human papillomavirus (HPV) and cervical cancer. Med Dosw Mikrobiol 2016;68(1):73–84.

[12] Bosch FX, de Sanjosé S. Chapter 1: Human papillomavirus and cervical cancer—burden and assessment of causality. J Natl Cancer Inst Monogr 2003;2003(31):3–13.

[13] Zaravinos A, et al. Molecular detection methods of human papillomavirus (HPV). Int J Biol Markers 2009;24(4):215–22.

[14] Smits HL, et al. Application of the NASBA nucleic acid amplification method for the detection of human papillomavirus type 16 E6-E7 transcripts. J Virol Methods 1995;54(1):75–81.

[15] Sotlar K, et al. Detection of high-risk human papillomavirus E6 and E7 oncogene transcripts in cervical scrapes by nested RT-polymerase chain reaction. J Med Virol 2004;74(1):107–16.

[16] Böhmer G, et al. No confirmed case of human papillomavirus DNA-negative cervical intraepithelial neoplasia grade 3 or invasive primary cancer of the uterine cervix among 511 patients. Am J Obstet Gynecol 2003;189(1):118–20.

[17] Jacobson D, et al. Cervical ectopy and the transformation zone measured by computerized planimetry in adolescents. Int J Gynaecol Obstet 1999;66(1):7–17.

[18] Sankaranarayanan R, Budukh AM, Rajkumar R. Effective screening programmes for cervical cancer in low-and middle-income developing countries. Bull World Health Organ 2001;79:954–62.

[19] Satterwhite CL, et al. Sexually transmitted infections among US women and men: prevalence and incidence estimates, 2008. Sex Transm Dis 2013;40(3):187–93.

[20] Das BC, et al. Prospects and prejudices of human papillomavirus vaccines in India. Vaccine 2008; 26(22):2669–79.

[21] Munoz N, et al. Chapter 1: HPV in the etiology of human cancer. Vaccine 2006;24(Suppl 3) S3/1-10.

[22] Rodríguez AC, et al. Longitudinal study of human papillomavirus persistence and cervical intraepithelial neoplasia grade 2/3: critical role of duration of infection. J Natl Cancer Inst 2010;102(5):315–24.

[23] Bharti AC, et al. Human papillomavirus and control of cervical cancer in India. Expert Rev Obstet Gynecol 2010;5(3):329–46.

[24] Doorbar JJCS. Molecular biology of human papillomavirus infection and cervical cancer. Clin Sci (Lond) 2006;110(5):525–41.

[25] Mantovani F, Banks L. The human papillomavirus E6 protein and its contribution to malignant progression. Oncogene 2001;20(54):7874–87.

[26] Munger K, et al. Biological activities and molecular targets of the human papillomavirus E7 oncoprotein. Oncogene 2001;20(54):7888–98.

[27] Hanahan D, Weinberg RA. Hallmarks of cancer: the next generation. Cell 2011;144(5):646–74.

[28] Bharti AC, et al. Anti-human papillomavirus therapeutics: facts & future. Indian J Med Res 2009;130(3):296.

[29] Stanley M. Immune responses to human papillomavirus. Vaccine 2006;24(Suppl 1):S16–22.

[30] Thierry F. Transcriptional regulation of the papillomavirus oncogenes by cellular and viral transcription factors in cervical carcinoma. Virology 2009;384(2):375–9.

[31] Grabowska AK, Riemer AB. The invisible enemy—how human papillomaviruses avoid recognition and clearance by the host immune system. Open Virol J 2012;6:249–56.

[32] Medzhitov R, Janeway Jr. CA. Innate immunity: impact on the adaptive immune response. Curr Opin Immunol 1997;9(1):4–9.

[33] Jo EK, et al. Molecular mechanisms regulating NLRP3 inflammasome activation. Cell Mol Immunol 2016;13(2):148–59.

[34] Daud II, et al. Association between toll-like receptor expression and human papillomavirus type 16 persistence. Int J Cancer 2011;128(4):879–86.

[35] Hasan UA, et al. TLR9 expression and function is abolished by the cervical cancer-associated human papillomavirus type 16. J Immunol 2007;178(5):3186–97.

[36] Tummers B, et al. The interferon-related developmental regulator 1 is used by human papillomavirus to suppress NFκB activation. Nat Commun 2015;6:6537.

[37] Oliveira LB, et al. Polymorphism in the promoter region of the Toll-like receptor 9 gene and cervical human papillomavirus infection. J Gen Virol 2013;94(Pt 8):1858–64.

[38] Kanodia S, Fahey LM, Kast WM. Mechanisms used by human papillomaviruses to escape the host immune response. Curr Cancer Drug Targets 2007;7(1):79–89.

[39] Um SJ, et al. Abrogation of IRF-1 response by high-risk HPV E7 protein in vivo. Cancer Lett 2002;179(2):205–12.

[40] Karim R, et al. Human papillomavirus deregulates the response of a cellular network comprising of chemotactic and proinflammatory genes. PLoS One 2011;6(3). e17848.

[41] Mota F, et al. The antigen-presenting environment in normal and human papillomavirus (HPV)-related premalignant cervical epithelium. Clin Exp Immunol 1999;116(1):33–40.

[42] Nunes RAL, et al. Innate immunity and HPV: friends or foes. Clinics (Sao Paulo) 2018;73(Suppl 1): e549s.

[43] Fonseca-Moutinho JA. Smoking and cervical cancer. ISRN Obstet Gynecol 2011;2011:847684.

[44] Lichtman MA. Obesity and the risk for a hematological malignancy: leukemia, lymphoma, or myeloma. Oncologist 2010;15(10):1083–101.

[45] Clarke MA, et al. Epidemiologic evidence that excess body weight increases risk of cervical cancer by decreased detection of precancer. J Clin Oncol 2018;36(12):1184–91.

[46] Thomas RJ, Kenfield SA, Jimenez A. Exercise-induced biochemical changes and their potential influence on cancer: a scientific review. Br J Sports Med 2017;51(8):640–4.

[47] Szender JB, et al. Impact of physical inactivity on risk of developing cancer of the uterine cervix: a case-control study. J Low Genit Tract Dis 2016;20(3):230–3.

[48] Barchitta M, et al. The association of dietary patterns with high-risk human papillomavirus infection and cervical cancer: a cross-sectional study in Italy. Nutrients 2018;10(4):469.

[49] Lee JK, et al. Mild obesity, physical activity, calorie intake, and the risks of cervical intraepithelial neoplasia and cervical cancer. PLoS One 2013;8(6):e66555.

[50] Reusser NM, et al. HPV carcinomas in immunocompromised patients. J Clin Med 2015;4(2):260–81.

[51] Ramirez PT, et al. Management of low-risk early-stage cervical cancer: should conization, simple trachelectomy, or simple hysterectomy replace radical surgery as the new standard of care? Gynecol Oncol 2014;132(1):254–9.

[52] Hu J, et al. Should ovaries be removed or not in early-stage cervical adenocarcinoma: a multicenter retrospective study of 105 patients. J Obstet Gynaecol 2017;37(8):1065–9.

[53] Zhou J, et al. Postoperative clinicopathological factors affecting cervical adenocarcinoma: stages I–IIB. Medicine 2018;97(2):e9323.

[54] Gray HJ. Primary management of early stage cervical cancer (IA1-IB) and appropriate selection of adjuvant therapy. J Natl Compr Canc Netw 2008;6(1):47–52.

[55] Fanfani F, et al. Radical hysterectomy after chemoradiation in FIGO stage III cervical cancer patients versus chemoradiation and brachytherapy: complications and 3-years survival. Eur J Surg Oncol 2016;42(10):1519–25.

[56] Shrivastava S, et al. Cisplatin chemoradiotherapy vs radiotherapy in FIGO stage IIIB squamous cell carcinoma of the uterine cervix: a randomized clinical trial. JAMA Oncol 2018;4(4):506–13.

[57] Bogani G, et al. The role of human papillomavirus vaccines in cervical cancer: prevention and treatment. Crit Rev Oncol Hematol 2018;122:92–7.

[58] Colombo N, et al. Cervical cancer: ESMO Clinical Practice Guidelines for diagnosis, treatment and follow-up. Ann Oncol 2012;23(Suppl_7):vii27–32.

[59] Melnikow J, et al. Cervical intraepithelial neoplasia outcomes after treatment: long-term follow-up from the British Columbia Cohort Study. J Natl Cancer Inst 2009;101(10):721–8.

[60] Prendiville W, Cullimore J, Norman S. Large loop excision of the transformation zone (LLETZ). A new method of management for women with cervical intraepithelial neoplasia. BJOG 1989;96(9):1054–60.

[61] Plante M, et al. Vaginal radical trachelectomy: a valuable fertility-preserving option in the management of early-stage cervical cancer. A series of 50 pregnancies and review of the literature. Gynecol Oncol 2005;98(1):3–10.

[62] Rogers L, et al. Radiotherapy and chemoradiation after surgery for early cervical cancer. Cochrane Database Syst Rev 2012;2012(5)Cd007583.

[63] Falcetta FS, et al. Adjuvant platinum-based chemotherapy for early stage cervical cancer. Cochrane Database Syst Rev 2016;**11**:Cd005342.

[64] Liu Y, et al. PD-1/PD-L1 inhibitors in cervical cancer. Front Pharmacol 2019;10:65.

[65] Lorusso D, et al. A systematic review comparing cisplatin and carboplatin plus paclitaxel-based chemotherapy for recurrent or metastatic cervical cancer. Gynecol Oncol 2014;133(1):117–23.

[66] Monk BJ, et al. Phase II, open-label study of pazopanib or lapatinib monotherapy compared with pazopanib plus lapatinib combination therapy in patients with advanced and recurrent cervical cancer. J Clin Oncol 2010;28(22):3562–9.

[67] Symonds RP, et al. Cediranib combined with carboplatin and paclitaxel in patients with metastatic or recurrent cervical cancer (CIRCCa): a randomised, double-blind, placebo-controlled phase 2 trial. Lancet Oncol 2015;16(15):1515–24.

[68] Tewari KS, et al. Bevacizumab for advanced cervical cancer: final overall survival and adverse event analysis of a randomised, controlled, open-label, phase 3 trial (Gynecologic Oncology Group 240). Lancet 2017;390(10103):1654–63.

[69] Krill LS, Tewari KS. Integration of bevacizumab with chemotherapy doublets for advanced cervical cancer. Expert Opin Pharmacother 2015;16(5):675–83.

[70] Minion LE, Tewari KS. Cervical cancer—state of the science: from angiogenesis blockade to checkpoint inhibition. Gynecol Oncol 2018;148(3):609–21.

[71] Pardoll DM. The blockade of immune checkpoints in cancer immunotherapy. Nat Rev Cancer 2012;12(4):252–64.

[72] Hodi FS, et al. Improved survival with ipilimumab in patients with metastatic melanoma. N Engl J Med 2010;363(8):711–23.

[73] Kurup SP, et al. Regulatory T cells impede acute and long-term immunity to blood-stage malaria through CTLA-4. Nat Med 2017;23(10):1220–5.

[74] Ribas A. Tumor immunotherapy directed at PD-1. N Engl J Med 2012;366(26):2517–9.

[75] Bagcchi S. Pembrolizumab for treatment of refractory melanoma. Lancet Oncol 2014;15(10). e419.

[76] Fife BT, Bluestone JA. Control of peripheral T-cell tolerance and autoimmunity via the CTLA-4 and PD-1 pathways. Immunol Rev 2008;224:166–82.

[77] Ku GY, et al. Single-institution experience with ipilimumab in advanced melanoma patients in the compassionate use setting: lymphocyte count after 2 doses correlates with survival. Cancer 2010;116(7):1767–75.

[78] Lheureux S, et al. Association of Ipilimumab with Safety and Antitumor Activity in women with metastatic or recurrent human papillomavirus–related cervical carcinoma. JAMA Oncol 2018;4(7):e173776.

[79] Papadopoulos KP, et al. A first-in-human study of REGN2810, a monoclonal, fully human antibody to programmed death-1 (PD-1), in combination with immunomodulators including hypofractionated radiotherapy (hfRT). J Clin Oncol 2016;34:3024.

[80] Frenel J-S, et al. Safety and efficacy of pembrolizumab in advanced, programmed death ligand 1–positive cervical cancer: results from the phase Ib KEYNOTE-028 trial. J Clin Oncol 2017;**35**(36):4035–41.

[81] Hollebecque A, et al. An open-label, multicohort, phase I/II study of nivolumab in patients with virus-associated tumors (CheckMate 358): efficacy and safety in recurrent or metastatic (R/M) cervical, vaginal, and vulvar cancers. J Clin Oncol 2017;35:5504.

[82] GlaxoSmithKline Vaccine HPV-007 Study Group, et al. Sustained efficacy and immunogenicity of the human papillomavirus (HPV)-16/18 AS04-adjuvanted vaccine: analysis of a randomised placebo-controlled trial up to 6.4 years. Lancet 2009;374(9706):1975–85.

[83] Kim JJ. Focus on research: weighing the benefits and costs of HPV vaccination of young men. N Engl J Med 2011;364(5):393–5.

[84] Lavanchy D. Worldwide epidemiology of HBV infection, disease burden, and vaccine prevention. J Clin Virol 2005;34:S1–3.

[85] Van Damme P, et al. Use of the nonavalent HPV vaccine in individuals previously fully or partially vaccinated with bivalent or quadrivalent HPV vaccines. Vaccine 2016;34(6):757–61.

CHAPTER 11

Identification of targeted molecules in cervical cancer by computational approaches

Manoj Kumar Gupta and Vadde Ramakrishna
Department of Biotechnology and Bioinformatics, Yogi Vemana University, Kadapa, Andhra Pradesh, India

Abstract

Cervical cancer is the second leading cause of cancer death in adult women. The three most widely employed techniques for the treatment of cervical cancer are radiotherapy, surgery, and hormone chemotherapy. Recently several biomarkers have also been identified using classical and high-throughput technologies. High-throughput technologies generate huge data, which in turn demand development of robust computational approaches for analysis of this big data in a more comprehensive way. This, in turn, will enable us to better understand mechanisms associated with many diseases, including cervical cancer. Considering this, in the present chapter, we present information about different computational approaches that have been employed to detect target molecules associated with cervical cancer. Information obtained revealed that to date limited computational studies have identified several cervical cancer-associated key hub genes (e.g., *BTD*, *PEG3*, *RPLP2*, and *SPON1*), long noncoding RNA (e.g., GOLGA2P5, EMX2OS, FLJ10038, FAM66C, ACVR2B-AS1, AMZ2P1, LINC00341, ZNF876P, MIR9-3HG, and ILF3-AS1), and miRNAs (e.g., *Hsa-mir-1273g*, *Hsa-mir-5095*, *Hsa-mir-5096*, and *Hsa-mir-1273f*) that play a key role in cervical cancer development. However, as there are only a few number of computational studies performed on cervical cancer datasets, there is still scope for developing more robust software/ algorithms and analyzing cervical cancer datasets. In the near future, the information in this chapter will be highly valuable for cancer biologists and immunologists toward cervical cancer treatment.

Keywords: Cervical cancer, Computational approach, Key genes, Drugs.

Abbreviations

circRNAs	circular RNA
GWAS	genome-wide association study
HPV	human papillomavirus
lncRNAs	long noncoding RNA
miRNA	micro RNA

1. Introduction

Cervical cancer is the second leading cause of cancer death in a young adult women. Cervical cancer affects women of different countries distinctly. Incidence of cervical cancer in women of high-income countries is lower than in low- and middle-income countries

G.P. Nagaraju, R.R. Malla (eds.)
A Theranostic and Precision Medicine Approach for Female-Specific Cancers
ISBN 978-0-12-822009-2
https://doi.org/10.1016/B978-0-12-822009-2.00011-X

[1]. This is mainly because of increased screening in women as well as human papillomavirus (HPV) vaccination in developed nations [2]. Irrespective of this, in 2018, ~570,000 new cases and ~311,000 deaths due to cervical cancer were reported globally [3]. In India, every year ~122,000 and ~67,000 women are diagnosed and die due to cervical cancer, respectively [4]. Hence there is always a demand for a more effective approach for cervical cancer treatment. To date, several techniques have been employed to treat cervical cancer based on disease state, histopathological tumor type, degree of tumor differentiation, patient age, and metastatic behavior. The three most widely employed techniques for the treatment of cervical cancer are radiotherapy, surgery, and hormone chemotherapy. Surgery is the only therapy that works effectively in the preinvasive and microinvasive stage (stage Ia). Radiation is widely employed to treat IIb, IIIa, IIIb, and IVa stages of cervical cancer. Radiation and surgery are employed in combination for treating Ib and IIa stages [5]. Additionally, several biomarkers have recently been identified, for example, methylation of tumor suppressor genes and type-specific viral load, using classical and high-throughput technologies. These biomarkers are widely employed for early detection as well as prevention of cervical cancer [5]. In the majority cases, HPV is the main reason for cervical cancer [6].

Earlier candidate gene studies have identified various cervical cancer-associated genes, namely, *IRF3, DMC1, EXO1, TLR2, FANCA, XRCC1, CYBA, ERAP1, EVER1/2, TP53, TAP2, IL17, TERT, LMP7, GTF2H4, DUT, SULF1, OAS3,* and *IFNG* [7–12]. In one genome-wide association study (GWAS), authors reported that knockdown of *ARRDC3* diminishes cell growth that subsequently prevents HPV 16 pseudovirions infection. This in turn supports that *ARRDC3* plays a key role during HPV infectious [13]. Several other GWAS studies have also established the relationship between cervical cancer and distinct HLA II alleles within the populations of Asia and Europe [14–17]. Few polymorphisms, namely, rs9277535 (*DPB1*), rs3077 (*DPA1*), rs3117027 (*DPB2*), and rs4282438 (*DPB2*), were also found to be associated with cervical cancer [18]. Other polymorphisms, namely, TNF-α-238G > A and TNF-α-308G > A, are also reported to decrease and increase the susceptibility of cervical cancer, respectively [19]. Haplotypes, namely, *HLA-B*0702-DRB1*1501/ HLADQB1*0602* and *HLA-B*1501/HLA-DRB1*1301/HLA-DQA1*0103/ HLA-DQB1*0603* are also reported to increase and decrease the risk of cervical cancer, respectively [12]. In 2016, Martínez-Nava et al. identified three polymorphisms, namely, *p21* (rs1801270) and *BRIP1* (rs11079454 and rs2048718), that reduce the risk of cervical cancer in women [20].

Recent advancement of high-throughput technologies generates huge data, which in turn demands development of robust computational approaches for analyzing the big data in a more comprehensive way. This will enable us to better understand mechanisms associated with many diseases, including cervical cancer [21–24]. Nevertheless, detecting protein-protein interaction and key genes via experimental requires huge amounts of capital and time. Computational approaches provide a unique way to address these problems in a short interval of time with less cost. Considering this, earlier, our laboratory employed different computational approaches to understand mechanisms associated with biliary stricture [25], Japanese encephalitis [26], diabetes [27], Alzheimer's disease, [24] and biotic and abiotic stresses in rice plants [28–30]. In this chapter, we present information about various cervical cancer-associated targeted molecules that were identified through computational approaches. In the near future, these target molecules may serve as biomarkers for early detection as well as prevention of cervical cancer in women.

2. Identification of target molecules

To date, several computational approaches have been developed to identify various cervical cancer-associated target molecules (Table 1).

Table 1 Cervical-associated target molecules identified via computational approaches.

Target molecules	Genes/miRNA/lncRNAs/cirRNAs	References
Genes	*MMP1, VIM, CDC45, CAT, BTD, PEG3, RPLP2, SPON1, SLC5A3, PRDX3, LSM3, AS1B, COPB2, PCNA, CDK1, CCNB1, BIRC5, MAD2L1, TOP2A, TSPO, FOS, CCND1, MCM2, PCNA & RNASEH2A, TNNI3, CCDC136, ABCG2, CYP26A1, TMEM233S, YT13, FOXC2, CXCL5, RRB1, ARRB2, CAV1, CFTR, EP300, ERBB3, HIF1A, INSR, JAK2, JUN, LYN, PML, RET, SMAD3, SR, RACGAP1, CHEK1, KIF11, KIF23, RRM2, CEP55,* and *ATAD2*	[31–37]
miRNAs	*miR-194-1, miR-204, miR-150, miR-193b-3p, miR-215-5p* and *miR-192-5p, miR-548d-5p, miR-5095, miR-548c-5p, miR-133a-3p, miR-215-5p, miR-944, miR-31-3p, miR-548d-5p, miR-11-3p, miR-491-3p, miR-107, miR-133a-5p, miR-133b, Hsa-mir-1273g, Hsa-mir-5095, Hsa-mir-5096,* and *Hsa-mir-1273f*	[38–42]
lncRNAs	EPB41L4A-AS1, LINC00649, HCP5, SNHG12, GOLGA2P10, ACVR2B-AS1, ATP1A1-AS1, LOH12CR2, A2M-AS1, FTX, MST1P2, GOLGA2P5, EMX2OS, FLJ10038, FAM66C, ACVR2B-AS1, AMZ2P1, LINC00341, ZNF876P, MIR9-3HG, ILF3-AS1, GOLGA2P5, MIR9-3HG ILF3-AS1, FAM66C, LINC00312, MIR9-3HG, SYS1-DBNDD2, and DDX12P	[43, 44]
cirRNAs	hsa_circRNA_103519, hsa_circRNA_101958, hsa_circRNA_000596, hsa_circRNA_400068, hsa_circRNA_104315, hsa_circ_0031027, hsa_circ_0070190, hsa_circ_0084927, hsa_circ_0043280, hsa_circ_0000745, and hsa_circ_0065898 and hsa_circ_0000077	[37, 45]

2.1 Key genes and proteins

Zhang and Zhao performed pathogenic network analysis to detect four cervical cancer-associated genes, namely, *MMP1, VIM, CDC45,* and *CAT* [31]. Functional analysis with a 21 gene expression dataset revealed four upregulated (*BTD, PEG3, RPLP2,* and *SPON1*) and five downregulated (*SLC5A3, PRDX3, LSM3, AS1B,* and *COPB2*) cervical cancer-associated genes [32]. Gene regulatory analysis via Cheng et al. suggests that DNA polymerases ("PLOA1/E2/E3/Q") and replicative helicase proteins (MCM6, MCM4, MCM2, MCM10, and MCM5) increased DNA replication during cervical cancer [46]. A group of transcriptional factors, kinases, and cyclins like CCNB2, CDK1, TFDP2, and CCNA2 were identified to enhance cell cycle shifts from G1 phase to S phase and from G2 phase to M phase during cervical cancer. Additionally, a group of motor proteins, namely, KIF4A, KIF11 and KIF14, along with PRC1, were identified to modulate cytokinesis throughout cervical cancer progression [46]. Another computational analysis identified six upregulated (*PCNA, CDK1, CCNB1, BIRC5, MAD2L1,* and *TOP2A*) and three downregulated (*TSPO, FOS,* and *CCND1*) cervical cancer-associated genes. These nine key genes were mainly associated with DNA replication, the p53 signaling pathway, oocyte meiosis, and cell cycle [33]. In 2016, Deng et al. identified 143 cervical cancer-associated differentially expressed genes using bioinformatics approaches [47]. Li et al. reported that three key genes, namely, *MCM2, PCNA,* and *RNASEH2A* and DNA replication play an important role in cervical cancer [34]. Another group of researchers performed bioinformatics analysis on the Cancer Genome Atlas database (https://portal.gdc.cancer.gov/) and reported that four hypermethylated genes (*TNNI3, CCDC136, ABCG2,* and *CYP26A1*) and four hypo-methylated genes (*TMEM233S, YT13, FOXC2,* and *CXCL5*) were positively and negatively associated with the overall survival of cervical cancer patients, respectively [35]. Hindumathi et al. identified 15 key cervical cancer-associated genes, namely, *RRB1, ARRB2, CAV1, CFTR, EP300, ERBB3, HIF1A, INSR, JAK2, JUN, LYN, PML, RET, SMAD3,* and *SRC* through gene ontology and graph theoretical approaches [36]. Yi et al. identified seven genes, namely, RACGAP1, CHEK1, KIF11, KIF23, RRM2, CEP55, and ATAD2, that play a key role in cervical cancer development [37].

2.2 Micro RNA

Micro RNAs (miRNAs) are ~22 long noncoding nucleotides and are highly conserved in nature. miRNAs are associated with numerous biological processes, including carcinogenesis. To date, several miRNAs have been recognized in nearby integration sites of HPV [48]. Liu et al. identified three cervical cancer-associated miRNAs, namely, *miR-194-1, miR-204,* and *miR-150* [38]. In 2018, Kori et al. identified several cervical cancer biomarkers, namely, metabolites (arachidonic acids), ephrin receptors (*EPHB2, EPHA4,* and *EPHA5*), endothelin receptors (*EDNRB* and *EDNRA*), nuclear receptors (*NR2C1,*

NR2C2, and NCOA3), miRNAs (miR-193b-3p, miR-215-5p, and miR-192-5p), transcription factors (ETS1, CUTL1, and E2F4) and proteins (CDK1, WNK1, GSK3B, CRYAB, KAT2B, and PARP1) using bioinformatics approaches [39]. In 2018, Rampogu et al. identified 272 cervical cancer-associated miRNAs using computational approaches. They also reported that three miRNAs, namely, miR-548d-5p, miR-5095, and miR-548c-5p, along with various genes, modulate cell cycle. Ten miRNAs, namely, miR-133a-3p, miR-215-5p, miR-944, miR-31-3p, miR-548d-5p, miR-11-3p, miR-491-3p, miR-107, miR-133a-5p, and miR-133b were found to be highly conserved. Another computational study identified that miR-21 and miR-29a are upregulated and downregulated during cervical cancer, respectively [40]. Another computational study identified that miR-21 is upregulated in cervical cancer patients [49]. In 2011, Reshmi et al. detected four novel miRNAs, namely, Hsa-mir-1273g, Hsa-mir-5095, Hsa-mir-5096, and Hsa-mir-1273f. They also reported that these miRNAs mainly reside in the promoter regions of host genes, for instance, SCP2 and BMP2K [41].

2.3 Long noncoding RNA

Long noncoding RNA (lncRNAs) are functional RNA molecules. They are usually longer than 200 nucleotides and are deprived of protein-coding capability [50]. They can modulate transcription factor activity and chromatin structure [51]. Recently, He et al. identified four cervical cancer-associated lncRNAs, namely, EPB41L4A-AS1, LINC00649, HCP5, and SNHG12, using computational approaches. These four lncRNAs are mainly associated with vacuolar transport, cell cycle arrest, and histone modification [43]. Another study identified five lncRNAs (GOLGA2P10, ACVR2B-AS1, ATP1A1-AS1, LOH12CR2, and A2M-AS1) that are differently expressed at various pathological stages, two lncRNAs (FTX and MST1P2) that are associated with different races of cervical cancer patients, 10 lncRNAs (GOLGA2P5, EMX2OS, FLJ10038, FAM66C, ACVR2B-AS1, AMZ2P1, LINC00341, ZNF876P, MIR9-3HG, and ILF3-AS1) that express based on tumor stage, four lncRNAs (GOLGA2P5, MIR9-3HG, ILF3-AS1, and FAM66C) that experiences abnormal expression based on the patient outcome assessment, four lncRNAs (LINC00312, MIR9-3HG, SYS1-DBNDD2, and DDX12P) that distinctly expressed based on the degree of HPV infection [44]. In 2018, Zhang et al. identified lncRNA, namely, NNT-AS1, using machine learning approaches. These lncRNAs initiate cell proliferation as well as invasion via the Wnt/β-catenin signaling pathway in cervical cancer [52].

2.4 Circular RNA

Circular RNA (circRNAs) are covalently closed, endogenous biomolecules present in eukaryotes. Their biogenesis is modulated via specific trans-acting factors and cis-acting elements. They are highly conserved and are associated with various biological processes and diseases, including cancer [53]. Earlier several studies have suggested that circRNAs function as miRNA sponges and regulate transcription. Few circRNAs are also reported to translate into proteins, thereby suggesting that circRNAs may also be associated with the expression of genes at various levels [54]. Yi et al. identified five cervical cancer–associated circRNAs, namely, hsa_circRNA_103519, hsa_circRNA_101958, hsa_circRNA_000596, hsa_circRNA_400068, and hsa_circRNA_104315, using computational approaches [37]. Another computational study identified seven cervical cancer–associated circRNAs, namely, hsa_circ_0031027, hsa_circ_0070190, hsa_circ_0084927, hsa_circ_0043280, hsa_circ_0000745, hsa_circ_0065898, and hsa_circ_0000077 [45]. These identified molecules, specifically proteins, have also been employed to identify the best possible drug against cervical cancer.

3. Identification of drug

In 2018, Rampogu et al. employed "pharmacophore guided molecular modeling approaches" and identified 51 compounds that are responsible for dose-dependent downregulation of *NOS2, PP2B*, and *IL-6* genes in the HeLa cells. These compounds are also associated with the expression of apoptotic genes, namely, caspases-3, Bax, and Bcl2, during cervical cancer [42]. Other computational studies identified that lissoclibadin 7, litospermic II acid, and kaempferol 3-ramnoglucoside have a better binding affinity with three cervical cancer–associated proteins, namely, HDAC4, HDAC 5, and HDAC10, respectively [55]. Ricci-López et al. identified three potential drugs, namely, ZINC111606147, ZINC96096545, and ZINC362643639, against HPV E6 protein [56]. Ramnath et al. reported that abietic acid may serve as a potential drug against cervical cancer and Alzheimer's disease [57]. The binding affinity of coumarin with caspase $3 <$ camptothecin $<$ epigallocatechin $<$ quercetin [58]. Daphnoretin [59] and ZINC14761180 [60] are also reported to have the best inhibitory property against E6 protein. Brazilein has a better binding affinity toward p53 and caspase 9 [61]. Chalcone 3 has a better cytotoxic effect against the HELA cell line [62]. Another drug, namely, 5-{(1-[(1-phenyl-1*H*-1,2,3-triazol-4-yl)methyl]-1*H*-indol-3-yl)methylene}pyrimidine-2,4,6-(1*H*,3*H*,5*H*)trione derivative (5e) is exhibting better anticancer property [63]. Thus, the computational approaches can be employed for the identification of both biomarkers as well as drug-associated cervical cancers.

4. Conclusion

In conclusion, cervical cancer is one of the most fatal cancers of women. Though radiotherapy, surgery, and hormone chemotherapy are widely used for cervical cancer treatment, survival rates are low. Recent advancement of high-throughput sequencing technologies have enabled early detection and prevention of cancer. However, these technologies also generate huge amounts of data, which in turn demand the development of robust computational approaches for analyzing this big data in a more comprehensive way. These computational approaches provide us a unique way to identify key genes and drugs using publicly available genomic and proteomic datasets in a short interval of time with less cost. To date, only limited studies using these computational approaches have identified numerous cervical cancer-associated genes, miRNAs, and proteins that serve as biomarkers that can be used in developing drugs against cervical cancer. Thus, there is still scope for developing more robust software/algorithms and analyzing cervical cancer datasets. In the near future, the information in this chapter will be highly valuable for cancer biologists and immunologists toward cervical cancer treatment.

Conflicts of Interest

None.

References

[1] Siegel RL, Miller KD, Jemal A. Cancer statistics, 2019. CA Cancer J Clin 2019;69:7–34. https://doi.org/10.3322/caac.21551.

[2] Lin M, Ye M, Zhou J, Wang ZP, Zhu X. Recent advances on the molecular mechanism of cervical carcinogenesis based on systems biology technologies. Comput Struct Biotechnol J 2019;17:241–50. https://doi.org/10.1016/j.csbj.2019.02.001.

[3] Bray F, Ferlay J, Soerjomataram I, Siegel RL, Torre LA, Jemal A. Global cancer statistics 2018: GLOBOCAN estimates of incidence and mortality worldwide for 36 cancers in 185 countries. CA Cancer J Clin 2018;68:394–424. https://doi.org/10.3322/caac.21492.

[4] Sreedevi A, Javed R, Dinesh A. Epidemiology of cervical cancer with special focus on India. Int J Womens Health 2015;7:405–14. https://doi.org/10.2147/IJWH.S50001.

[5] Šarenac T, Mikov M. Cervical cancer, different treatments and importance of bile acids as therapeutic agents in this disease. Front Pharmacol 2019;10: https://doi.org/10.3389/fphar.2019.00484.

[6] Chan CK, Aimagambetova G, Ukybassova T, Kongrtay K, Azizan A. Human papillomavirus infection and cervical cancer: epidemiology, screening, and vaccination—review of current perspectives. J Oncol 2019; https://doi.org/10.1155/2019/3257939.

[7] Wang SS, Bratti MC, Rodríguez AC, Herrero R, Burk RD, Porras C, et al. Common variants in immune and DNA repair genes and risk for human papillomavirus persistence and progression to cervical cancer. J Infect Dis 2009;199:20–30. https://doi.org/10.1086/595563.

[8] Klug SJ, Ressing M, Koenig J, Abba MC, Agorastos T, Brenna SM, et al. TP53 codon 72 polymorphism and cervical cancer: a pooled analysis of individual data from 49 studies. Lancet Oncol 2009;10:772–84. https://doi.org/10.1016/S1470-2045(09)70187-1.

[9] Mehta AM, Jordanova ES, van Wezel T, Uh H-W, Corver WE, Kwappenberg KMC, et al. Genetic variation of antigen processing machinery components and association with cervical carcinoma. Genes Chromosomes Cancer 2007;46:577–86. https://doi.org/10.1002/gcc.20441.

[10] Rafnar T, Sulem P, Stacey SN, Geller F, Gudmundsson J, Sigurdsson A, et al. Sequence variants at the TERT-CLPTM1L locus associate with many cancer types. Nat Genet 2009;41:221–7. https://doi.org/10.1038/ng.296.

[11] Hardikar S, Johnson LG, Malkki M, Petersdorf EW, Galloway DA, Schwartz SM, et al. A population-based case–control study of genetic variation in cytokine genes associated with risk of cervical and vulvar cancers. Gynecol Oncol 2015;139:90–6. https://doi.org/10.1016/j.ygyno.2015.07.110.

[12] Leo PJ, Madeleine MM, Wang S, Schwartz SM, Newell F, Pettersson-Kymmer U, et al. Defining the genetic susceptibility to cervical neoplasia—a genome-wide association study. PLoS Genet 2017;13: https://doi.org/10.1371/journal.pgen.1006866.

[13] Takeuchi F, Kukimoto I, Li Z, Li S, Li N, Hu Z, et al. Genome-wide association study of cervical cancer suggests a role for ARRDC3 gene in human papillomavirus infection. Hum Mol Genet 2019;28:341–8. https://doi.org/10.1093/hmg/ddy390.

[14] Lehoux M, D'Abramo CM, Archambault J. Molecular mechanisms of human papillomavirus-induced carcinogenesis. Public Health Genomics 2009;12:268–80. https://doi.org/10.1159/000214918.

[15] Burk RD, Harari A, Chen Z. Human papillomavirus genome variants. Virology 2013;445:232–43. https://doi.org/10.1016/j.virol.2013.07.018.

[16] Skinner SR, Wheeler CM, Romanowski B, Castellsagué X, Lazcano-Ponce E, Rosario-Raymundo MRD, et al. Progression of HPV infection to detectable cervical lesions or clearance in adult women: analysis of the control arm of the VIVIANE study. Int J Cancer 2016;138:2428–38. https://doi.org/10.1002/ijc.29971.

[17] Mirabello L, Clarke MA, Nelson CW, Dean M, Wentzensen N, Yeager M, et al. The intersection of HPV epidemiology. Genomics and mechanistic studies of HPV-mediated carcinogenesis. Viruses 2018;10:80. https://doi.org/10.3390/v10020080.

[18] Cheng L, Guo Y, Zhan S, Xia P. Association between HLA-DP gene polymorphisms and cervical cancer risk: a meta-analysis. Biomed Res Int 2018; https://doi.org/10.1155/2018/7301595.

[19] Pan F, Tian J, Ji C-S, He Y-F, Han X-H, Wang Y, et al. Association of TNF-α-308 and -238 polymorphisms with risk of cervical cancer: a meta-analysis. Asian Pac J Cancer Prev 2012;13:5777–83. https://doi.org/10.7314/apjcp.2012.13.11.5777.

[20] Martínez-Nava GA, Fernández-Niño JA, Madrid-Marina V, Torres-Poveda K. Cervical cancer genetic susceptibility: a systematic review and meta-analyses of recent evidence. PLoS One 2016;11: https://doi.org/10.1371/journal.pone.0157344.

[21] Gupta MK, Sarojamma V, Reddy MR, Shaik JB, Vadde R. Computational biology: toward early detection of pancreatic cancer. Crit Rev Oncog 2019;24: https://doi.org/10.1615/CritRevOncog.2019031335.

[22] Gupta MK, Donde R, Gouda G, Vadde R, Behera L. De novo assembly and characterization of transcriptome towards understanding molecular mechanism associated with MYMIV-resistance in Vigna mungo – a computational study. BioRxiv 2019;844639: https://doi.org/10.1101/844639.

[23] Gupta MK, Vadde R. Insights into the structure-function relationship of both wild and mutant zinc transporter ZnT8 in human: a computational structural biology approach. J Biomol Struct Dyn 2019;1–22. https://doi.org/10.1080/07391102.2019.1567391.

[24] Gupta MK, Vadde R. In silico identification of natural product inhibitors for γ-secretase activating protein, a therapeutic target for Alzheimer's disease. J Cell Biochem 2018; https://doi.org/10.1002/jcb.28316.

[25] Gupta MK, Behara SK, Vadde R. In silico analysis of differential gene expressions in biliary stricture and hepatic carcinoma. Gene 2017;597:49–58.

[26] Gupta MK, Behera SK, Dehury B, Mahapatra N. Identification and characterization of differentially expressed genes from human microglial cell samples infected with Japanese encephalitis virus. J Vector Borne Dis 2017;54:131–8.

[27] Gupta MK, Vadde R. Identification and characterization of differentially expressed genes in type 2 diabetes using in silico approach. Comput Biol Chem 2019; https://doi.org/10.1016/j.compbiolchem.2019.01.010.

[28] Gupta MK, Vadde R, Donde R, Gouda G, Kumar J, Nayak S, et al. Insights into the structure–function relationship of brown plant hopper resistance protein, Bph14 of rice plant: a computational structural biology approach. J Biomol Struct Dyn 2018;1–17. https://doi.org/10.1080/07391102.2018.1462737.

[29] Donde R, Gupta MK, Gouda G, Kumar J, Vadde R, Sahoo KK, et al. Computational characterization of structural and functional roles of DREB1A, DREB1B and DREB1C in enhancing cold tolerance in rice plant. Amino Acids 2019;51:839–53. https://doi.org/10.1007/s00726-019-02727-0.

[30] Gupta MK, Vadde R, Gouda G, Donde R, Kumar J, Behera L. Computational approach to understand molecular mechanism involved in BPH resistance in Bt-rice plant. J Mol Graph Model 2019;88:209–20. https://doi.org/10.1016/j.jmgm.2019.01.018.

[31] Zhang Y-X, Zhao Y-L. Pathogenic network analysis predicts candidate genes for cervical cancer. Comput Math Methods Med 2016; https://doi.org/10.1155/2016/3186051.

[32] Tan MS, Chang S-W, Cheah PL, Yap HJ. Integrative machine learning analysis of multiple gene expression profiles in cervical cancer. Peer J 2018;6:e5285 https://doi.org/10.7717/peerj.5285.

[33] Wu X, Peng L, Zhang Y, Chen S, Lei Q, Li G, et al. Identification of key genes and pathways in cervical cancer by bioinformatics analysis. Int J Med Sci 2019;16:800–12. https://doi.org/10.7150/ijms.34172.

[34] Li X, Tian R, Gao H, Yan F, Ying L, Yang Y, et al. Identification of significant gene signatures and prognostic biomarkers for patients with cervical cancer by integrated bioinformatic methods. Technol Cancer Res Treat 2018; https://doi.org/10.1177/1533033818767455.

[35] Xie F, Dong D, Du N, Guo L, Ni W, Yuan H, et al. An 8-gene signature predicts the prognosis of cervical cancer following radiotherapy. Mol Med Rep 2019;20:2990–3002. https://doi.org/10.3892/mmr.2019.10535.

[36] Hindumathi V, Kranthi T, Rao SB, Manimaran P. The prediction of candidate genes for cervix related cancer through gene ontology and graph theoretical approach. Mol Biosyst 2014;10:1450–60. https://doi.org/10.1039/C4MB00004H.

[37] Yi Y, Liu Y, Wu W, Wu K, Zhang W. Reconstruction and analysis of circRNA-miRNA-mRNA network in the pathology of cervical cancer. Oncol Rep 2019;41:2209–25. https://doi.org/10.3892/or.2019.7028.

[38] Liu H, Liu L, Zhu H. The role of significantly deregulated microRNAs in recurrent cervical cancer based on bioinformatic analysis of the cancer genome atlas data. J Comput Biol 2019;26:387–95. https://doi.org/10.1089/cmb.2018.0241.

[39] Kori M, Arga KY. Potential biomarkers and therapeutic targets in cervical cancer: insights from the meta-analysis of transcriptomics data within network biomedicine perspective. PLoS One 2018;13: e0200717 https://doi.org/10.1371/journal.pone.0200717.

[40] Pardini B, De Maria D, Francavilla A, Di Gaetano C, Ronco G, Naccarati A. MicroRNAs as markers of progression in cervical cancer: a systematic review. BMC Cancer 2018;18:696. https://doi.org/10.1186/s12885-018-4590-4.

[41] Reshmi G, Chandra SSV, Babu VJM, Babu PSS, Santhi WS, Ramachandran S, et al. Identification and analysis of novel microRNAs from fragile sites of human cervical cancer: computational and experimental approach. Genomics 2011;97:333–40. https://doi.org/10.1016/j.ygeno.2011.02.010.

[42] Rampogu S, Ravinder D, Pawar SC, Lee KW. Natural compound modulates the cervical cancer microenvironment—a pharmacophore guided molecular modelling approaches. J Clin Med 2018;7: https://doi.org/10.3390/jcm7120551.

[43] He M, Lin Y, Xu Y. Identification of prognostic biomarkers in colorectal cancer using a long non-coding RNA-mediated competitive endogenous RNA network. Oncol Lett 2019;17:2687–94. https://doi.org/10.3892/ol.2019.9936.

[44] Wu W-J, Shen Y, Sui J, Li C-Y, Yang S, Xu S-Y, et al. Integrated analysis of long non-coding RNA competing interactions revealed potential biomarkers in cervical cancer: based on a public database. Mol Med Rep 2018;17:7845–58. https://doi.org/10.3892/mmr.2018.8846.

[45] Gong J, Jiang H, Shu C, Hu M, Huang Y, Liu Q, et al. Integrated analysis of circular RNA-associated ceRNA network in cervical cancer: observational study. Medicine (Baltimore) 2019;98:e16922 https://doi.org/10.1097/MD.0000000000016922.

[46] Cheng J, Lu X, Wang J, Zhang H, Duan P, Li C. Interactome analysis of gene expression profiles of cervical cancer reveals dysregulated mitotic gene clusters. Am J Transl Res 2017;9:3048–59.

[47] Deng S-P, Zhu L, Huang D-S. Predicting hub genes associated with cervical cancer through gene co-expression networks. IEEE/ACM Trans Comput Biol Bioinform 2016;13:27–35. https://doi.org/10.1109/TCBB.2015.2476790.

[48] Guerrero Flórez M, Guerrero Gómez OA, Mena Huertas J, Yépez Chamorro MC. Mapping of micro-RNAs related to cervical cancer in Latin American human genomic variants. F1000Research 2018;6:946. https://doi.org/10.12688/f1000research.10138.2.

[49] Liolios T, Kastora SL, Colombo G. MicroRNAs in female malignancies. Cancer Inform 2019;18: https://doi.org/10.1177/1176935119828746 1176935119828746.

[50] Cao H, Wahlestedt C, Kapranov P. Strategies to annotate and characterize long noncoding RNAs: advantages and pitfalls. Trends Genet 2018;34:704–21. https://doi.org/10.1016/j.tig.2018.06.002.

[51] Guglas K, Bogaczyńska M, Kolenda T, Ryś M, Teresiak A, Bliźniak R, et al. lncRNA in HNSCC: challenges and potential. Contemp Oncol 2017;21:259–66. https://doi.org/10.5114/wo.2017.72382.

[52] Zhang X, Wang J, Li J, Chen W, Liu C. CRlncRC: a machine learning-based method for cancer-related long noncoding RNA identification using integrated features. BMC Med Genomics 2018;11:120. https://doi.org/10.1186/s12920-018-0436-9.

[53] Kristensen LS, Andersen MS, Stagsted LVW, Ebbesen KK, Hansen TB, Kjems J. The biogenesis, biology and characterization of circular RNAs. Nat Rev Genet 2019;20:675–91. https://doi.org/10.1038/s41576-019-0158-7.

[54] Bach D-H, Lee SK, Sood AK. Circular RNAs in cancer. Mol Ther Nucleic Acids 2019;16:118–29. https://doi.org/10.1016/j.omtn.2019.02.005.

[55] Tambunan USF, Parikesit AA, Nasution MAF, Hapsari A, Kerami D. Exposing the molecular screening method of Indonesian natural products derivate as drug candidates for cervical cancer. Iran J Pharm Res 2017;16:1113–27.

[56] Ricci-López J, Vidal-Limon A, Zunñiga M, Jiménez VA, Alderete JB, Brizuela CA, et al. Molecular modeling simulation studies reveal new potential inhibitors against HPV E6 protein. PLoS One 2019;14:e0213028 https://doi.org/10.1371/journal.pone.0213028.

[57] Ramnath MG, Thirugnanasampandan R, NagaSundaram N, Bhuvaneswari G. Molecular docking and dynamic simulation studies of terpenoids of I. wightii (Bentham) H. Hara against acetylcholinesterase and histone deacetylase3 receptors. Curr Comput Aided Drug Des 2018;14:234–45. https://doi.org/10.2174/1573409914666180321111925.

[58] Ashwini S, Varkey SP, Shantaram M. In silico docking of polyphenolic compounds against caspase 3-HeLa cell line protein. Int J Drug Dev Res 2017;9:28–32.

[59] Mamgain S, Sharma P, Pathak RK, Baunthiyal M. Computer aided screening of natural compounds targeting the E6 protein of HPV using molecular docking. Bioinformation 2015;11:236. https://doi.org/10.6026/97320630011236.

[60] Kumar A, Rathi E, Kini SG. E-pharmacophore modelling, virtual screening, molecular dynamics simulations and in-silico ADME analysis for identification of potential E6 inhibitors against cervical cancer. J Mol Struct 2019;1189:299–306. https://doi.org/10.1016/j.molstruc.2019.04.023.

[61] Laksmiani NPL, Astuti NMW, Arisanti CIS, Paramita NLPV. Ethyl acetate fraction of secang as anti cervical cancer by inducing p53 and caspase 9. IOP Conf Ser Earth Environ Sci 2018;207:012065. https://doi.org/10.1088/1755-1315/207/1/012065.

[62] Tantawy MA, Sroor FM, Mohamed MF, El-Naggar ME, Saleh FM, Hassaneen HM, et al. Molecular docking study, cytotoxicity, cell cycle arrest and apoptotic induction of novel chalcones incorporating thiadiazolyl isoquinoline in cervical cancer. Anticancer Agents Med Chem 2019; https://doi.org/10.2174/1871520619666191024121116.

[63] Kumar A, Sathish Kumar B, Sreenivas E, Subbaiah T. Synthesis, biological evaluation, and molecular docking studies of novel 1,2,3-triazole tagged 5-[(1H-Indol-3-yl)methylene]pyrimidine-2,4,6 (1H,3H,5H)trione derivatives. Russ J Gen Chem 2018;88:587–95. https://doi.org/10.1134/S1070363218030313.

CHAPTER 12

Cervical cancer metabolism: Major reprogramming of metabolic pathways and cellular energy production

K. Vijaya Rachel and Nagarjuna Sivaraj

Department of Biochemistry and Bioinformatics, GITAM Deemed to be University, Visakhapatnam, Andhra Pradesh, India

Abstract

A few important changes in tumor cellular bioenergetics include elevated glycolysis, enhanced gluta-minolytic flux, amino acid upregulation and lipid metabolism, mitochondrial biogenesis development, pentose phosphate pathway (PPP) stimulation, and biosynthesis of macromolecules. Cervical cancer (CC) cells are also found to reprogram cellular metabolic pathways. Cancer cells are benefitted due to increased hexokinase 2 expression by providing metabolites for growth, supplying intermediates for tricarboxylic acid, assisting autophagy in response to deprivation of glucose, and binding to voltage-dependent anion channel. Oncogenic signals like protein kinase B mechanistic target of rapamycin-related pathways were found to be activated by human papillomavirus (HPV) E6/E7 and they have put out the action of p53 and pRb tumor-suppressive paths, suggesting that HPV E6/E7 could regulate cellular glucose breakdown of CC. Variations in cancer cells' metabolism are related with overexpression of oncogenes (c-Myc) and transcription factors (hypoxia inducible factor 1a). The expression of 6-phospho glyceraldehyde dehydrogenase and PPP activity to phosphorylation of c-MET is being investigated. It is also known that CC cells can turn out lipoprotein A under certain conditions. Metabolic reprogramming also aids in therapeutic interventions. For cervical and other cancers, the maximum standardized uptake value is an additional prognostic marker along with lymph node grade, cancer stage, and tumor size.

Keywords: Cervical cancer, Hexokinase 2, HPV E6/E7, Pentose phosphate pathway, LPA, SUV_{max}.

Abbreviations

5-FU	fluorouracil
ADC	apparent diffusion coefficient
ADP	adenosine diphosphate
ATP	adenosine triphosphate
CC	cervical cancer
c-Myc	cytoplasmic proto-oncogene regulator
DWI	diffusion–weighted imaging
HGF	hepatocyte growth factor
HPV	human papillomavirus
IL-2	interleukin-2
LPA	lipoprotein A
MCF7	Michigan Cancer Foundation-7

G.P. Nagaraju, R.R. Malla (eds.)
A Theranostic and Precision Medicine Approach for Female-Specific Cancers
ISBN 978-0-12-822009-2
https://doi.org/10.1016/B978-0-12-822009-2.00012-1

Met	metformin
MRS	magnetic resonance spectroscopy
NADPH	nicotinamide adenine dinucleotide phosphate hydrogen
p53	tumor protein 53
PMA	phorbol 12-myristate 13-acetate
PPP	pentose phosphate pathway
pRB	retinoblastoma protein
SUVmax	maximum standardized uptake value
TCA	tricarboxylic acid
VDAC	voltage-dependent anion channel

1. Introduction

Cervical cancer (CC) is a female-specific cancer [1]. As in all types of cancers, CC is also a consequence of uncontrolled growth of cells in the cervix camera or in any remote organ [2]. In 2018, nearly 570,000 cases of CC were reported, of which there were 311,000 deaths. After breast, colorectal, and lung cancer (2.1 million, 0.8 million, and 0.7 million cases), CC is the fourth most common cancer in women. A wide variance was observed among different nations with age-standardized incidence of CC of 13.1 per 100,000 women around the world. CC was found to be the leading cause of death among women in western, eastern, middle, and South African countries. With nearly 6.5% of women developing CC before 75 years of age, the highest incidence was estimated in Eswatini. China and India contribute to more than a third of global CC burden with nearly 106,000 cases in China and 97,000 cases in India leading to 48,000 deaths and 60,000 deaths in China and India, respectively. Average age at diagnosis of CC was 53 years globally; average age is 44 years in Vanuatu and 68 years in Singapore. Global average age at death from CC is 59 years; average age of death is 45 years in Vanuatu and 76 years in Martinique. Among 185 countries assessed, 146 (79%) of women younger than 45 years were affected with CC [3]. CC is the third most widespread cancer among women globally, accounting for 275,000 deaths in 2008, 88% of which occurred in developing countries; 159,800 in Asia alone [4]. Although the rate of CC has progressively decreased due to easy accessibility to screening and vaccines, it still has devastating effects in developing countries, including India [5]. The majority of CCs are caused by human papillomavirus (HPV). Other risk factors include multiple sexual partners, prolonged use of birth control pills, and smoking [6]. Cancer is a multifactorial disease connecting to a number of hallmarks. Reprogramming of energy metabolism by cancer cells causes elevated energy mediators on the condition that cell growth will benefit cell survival, metastasis, migration, and resistance to chemotherapy and radiotherapy [7–9]. A few important changes in tumor cellular bioenergetics include elevated glycolysis, enhanced glutaminolytic flux, amino acid upregulation and lipid metabolism, mitochondrial biogenesis development, pentose phosphate pathway (PPP) stimulation, and macromolecule biosynthesis [1–9].

2. Glycolysis

Even with the availability of HPV vaccines, CC is a major cause of death in women worldwide At present, cytostatic drugs for chemotherapy, mainly cisplatin (*cis*-dichloro diamine platinum (II)), are quite primitive, despite less specificity and significant toxicity to patients [10]. Aerobic glycolysis is metabolically distinctive of cancer cells.

Studies reporting activity in CC and increased hexokinase 2 expression likened to normal cervical epithelium date back to the 1970s. Normal cervical tissues showed low or no hexokinase 2 signals in recent immunohistochemistry analyses, while nearly 60% of the CC specimens ($n = 197$) were stained positive for hexokinase 2. This metabolic presentation could also be of clinical relevance, as increased hexokinase 2 expression is a negative prognostic marker for numerous solid tumors. High hexokinase 2 levels are found to be associated with resistance to radiotherapy, which is in agreement with the argument that enhanced glycolysis can confer radiation resistance in cancer cells. Due to its elevated expression in many tumors and its tumor-promoting potential, hexokinase 2 is understood to be a novel target for cancer therapy.

According to some studies, cancer cells benefit from increased hexokinase 2 expression by different means. Energy along with necessary metabolites for growth of tumor cells are provided by hexokinase 2 stimulation aerobically. Additionally, an important role is played by hexokinase 2 supplying intermediates for the tricarboxylic acid (TCA) cycle, glutamine-derived carbon utilization in anaplerosis, contributing to preservation of cellular energy homeostasis through assisting autophagy in response to deprivation of glucose. Further, hexokinase 2 is known to bind to the voltage-dependent anion channel (VDAC), a pore-forming membrane protein located on the outer mitochondrial membrane. Interaction of the VDAC with hexokinase 2 is of pivotal importance as it couples glycolysis to oxidative phosphorylation. Furthermore, the hexokinase 2-VDAC interaction can choke apoptosis by deterring proapoptotic proteins targeting the outer mitochondrial membrane thus meddling with emergence of mitochondrial permeability transition pores. Recent reports establish potentiality of E6/E7 in stimulating expression of hexokinase 2 offer a direct link between HPV oncogenes and expression of crucial cellular enzymes accountable for metabolic reprogramming and resistance to apoptosis in cancer cells. This scenario is concomitant to reduction in therapeutic sensitivity and increased oncogenicity [11].

Variations in cancer cells' metabolism are related to overexpression of oncogenes (c-Myc) and transcription factors (hypoxiainducible factor 1a). They help keep cancerous cells away from apoptosis persuaded by reactive oxygen species. Various studies were dedicated to the effect of metformin (Met) on metabolism of metastatic tumor cells of the cervix. They recorded apoptosis of violent CC cells through Met-inhibited glycolytic phenotypes by regulating the expression of oncogenes and their downstream proteins. In those studies, Met synchronized mitochondrial metabolism, particularly via supplementation of the TCA cycle with $C_3H_4O_3$ (keto acid pyruvate) and $C_5H_{10}N_2O_3$-glutamine. Epithelial and mesenchymal markers of tumor cells, when targeted by Met, increased persistent characteristics of CC cells [12–14].

3. Pentose phosphate pathway

Metabolites required for anabolism are principally supplied by the PPP and are associated with glycolysis along with other pathways and hence can be designated to have a definite role in tumor cell production [15]. The PPP utilizes glucose-6-phosphate as fuel for sequential oxidative and nonoxidative sites of the pathway. The oxidative PPP forms ribulose-5-phosphate and, as elevation yield, one CO_2, three protons, and two nicotinamide adenine dinucleotide phosphate hydrogen (NADPH) pathways per molecule of glucose-6-phosphate utilization. Carbon dioxide and protons contribute toward tumor acidification [16–18], NADPH being an essential cofactor for glutathione peroxidase and for synthesis of fatty acids. On the nonoxidative front of the PPP, ribulose-5-phosphate is utilized for synthesis of nucleic acid. Numerous researchers have identified the necessity of PPP in giving tumor cell growth devoid of anchorage [19, 20]. Pairing of glycolysis and PPP is delimited by pyruvate kinase M, an enzyme of glycolysis, converting phosphoenolpyruvate and adenosine diphosphate to pyruvate and adenosine triphosphate. Differentiated cells mainly express pyruvate kinase muscle isoenzyme M1 and malignant cells express embryonic pyruvate kinase muscle isoenzyme M2 isoform consequential from alternate splicing. Expression of specific ribonucleoprotein is induced pyruvate kinase muscle isoenzyme and pyruvate kinase muscle isoenzyme 2, which in turn are induced by HIF1 and c-Myc, respectively [21]. Even though the involvement of the PPP to conflict with anoikis is well predictable, how it influences tumor cell movement and attack had not been well established until recently. Investigational confirmation points to the silencing of the third enzyme of the PPP, 6-phosphogluconate dehydrogenase, which was shown to reduce in vitro metastasis of lung carcinoma cells when stimulated by hepatocyte growth factor (HGF) [22]. The mechanism underlying this phenomena is decreasing phosphorylation of tyrosine/activation of HGF receptor c-MET. The molecular mechanism from the expression of 6-phospho-glyceraldehyde dehydrogenase and PPP activity to phosphorylation of c-MET is still to be recognized.

4. Fatty acid oxidation

Zhongzhou Shen et al. [22] presented information that demonstrates that ovarian and CC cells can turn out lipoprotein A (LPA) under certain conditions. They demonstrated that malignant cells can generate LPA [23]. The data show that ovarian cancer cells secrete LPA, but breast cancer (metastatic breast cancer-M.D. Anderson and Michigan Cancer Foundation-7 (MCF7) and leukemia (Jurkat and K562) cells do not. Inhibitor studies suggest contribution of a cytosolic phospholipase A2 or an Ca^{2+} independent phospholipase A2 is elevated in CC and ovarian cancers, but do not regulate the probable association of a secretory phospholipase A2 in LPA secretion of tumor cells. It is probable that under varied conditions these leukemia and breast cells might secret LPA into the culture medium. Another study states that nonproduction of LPA in leukemia and breast cells is not due to the lack of sensitivity of these cells to phorbol 12-myristate 13-acetate (PMA). Jurkat cells are known to proliferate and generate interleukin-2 (IL-2) under the stimulus of PMA [24]. It is evident that K562 cells distinguish in reaction to PMA [25, 26]. Additionally, PMA has been revealed to encourage aromatase activity in MCF7 cells [27]. PMA also encourages a dose-dependent augment in prostaglandin E2 production, inhibits cell growth, and arouses aromatase action in 231 cells [28, 29]. The stimuli that trigger these differences continue to be unknown. The same type of specificity is exhibited in vitro by various malignant cells, evidenced by computing plasma LPA [30]. There outcomes recommend that cancerous ovarian epithelial cells may characterize a significant basis of the high LPA detected in patients with ovarian cancers [31]. It was reported that once the growth factors are exhausted the ovarian cancer cells cease in vitro secretion of LPA. Several factors like growth factors and cytokines, transforming growth factor b, counting epidermal growth factor, amphiregulin, insulin-like growth factor, IL-1, basic fibroblast growth factor, and tumor necrosis factor alpha act as in vitro triggers to turn on phospholipase2 in cell systems [32–34] and ovarian cancer cells are no exception in physiological and pathological circumstances [35–37]. Research is being done to classify these physiological stimuli for assembly of LPA in ovarian cancer cells. Animal models are essential to endorse source of elevated plasma LPA secreted by a tumor tissue/primary tumor culture for the studies carried out under pathological and physiological conditions (Fig. 1).

5. Cancer metabolism and diagnostic imaging

5.1 The combination of magnetic resonance imaging and magnetic resonance spectroscopic imaging

Restricted diffusion is demonstrated by malignant cervical tissue and hence lessen apparent diffusion coefficient (ADC) values in comparison to normal tissue. Diffusion-weighted imaging (DWI) and ADC maps allocate separation of malignant from benign zones of the cervix with great sensitivity and precision. Discrimination of histological type and grade is possible by ADC facility in the launch data [38]. Payne et al. [39] demonstrated statistically noteworthy differentiation in ADC values of healthy and cancerous cervixes (1.7×10^{-3} mm²/s vs 1.1×10^{-3} mm²/s; $P < 0.001$). Higher ADC values are observed in well-differentiated tumors compared to incompetently distinguished tumors (1.2×10^{-3} mm²/s vs 1.1×10^{-3} mm²/s; $P = 0.01$). Similar correlation of ADC and tumor grade was reported by another research group [40]. However, heterogeneity

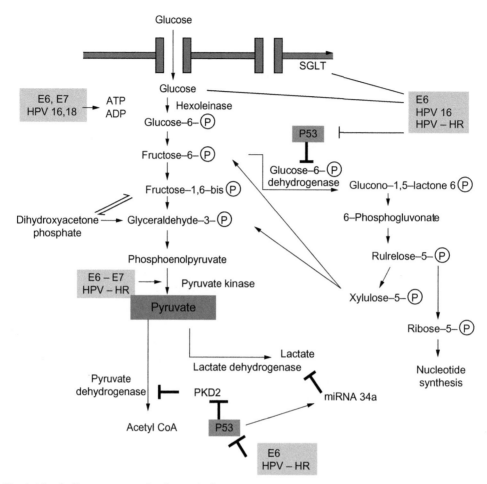

Fig. 1 Metabolic reprogramming in cervical cancer.

has been experiential in ability to discriminate histological subtype. While one analysis [41] recognized ADC of squamous carcinomas to be lower than adenocarcinomas, another investigation described no variance in ADC among different histological subtypes [42]. Studies are needed to authenticate these results, however. The available information reveals great sensitivity and precision of DWI in recognizing primary tumor and secondary nodal deposits. Its comparative or incremental performance over T2 weighted imaging has not been methodically assessed. While the vast majority of studies have imaged early stage tumors with ADC values ranging between 0.86 and 1.38×10^{-3} mm^2/s, there is a distinct probability that big infiltrative tumors may be accompanied with higher average ADC values due to intratumoral necrosis and consequential increase in extracellular volume.

Not many studies have investigated the use of in vivo and/or high field ex vivo magnetic resonance spectroscopy (MRS)/magic-angle spinning for diagnosing CC. Delikatny [43] evaluated 159 CC samples (40 invasive and 119 preinvasive). While a high resolution lipid peak (1.3 ppm) was observed in patients with invasive cancer, the preinvasive specimens demonstrated little or no lipid spectra, however, they had a strong unresolved resonance between 3.8 and 4.2 ppm. Mahon et al. evaluated the performance of MRS and allied ex vivo spectroscopy with histopathology. Investigators anticipated that important in-phase triglycerides may be used for in vivo detection of malignancy; however, the analysis at times may be limited by out-of-phase lipid signals [44, 45].

5.2 Therapeutic implications

According to Kidd et al. [45] general endurance is a crucial endpoint strategy in CC. Usually, tumor staging is a forecaster for general survival. Latest studies have validated that lymph node involvement in CC is a reliable forecaster and is indicative of cancer at Stage II. Other studies revealed maximum standardized uptake value (SUV_{max}) as an additional prognostic marker for endurance rather than tumor staging. One study examined three distinct result strata of patients based on SUV_{max} for general endurance. They demonstrated that for CC and other cancers SUV_{max} is an additional prognostic consequence than lymph node grade, cancer stage, or tumor size. While primary tumor SUV_{max} is enhanced in prophesying the consequences over earlier criteria. This criteria aids in describing SUV_{max} as a responsive gauge for prediction. In adding to their study, the researchers suggest categorizing patients based on SUV_{max} analysis into three different result confederacies. SUV_{max} can be a quantitative biomarker in foreseeing the effect of the marker on a particular patient before the management is commenced. Primary tumor SUV_{max} can thus be a responsive and measureable biomarker for CC outcome.

5.3 Inhibition of E6/E7-mediated glycolysis sensitizes 5-FU-resistant cell

Most of the evidences advocate the dependence of cancer cells on glycolysis of anaerobic mode as energy source and other metabolic intermediates for the development of tumors [46–49]. Amplified lactate dehydrogenase activity was found to be involved in altering cellular glucose metabolism and carcinoma of the cervix [50]. Oncogenic signals like protein kinase B mechanistic target of rapamycin-related pathways were found to be activated by HPV E6/E7 [51] and they have put out the action of p53 and pRb tumor-suppressive paths [52]. This evidence points to the conclusion that HPV E6/E7 could regulate cellular glucose breakdown of CC. CC through HPV septicity were found to depend more on glycolysis, which is evidenced by enhanced activities of hexokinase 2, lactate dehydrogenase, and pyruvate kinase muscle isoenzyme M2 [53]. Overexpression of HPV-16 E6/E7 stimulated breakdown of glucose. Knockdown E6/E7 extensively concealed the rate of glucose breakdown in cancer cells in the cervix.

Various investigators reported high glycolysis of cancer cells contributing to acquired fluorouracil (5-FU) resistance, [54, 55] suggesting that targeting the glycolysis pathway may be an efficient approach against chemoresistance. Additionally, inhibiting glycolysis will overcome obtained 5-FU resistance in E6/E7 ectopic overexpressing CC cells, signifying that dysregulated glycolysis acts as a downstream cellular route of E6/E7 oncoproteins. Primary toward increasing E6/E7 inhibition as a new approach against high glycolytic pathway for sensitization of CC cells to 5-FU, targeting cellular glucose metabolism might be a suitable beneficial approach to decrease or eradicate chemoresistant tumors.

References

[1] Elfström KM, Arnheim-Dahlström L, von Karsa L, Dillner J. CC screening in Europe: quality assurance and organisation of programmes, Eur J Cancer 2015;51(8):950–68. https://doi.org/10.1016/j.ejca.2015.03.008. Epub 2015 Mar 25, 25817010.

[2] Sarkar K, Bhattacharya S, Bhattacharyya S, Chatterjee S, Mallick AH, Chakraborti S, Chatterjee D, Bal B. Oncogenic human papilloma virus and cervical pre-cancerous lesions in brothel-based sex workers in India, J Infect Public Health 2008;1(2):121–8. https://doi.org/10.1016/j.jiph.2008.09.01. Epub 2008 Nov 12, 20701853.

[3] Arbyn M, Weiderpass E, Bruni L, de Sanjosé S, Saraiya M, Ferlay J, Bray F. Estimates of incidence and mortality of CC in 2018: a worldwide analysis. Lancet Glob Health 2020;8:e191–203.

[4] World Health Organization. Comprehensive CC control integrating health care for sexual and reproductive health care for sexual and reproductive health and chronic disease a guide to practice. Geneva: WHO Library Cataloguing; 2006, p. 3.

[5] Kaarthigeyan K. CC in India and HPV vaccination, Indian J Med Paediatr Oncol 2012;33(1):7–12. https://doi.org/10.4103/0971-5851.96961. 22754202 PMC3385284.

[6] Panatto D, Amicizia D, Trucchi C, Casabona F, Lai PL, Bonanni P, Boccalini S, Bechini A, Tiscione E, Zotti CM, Coppola RC, Masia G, Meloni A, Castiglia P, Piana A, Gasparini R. Sexual behaviour and risk factors for the acquisition of human papillomavirus infections in young people in Italy: suggestions for future vaccination policies, BMC Public Health 2012;12:623. https://doi.org/10.1186/1471-2458-12-623. 22871132 PMC3490840.

[7] Marx J. Cancer research. Obstacle for promising cancer therapy. Science 2002;295(5559):1444. https://doi.org/10.1126/science.295.5559.1444a.

[8] Meehan K, Vella LJ. The contribution of tumour-derived exosomes to the hallmarks of cancer. Crit Rev Clin Lab Sci 2016;53(2):121–31. https://doi.org/10.3109/10408363.2015.1092496.

[9] Zhou S, Huang C, Wei Y. The metabolic switch and its regulation in cancer cells. Sci China Life Sci 2010;53(8):942–58. https://doi.org/10.1007/s11427-010-4041-1.

[10] Huang YT, Wang CC, Tsai CS, Lai CH, Chang TC, Chou HH, Lee SP, Hong JH. Clinical behaviors and outcomes for adenocarcinoma or adenosquamous carcinoma of cervix treated by radical hysterectomy and adjuvant radiotherapy or chemoradiotherapy, Int J Radiat Oncol Biol Phys 2012;84 (2):420–7. https://doi.org/10.1016/j.ijrobp.2011.12.013. Epub 2012 Feb 24, 22365621.

[11] Hoppe-Seyler K, Honegger A, Bossler F, Sponagel J, Bulkescher J, Lohrey C. Viral E6/E7 oncogene and cellular hexokinase 2 expressiom in HPV-positive cancer cell lines. Oncotarget 2017;8 (63):106342–51.

[12] Kim MY, Kim YS, Kim M, Choi MY, Roh GS, Lee DH, Kim HJ, Kang SS, Cho GJ, Shin JK, Choi WS. Metformin inhibits CC cell proliferation via decreased AMPK O-GlcNAcylation, Anim Cells Syst (Seoul) 2019;23(4):302–9. https://doi.org/10.1080/19768354.2019.1614092. 31489252 PMC6711131.

[13] Tseng CH. Metformin use and CC risk in female patients with type 2 diabetes, Oncotarget 2016;7 (37):59548–55. https://doi.org/10.18632/oncotarget.10934. 27486978 PMC5312330.

[14] Jin L, Alesi GN, Kang S. Glutaminolysis as a target for cancer therapy, Oncogene 2016;35(28):3619–25. https://doi.org/10.1038/onc.2015.447. Epub 2015 Nov 23, 26592449 PMC5225500.

[15] Lunt SY, Vander Heiden MG. Aerobic glycolysis: meeting the metabolic requirements of cell proliferation, Annu Rev Cell Dev Biol 2011;27:441–64. https://doi.org/10.1146/annurev-cellbio-092910-154237. Review, 21985671.

[16] Newell K, Franchi A, Pouysségur J, Tannock I. Studies with glycolysis-deficient cells suggest that production of lactic acid is not the only cause of tumor acidity, Proc Natl Acad Sci U S A 1993;90 (3):1127–31. 8430084 PMC45824.

[17] Shan L. Cy5.5-labeled pH low insertion peptide (pHLIP), In: Molecular imaging and contrast agent database (MICAD) [Internet]. Bethesda, MD: National Center for Biotechnology Information (US); 2009. p. 2004–13. Aug 8 [updated 2009 Nov 12]. Available from: http://www.ncbi.nlm.nih.gov/books/NBK23623/ PubMed PMID: 20641819.

[18] Helmlinger G, Sckell A, Dellian M, Forbes NS, Jain RK. Acid production in glycolysis-impaired tumors provides new insights into tumor metabolism, Clin Cancer Res 2002;8(4):1284–91. 11948144.

[19] Yamamoto M, Inohara H, Nakagawa T. Targeting metabolic pathways for head andneck cancers therapeutics, Cancer Metastasis Rev 2017;36(3):503–14. https://doi.org/10.1007/s10555-017-9691-z. Review, 28819926.

[20] Lu J, Tan M, Cai Q. The Warburg effect in tumor progression: mitochondrial oxidative metabolism as an anti-metastasis mechanism, Cancer Lett 2015;356(2 Pt A):156–64. https://doi.org/10.1016/j.canlet.2014.04.001. Epub 2014 Apr 13. Review, 24732809 PMC4195816.

[21] Chan B, VanderLaan PA, Sukhatme VP. 6-Phosphogluconate dehydrogenase regulates tumor cell migration *in vitro* by regulating receptor tyrosine kinase c-Met. Biochem Biophys Res Commun 2013;439(2):247–51. https://doi.org/10.1016/j.bbrc.2013.08.048.

[22] Shen Z, Belinson J, Morton RE, Xu Y, Xu Y. Phorbol 12-myristate 13-acetate stimulates lysophosphatidic acid secretion from ovarian and CC cells but not from breast or leukemia cells, Gynecol Oncol 1998;71(3):364–8. ISSN 0090-8258. https://doi.org/10.1006/gyno.1998.5193.

[23] Kostenis E. Novel clusters of receptors for sphingosine-1-phosphate, sphingosylphosphorylcholine, and (lyso)-phosphatidic acid: new receptors for "old" ligands, J Cell Biochem 2004;92(5):923–36. Review, 15258916.

[24] Murray NR, Baumgardner GP, Burns DJ, Fields AP. Protein kinase C isotypes in human erythroleukemia (K562) cell proliferation and differentiation. Evidence that beta II protein kinase C is required for proliferation, J Biol Chem 1993;268(21):15847–53.8340409.

[25] Chang MS, Chen BC, Yu MT, Sheu JR, Chen TF, Lin CH. Phorbol 12-myristate 13-acetate upregulates cyclooxygenase-2 expression in human pulmonary epithelial cells via Ras, Raf-1, ERK, and NF-kappaB, but not p38 MAPK, pathways, Cell Signal 2005;17(3):299–310.15567061.

[26] Starzec AB, Spanakis E, Nehme A, Salle V, Veber N, Mainguene C, Planchon P, Valette A, Prevost G, Israel L. Proliferative responses of epithelial cells to 8-bromo-cyclic AMP and to a phorbol ester change during breast pathogenesis. J Cell Physiol 1994;161:31–8.

[27] Brueggemeier RW, Díaz-Cruz ES. Relationship between aromatase and cyclooxygenases in breast cancer: potential for new therapeutic approaches, Minerva Endocrinol 2006;31(1):13–26. Review, 16498361.

[28] Orsó E, Schmitz G. Lipoprotein(a) and its role in inflammation, atherosclerosis and malignancies, Clin Res Cardiol Suppl 2017;12(Suppl 1):31–7. https://doi.org/10.1007/s11789-017-0084-1. 28188431 PMC5352764.

[29] Westermann AM, Havik E, Postma FR, Beijnen JH, Dalesio O, Moolenaar WH, et al. Malignant effusions contain lysophosphatidic acid (LPA)-like activity. Ann Oncol 1998;9:437–42. https://doi.org/10.1023/A:1008217129273.

[30] Xu Y, Shen Z, Wiper DW, Wu M, Morton RE, Elson P, et al. Lysophosphatidic acid as a potential biomarker for ovarian and other gynecologic cancers. JAMA 1998;280:719–23. https://doi.org/10.1001/jama.280.8.719.

[31] Eder AM, Sasagawa T, Mao M, Aoki J, Mills GB. Constitutive and lysophosphatidic acid (LPA)-induced LPA production: role of phospholipase D and phospholipase A2. Clin Cancer Res 2000;6:2482–91.

[32] Hoogendam JP, Klerkx WM, de Kort GA, Beepat S, Zweemer RP, Sie-Go DM, et al. The influence of the b-value combination on apparent diffusion coefficient based differentiation between malignant and benign tissue in CC. J Magn Reson Imaging 2010;32:376–82.

[33] Xue HD, Li S, Sun F, Sun HY, Jin ZY, Yang JX, et al. Clinical application of body diffusion weighted MR imaging in the diagnosis and preoperative N staging of CC. Chin Med Sci J 2008;23:133–7.

[34] Chen YB, Hu CM, Chen GL, Hu D, Liao J. Staging of uterine cervical carcinoma: whole-body diffusion-weighted magnetic resonance imaging. Abdom Imaging 2011;36:619–26.

[35] Schalkwijk CG, van der Heijden MA, Bunt G, Maas R, Tertoolen LG, van Bergen Henegouwen PM, Verkleij AJ, van den Bosch H, Boonstra J. Maximal epidermal growth-factor-induced cytosolic phospholipase A2 activation in vivo requires phosphorylation followed by an increased intracellular calcium concentration, Biochem J 1996;313(Pt 1):91–6. https://doi.org/10.1042/bj3130091. 8546715 PMC1216914.

[36] Sato T, Nakajima H, Fujio K, Mori Y. Enhancement of prostaglandin E2 production by epidermal growth factor requires the coordinate activation of cytosolic phospholipase A2 and cyclooxygenase 2 in human squamous carcinoma A431 cells. Prostaglandins 1997;53(5):355–69. https://doi.org/10.1016/0090-6980(97)00036-1.

[37] MacGregor DR, Gould P, Foreman J, Griffiths J, Bird S, Page R, Stewart K, Steel G, Young J, Paszkiewicz K, Millar AJ, Halliday KJ, Hall AJ, Penfield S. High expression of osmotically responsive genes1 is required for circadian periodicity through the promotion of nucleo-cytoplasmic mRNA export in Arabidopsis, Plant Cell 2013;25(11):4391–404. https://doi.org/10.1105/tpc.113.114959. Epub 2013 Nov 19, 24254125 PMC3875725.

[38] Kilickesmez O, Bayramoglu S, Inci E, Cimilli T, Kayhan A. Quantitative diffusion-weighted magnetic resonance imaging of normal and diseased uterine zones. Acta Radiol 2009;50:340–7.

[39] Payne GS, Schmidt M, Morgan VA, Giles S, Bridges J, Ind T, et al. Evaluation of magnetic resonance diffusion and spectroscopy measurements as predictive biomarkers in stage 1 CC. Gynecol Oncol 2010;116:246–52.

[40] Liu Y, Bai R, Sun H, Liu H, Wang D. Diffusion-weighted magnetic resonance imaging of CC. J Comput Assist Tomogr 2009;33:858–62.

[41] Delikatny EJ, Russell P, Hunter JC, Hancock R, Atkinson AH, Van Haften-Day C, et al. Proton MR and human cervical neoplasia: ex vivo spectroscopy allows distinction of invasive carcinoma of the cervix from carcinoma in situ and other preinvasive lesions. Radiology 1993;188:791–6.

[42] Mahon MM, Cox IJ, Dina R, Soutter WP, McIndoe GA, Williams AD, et al. (1)H magnetic resonance spectroscopy of preinvasive and invasive CC: in vivo-ex vivo profiles and effect of tumor load. J Magn Reson Imaging 2004;19:356–64.

[43] Manoharan D, Das CJ, Aggarwal A, Gupta AK. Diffusion weighted imaging in gynecological malignancies—present and future, World J Radiol 2016;8(3):288–97. https://doi.org/10.4329/wjr.v8.i3.288. 27027614 PMC4807338.

[44] Oh JW, Rha SE, Oh SN, Park MY, Byun JY, Lee A. Diffusion-weighted MRI of epithelial ovarian cancers: correlation of apparent diffusion coefficient values with histologic grade and surgical stage, Eur J Radiol 2015;84(4):590–5. https://doi.org/10.1016/j.ejrad.2015.01.005. Epub 2015 Jan 16, 25623826.

[45] Kidd EA, Siegel BA, Dehdashti F, Grigsby PW. The standardized uptake value for F-18 fluorodeoxyglucose is a sensitive predictive biomarker for CC treatment response and survival. Cancer 2007;110(8):1738–44. https://doi.org/10.1002/cncr.22974.

[46] Pavlova NN, Thompson CB. The emerging hallmarks of cancer metabolism. Cell Metab 2016;23:27–47. https://doi.org/10.1016/j.cmet.2015.12.006.

[47] Kalyanaraman B. Teaching the basics of cancer metabolism: developing antitumor strategies by exploiting the differences between normal and cancer cell metabolism. Redox Biol 2017;12:833–42. https://doi.org/10.1016/j.redox.2017.04.018.

[48] Otto AM. Warburg effect(s)—a biographical sketch of Otto Warburg and his impacts on tumor metabolism. Cancer Metab 2016;4:5. https://doi.org/10.1186/s40170-016-0145-9.

[49] Zhang R, Su J, Xue SL, et al. HPV E6/p53 mediated down-regulation of miR-34a inhibits Warburg effect through targeting LDHA in CC. Am J Cancer Res 2016;6:312–20.

[50] Zeng Q, Chen J, Li Y, et al. LKB1 inhibits HPV-associated cancer progression by targeting cellular metabolism. Oncogene 2017;36:1245–55. https://doi.org/10.1038/onc.2016.290.

[51] Pastrez PRA, Mariano VS, Da Costa AM, et al. The relation of HPV infection and expression of p53 and p16 proteins in esophageal squamous cells carcinoma. J Cancer 2017;8:1062–70. https://doi.org/10.7150/jca.17080.

[52] Hoppe-Seyler K, Honegger A, Bossler F, et al. Viral E6/E7 oncogene and cellular hexokinase 2 expression in HPV-positive cancer cell lines. Oncotarget 2017;8:106342–51. https://doi.org/10.18632/oncotarget.22463.

[53] Wang T, Ning K, Sun X, Zhang C, Jin LF, Hua D. Glycolysis is essential for chemoresistance induced by transient receptor potential channel C5 in colorectal cancer. BMC Cancer 2018;18:207. https://doi.org/10.1186/s12885-018-4242-8.

[54] Grasso C, Jansen G, Giovannetti E. Drug resistance in pancreatic cancer: impact of altered energy metabolism. Crit Rev Oncol Hematol 2017;114:139–52. https://doi.org/10.1016/j.critrevonc.2017.03.026.

[55] He J, Xie G, Tong J, et al. Overexpression of microRNA-122 resensitizes 5-FU-resistant colon cancer cells to 5-FU through the inhibition of PKM2 *in vitro* and in vivo. Cell Biochem Biophys 2014;70:1343–50. https://doi.org/10.1007/s12013-014-0062.

CHAPTER 13

Pharmacoeconomics and cost-effectiveness of treatments related to breast and cervix cancers

Mohan Krishna Ghanta[a], Santosh C. Gursale[b], Narayan P. Burte[c], and L.V.K.S. Bhaskar[d]

[a]Department of Pharmacology, SRMC & RI, Sri Ramachandra Institute of Higher Education and Research, Chennai, Tamil Nadu, India
[b]Department of Pharmacology, BKL Walawalkar Rural Medical College, Sawarde, Ratnagiri, Maharashtra, India
[c]Department of Pharmacology, Viswabharathi Medical College, Kurnool, Andhra Pradesh, India
[d]Department of Zoology, Guru Ghasidas Vishwavidyalaya, Bilaspur, Chhattisgarh, India

Abstract

Breast and cervix cancers are major diseases among women with increasing incidence worldwide. Apart from disease burden, costs related to treatments and diagnoses of breast and cervical cancers put a major stress on the health and economic status of patients as well as the countries they are from. In most low- and middle-income countries, treatments of these diseases cause a larger economic burden on patients. The introduction of pharmacoeconomics reduces the economic burden of disease management on both patient and governments. This chapter discusses the worldwide cost burden and cost-effective treatments of breast and cervix cancer, which may be a pharmacoeconomic input.

Keywords: Cervical cancer, Breast cancer, Pharmacoeconomic concepts, Cost-effectiveness treatments, Treatment cost burden.

Abbreviations

BRCA1	human tumor suppressor gene
CBA	cost-benefit analysis
CEA	cost-effective analysis
CMA	cost-minimization analysis
CRT	chemoradiation therapy
CT	computerized tomography
CUA	cost-utility analysis
ER	estrogen receptors
GDP	gross domestic product
HER2	human epidermal growth factor receptor 2
HPV	human papillomavirus
INR	Indian rupee
IV	intravenous
NAC	neoadjuvant chemotherapy
QALY	quality adjusted life years
TNM	tumor node metastasis
USD	United States of America Dollar

G.P. Nagaraju, R.R. Malla (eds.)
A Theranostic and Precision Medicine Approach for Female-Specific Cancers
ISBN 978-0-12-822009-2
https://doi.org/10.1016/B978-0-12-822009-2.00013-3

1. Introduction

Women make up 49.5% of total world population. A recent study revealed cancer as the second leading cause of mortality in females worldwide. The cancer burden among women is growing, thus it is important to address as it not only affects the patient but her family and caregivers as well. Female-specific cancer causes an economic burden to societies, families, and individuals. Among total prevalent cancers in women, breast, uterine, cervix, ovary, lung, liver, and colorectal cancers constitute 60%. This chapter details pharmacoeconomics and its global status as well as cost burden of and cost-effective therapies for breast and cervix cancers [1–3].

2. Pharmacoeconomics

Pharmacoeconomics deals with health economics. It links clinical treatments and procedural outcomes with economic cost measures [4]. Pharmacoeconomics is defined as the description and analysis of costs of drug therapy to the healthcare system and society [5]. Pharmacoeconomics is also known as "economic evaluation of pharmacotherapy" [6].

The role of pharmacoeconomics is to secure the gross domestic product (GDP) of a country against healthcare policies, to rationalize the use and expenses of medicines to clinical efficacy, to allocate resources to health systems, to identify inefficient and most efficient treatment options, to decide pharmaceutical benefits, to provide decision making in pricing and reimbursement of medicines, to regulate single exit price for medicines, to preserve the value of health services, to enhance the physician–patient relationship, to regulate maximum reimbursement pricing of medicines, to evaluate medicines for preference to formularies, to regulate reimbursement policies, and to regulate the addition of medicines to the essential medicines list based on their cost-effectiveness.

Various methodologies have been developed to quantify the pharmacoeconomic status of pharmacotherapies. These include cost-minimization analysis (CMA), cost-benefit analysis (CBA), cost-effective analysis (CEA), cost–utility analysis (CUA), and sensitivity analysis. CMA is done between two drugs having similar clinical outcomes to obtain an optimal therapeutic medicine based on cost. CMA is applied for making formulary decisions and in developing guidelines of pharmacotherapy. CBA evaluates the costs associated with treatments and financial benefits. This method is applied for making policy decisions on healthcare programs. CEA evaluates the cost of treatment and its clinical efficacy to patients, like monetary units per physical unit of clinical efficacy in an individual patient. This method is applied for determining healthcare interventions for individual patients [7] as well as aiding formulary decisions. CUA evaluates healthcare interventions and quality of life in terms of monetary unit per unit of health (quality adjusted life years, QALY). This is applied for comparing different interventions used in chronic diseases [8].

3. Breast cancer

Breast cancer is a common cancer in women and accounts for more than 25% of the incidence rate among all other cancers in women [9]. Worldwide, there were 1,671,100 new cases of breast cancer and 521,900 deaths due to breast cancer in 2012 [1]. Breast cancer incidence is high in Australia, New Zealand, Europe, and North America. Five-year survival rates were 85% in the United States, Australia, Brazil, Israel, and Europe, and 60% in India, South Africa, and Algeria [10].

3.1 Pathological classification and diagnosis of breast cancer

Based on hormone receptors' expression, breast carcinoma is classified as either estrogen receptor (ER)-positive, human epidermal growth factor receptor 2 (HER2)-positive, or triple negative. ER-positive carcinoma is not associated with HER2 receptors, HER2-positive carcinoma may or may not be ER-positive, and triple negative carcinoma doesn't involve ER, progesterone receptor (PR), or HER2. Another alternative classification of breast cancers based on molecular subtyping reveals four types of breast cancers, namely, luminal A, which is ER-positive (lower grade) and HER2 negative; luminal B, which is ER-positive (higher grade) and HER2 positive; HER2-enriched, which is overexpressed with HER2 and ER-negative; and basal-like breast carcinoma, which is both ER- and HER2-negative [11].

Morphologically, breast cancers are classified as noninvasive and invasive. Noninvasive breast cancer includes ductal carcinoma in situ and lobular carcinoma in situ. Ductal carcinoma in situ enlarges affected lobules and lobular carcinoma in situ deforms lobules into tubular spaces. Invasive breast carcinomas include invasive ductal carcinoma, invasive lobular carcinoma, carcinoma with medullary features, mucinous carcinoma, tubular carcinoma, and other types. Invasive ductal carcinoma microscopically ranges from fully developed tubules with low grade nuclei to tumors with layers of anaplastic cells, and appears as a hard, palpable, irregular mass. This type of cancer is mostly ER-positive, but can be HER2-positive as well as negative for both ER and HER2. Invasive lobular carcinoma is morphologically similar to lobular carcinoma in situ. Stromal invasion presents as "single-file" and ER-positive. HER2 upregulation is rare. Carcinoma with medullary features presents as rounded masses with layers of large anaplastic cells and lymphocyte infiltration. This is common with germline *BRCA1* mutation. Mucinous or colloid carcinoma presents as soft and gelatinous tumors that are ER-positive and HER2-negative. Tubular carcinoma presents as small irregular tumors with developed tubules and low-grade nuclei that are ER-positive and HER2-negative. Inflammatory carcinoma clinically presents as an enlarged erythematous breast without palpable tumors or masses [12–14]. This cancer is both ER-positive and HER2-positive. Clinical staging of disease is based on tumor node metastasis (TNM) classification (Table 1).

Table 1 TNM staging of breast cancer [15].

Staging	Breast cancer
Tumor	
TX	Assessment of the primary tumor cannot be made
T0	No evidence of the primary tumor
T1	Tumor size <20 mm
T2	Tumor size 20–50 mm
T3	Tumor size >50 mm
T4	Tumor invasion into the chest wall and or to the skin
Tis	Carcinoma in situ
Tis (DCIS)	Ductal carcinoma in situ
Tis (LCIS)	Lobular carcinoma in situ
Node	
NX	Assessment of regional lymph nodes cannot be made
N0	No regional lymph node involvement
N1	Involvement of ipsilateral axillary lymph nodes and are movable
N2	Involvement of ipsilateral axillary lymph nodes and are adhered to surrounding structures
N3	Involvement of clavicular, mammary, and axillary lymph nodes of same side
Metastasis	
M0	No clinical or radiographic evidence of distant metastases
M1	Distant metastases diagnosed by clinical or radiological findings

Diagnosis of breast cancer is based on clinical examination, mammography, ultrasonography, biopsy either with fine needle aspiration for cytology or core biopsy for histology, computerized tomography (CT) of thorax, abdomen, isotope bone scan, and molecular subtyping [13].

3.2 Treatments and cost burden of breast cancer

The mainstay of treatment for breast cancer patients is surgical management, which includes lumpectomy or mastectomy. Adjuvant radiotherapy and chemotherapy is suggested for patients with risk of recurrence. Chemotherapy is advised for ER-negative breast cancer patients or those with involvement of axillary lymph nodes. Management of metastasis includes radiotherapy and endocrine therapy. Current treatment based on breast cancer staging with molecular subtyping is described in Table 2. Standard chemotherapy regimens contain anthracycline and taxane. The anthracycline cyclophosphamide-taxane regimen includes four cycles of doxorubicin and cyclophosphamide followed by four cycles of paclitaxel [20]. Other regimens

Table 2 Current treatment based on breast cancer staging.

Staging	Treatment
HR-positive tumors	Anastrozole/letrozole/tamoxifen [16]
HER2-positive tumors	Trastuzumab, pertuzumab, and lapatinib; stage II/III—trastuzumab along with anthracycline cyclophosphamide-taxane regimen [17, 18]
Triple negative tumors	Single or combination regimens of chemotherapeutic drugs [19]

include anthracycline and cyclophosphamide followed by paclitaxel for 12 weeks or docetaxel thrice-weekly for four cycles [21, 22]. Associated adverse events with breast cancer treatment include cognitive impairment, osteoporosis, chronic fatigue, and vaginal dryness [23].

A European study revealed the total cost of breast cancer treatment including posttreatment follow-up. The diagnostic expenditures (mean cost) were €414. The treatment mean cost was €8780. The follow-up mean cost was €2351. The total direct mean cost was €10,970. The follow-up cost was highly variable. A study from Punjab, India, reported that the total cost for diagnosis or screening was 4,147,700 INR (Indian rupee), total medical cost was 79,080,682 INR, total direct cost was 82,639,662 INR, and total indirect cost was 21,594,507.97 INR [24]. A meta-analysis reported the cumulative treatment cost of breast cancer by stage. The average cost for stage I breast cancer treatment was $20,852 USD, stage II was $29,430 USD, stage III was $40,211 USD, and stage IV was $41,560 USD [25]. This cost was higher in comparison to treatment costs in Canada and Italy [26, 27].

3.3 Cost-effective therapies

With an increasing burden of breast cancer disease and its treatment cost, introduction of cost-effective therapies and diagnostic models may prevent cancer mortality. As a low economic diagnostic model, low axillary lymph node sampling technology proved better in comparison to sentinel node biopsy [28, 29]. Lairson et al. performed CEA of breast cancer chemotherapy, comparing anthracycline-based regimens to non-anthracycline-based regimens. They found the anthracycline-based regimen to have 12.05 QALY with total treatment cost of $119,055 and the non-anthracycline-based regimen to have 9.56 QALY with total treatment cost of $86,383. In this comparative study, anthracycline-based regimens were found to be cost-effective in comparison to non-anthracycline-based regimens of breast cancer chemotherapy [30]. Shorter duration of treatment with adjuvant trastuzumab showed noninferiority to 1-year treatment [31]. A CMA study in Spain reported trastuzumab subcutaneous to be cost-effective compared to trastuzumab IV [32]. Preoperative progesterone treatment was found to be a cost-effective intervention in node-positive breast cancer patients [33]. Palbociclib was found to be cost-effective in treatment of advanced breast cancer, with a treatment cost of $768,498/QALY in adjuvant with letrozole and $918,166/QALY in adjuvant with fulvestrant [34]. Metronomic chemo-therapy was suggested as another cost-effective treatment for breast cancer [35].

4. Cervix cancer

Cervical cancer ranks fourth in both incidence and mortality among all cancers affecting women worldwide and second in developing countries. There were 527,600 cases of cervical cancer and 265,700 deaths worldwide in 2012. The incidence is highest in Sub-Saharan Africa, Asia, and South America. Burden of cervical cancer is greater in India (90%) when compared to other developing countries. The 5-year survival rate is 60%–70% in high-income countries and 46% in India among low- and middle-income countries. Cervical cancer is commonly diagnosed with abnormal smear test, clinical presentation of vaginal bleeding, and symptoms related to bladder, rectum, and pelvis involvement. Distant metastasis is commonly seen in the bones and lungs [13].

Pathologically, cervical cancers may be squamous cell carcinoma, adenocarcinoma, mixed adenosquamous carcinoma, or small cell neuroendocrine carcinoma. Various investigational procedures for cervical cancer patients include smear or cone biopsy, human papillomavirus (HPV) testing, cystoscopy, flexible sigmoidoscopy, magnetic resonance imaging, chest X-ray, and CT of the abdomen and pelvis [13]. Staging of cervical cancer is discussed in Table 3 [37].

4.1 Treatments and cost burden of cervical cancer

Surgery, radiotherapy, and chemotherapy are the common modalities of cervical cancer treatment. Management of cervical cancer is decided by the stage of disease (Table 3). Liu et al. [38] revealed the cost burden of cervical cancer treatment. Mean total cost of treatment during the prediagnosis phase was $3155 (Canadian dollars) and $17,938 during the initial phase of treatment. The cost of treatments during the last year of life was $58,319 [38]. A study in Brazil revealed that more cost (direct cost) was seen in radiotherapy $199,794.32, and next cost burden was seen with chemotherapy $143,268.17. The cost of surgical management was $44,431.43. The diagnostic methods direct cost was calculate by summing laboratory test cost and imaging studies ($19,714.81). The total cost burden of cervical cancer on the patient was $523,218.22. A few studies from the United States have revealed the cost of cervical cancer management using healthcare insurance claims. The cervical cancer management mean cost during 1990–91 was $30,136 [39]. McCrory et al. evaluated the mean cost of cervical cancer management by stage. It was $17,645, $27,069, $40,280 for stage I, II/III, and IV, respectively [40]. Insinga et al., in their retrospective study, estimated the mean mortality-adjusted cost of $29,649 [41]. Cromwell et al. calculated the overall mean cost charges on patient from diagnosis to death at $19,153±$3484 [42]. Another study estimated the direct cost of cervical cancer management in Ontario, Canada during the first 5 years after diagnosis. The cost burden

Table 3 Staging (FIGO) and management of cervical cancer.

Cancer stage			Description	Management
Stage 0			Intraepithelial cancer of cervix or carcinoma in situ	Chemotherapy
Stage I			Cancer confined to cervix	
	Ia		Invasive carcinoma (diagnosis only by microscopy)	
		Ia1	Microinvasive or stromal invasion less than 3 mm in depth and less than 7 mm in extension	
		Ia2	Stromal invasion greater than 3 mm in depth and 5–7 mm in extension	
	Ib		Clinically visible lesion confined to cervix	Surgical management and radiotherapy
		Ib1	Tumor/lesion <4 cm	
		Ib2	Tumor/lesion >4 cm	
Stage II			Cancer beyond the cervix, but not to pelvis side walls or to lower third of vagina	
	IIa		Carcinoma up to upper two-thirds of vagina	
		IIa1	Visible tumor/lesion <4 cm	
		IIa2	Visible tumor/lesion >4 cm	
	IIb		Lesion with parametrial invasion	Only radiotherapy
Stage III	IIIa		Lesion beyond lower one third of vagina with no extension to pelvic sidewall	
	IIIb		Lesion/tumor extending to pelvic sidewall	
Stage IV			Lesion/tumor extending beyond true pelvis or biopsy showing involvement of bladder or rectum mucosa	
	IVa		Rectal or bladder invasion	
	IVb		Distant metastasis	Cisplatin-based chemotherapy and radiotherapy confined to affected sites [13, 36]

per patient during years 1–5 was $39,187, $14,425, $11,280, $8444, and $5480, respectively [43]. Estimation of treatment cost in Ontario during 2013 reported by Oliveira et al. was $4272, which was much lower than that reported during 2016 ($13,697) [43, 44]. Subramanian et al. calculated the cost of cervical cancer management for the initial 6 months to be $3807 for in situ stage, $23,187 for local cancer, $35,853 for regional cancer, and $45,028 for distant cancers [45].

4.2 Cost-effective therapies

Cost-effectiveness based on thresholds of GDP per capita was criticized [46] and also withdrawn from World Health Organization recommendations. Alternatives like comparison of incremental cost-effectiveness between new interventions and existing interventions and estimating the willingness to pay for health benefits was suggested. Prevention of HPV infection resulted in reduction of health and cost burden of cervical cancer patients. HPV vaccination of younger adolescents proved cost-effective by many cost-effectiveness studies from different countries [47]. Use of bevacizumab was 13 times costlier than chemotherapy alone in advanced cervical cancer patients [48]. Cervical cancer screening with acetic acid was the cost-effective screening method with an incremental cost-effectiveness of $87 per life-year gained [49]. A neoadjuvant chemotherapy (NAC) regimen consists of intravenous (IV) infusion of cisplatin $80 \, mg/m^2$, IV bolus of bleomycin 30 mg, and IV infusion of vincristine $1.4 \, mg/m^2$ on the first day of every 21-day cycle for three cycles. Chemoradiation therapy (CRT) consists of IV infusion of cisplatin $40 \, mg/m^2$ weekly for 5 weeks starting on day 1 of external beam radiotherapy. The NAC regimen was found to be noninferior to CRT with advantages of less gastrointestinal and hematological toxicities. But the cost-effectiveness of these two regimens was not studied [50].

5. Conclusion

This chapter elaborated the high economic burden of breast and cervix cancer treatments on both patient and countries. This may further provide knowledge helpful in delivering cost-effective treatments using pharmacoeconomic tools that are currently poorly implemented in many regions of the world, particularly in low- and middle-income countries. This chapter concludes that more research is required nationally on cost-effectiveness, impacts on budget, and cost and burden of cancers.

References

[1] Torre LA, Islami F, Siegel RL, Ward EM, Jemal A. Global cancer in women: burden and trends. Cancer Epidemiol Biomarkers Prev 2017;26(4):444–57.

[2] United Nations. World population prospects: the 2015 revision. U N Econ Soc Aff 2015;33(2):1–66.

[3] World Health Organization. Women and health: today's evidence tomorrow's agenda. World Health Organization; 2009.

[4] Bakst A. Pharmacoeconomics and the formulary decision-making process. Hosp Formul 1995;30(1): 42–50.

[5] Richardson J. Cost utility analysis: what should be measured? Soc Sci Med 1994;39(1):7–21.

[6] Arenas-Guzman R, Tosti A, Hay R, Haneke E. Pharmacoeconomics—an aid to better decision-making. J Eur Acad Dermatol Venereol 2005;19(s1):34–9.

[7] Hill SR. Cost-effectiveness analysis for clinicians. BMC Med 2012;10(1):10.

[8] Reimbursement decisions and the implied value of life: cost effectiveness analysis and decisions to reimburse pharmaceuticals in Australia 1993–1996. George B, Harris A, Mitchell A, editors. Economics and

Health: 1997 proceedings of the nineteenth Australian conference of health economists. Sydney, Australia: School of Health Services Management, University of New South Wales; 1998.

[9] Ferlay J, Soerjomataram I, Dikshit R, Eser S, Mathers C, Rebelo M, et al. Cancer incidence and mortality worldwide: sources, methods and major patterns in GLOBOCAN 2012. Int J Cancer 2015;136 (5):E359–86.

[10] Allemani C, Weir HK, Carreira H, Harewood R, Spika D, Wang XS, et al. Global surveillance of cancer survival 1995–2009: analysis of individual data for 25,676,887 patients from 279 population-based registries in 67 countries (CONCORD-2). Lancet 2015;385(9972):977–1010.

[11] Kumar V, Abbas AK, Aster JC. Robbins basic pathology e-book. Elsevier Health Sciences; 2017.

[12] Sinn H-P, Kreipe H. A brief overview of the WHO classification of breast tumors, 4th edition, focusing on issues and updates from the 3rd edition. Breast Care 2013;8(2):149–54.

[13] Ralston SH, Penman ID, Strachan MW, Hobson R. Davidson's principles and practice of medicine E-book. Elsevier Health Sciences; 2018.

[14] Eliyatkın N, Yalçın E, Zengel B, Aktaş S, Vardar E. Molecular classification of breast carcinoma: from traditional, old-fashioned way to a new age, and a new way. J Breast Health 2015;11(2):59–66.

[15] Koh J, Kim MJ. Introduction of a new staging system of breast cancer for radiologists: an emphasis on the prognostic stage. Korean J Radiol 2019;20(1):69–82.

[16] Davies C, Godwin J, Gray R, Clarke M, Cutter D, Darby S, et al. Relevance of breast cancer hormone receptors and other factors to the efficacy of adjuvant tamoxifen: patient-level meta-analysis of randomised trials. Lancet 2011;378(9793):771–84.

[17] Gianni L, Pienkowski T, Im YH, Roman L, Tseng LM, Liu MC, et al. Efficacy and safety of neoadjuvant pertuzumab and trastuzumab in women with locally advanced, inflammatory, or early HER2-positive breast cancer (NeoSphere): a randomised multicentre, open-label, phase 2 trial. Lancet Oncol 2012;13(1):25–32.

[18] Tolaney SM, Barry WT, Dang CT, Yardley DA, Moy B, Marcom PK, et al. Adjuvant paclitaxel and trastuzumab for node-negative, HER2-positive breast cancer. N Engl J Med 2015;372(2):134–41.

[19] Gogate A, Rotter JS, Trogdon JG, Meng K, Baggett CD, Reeder-Hayes KE, et al. An updated systematic review of the cost-effectiveness of therapies for metastatic breast cancer. Breast Cancer Res Treat 2019;174(2):343–55.

[20] Citron ML, Berry DA, Cirrincione C, Hudis C, Winer EP, Gradishar WJ, et al. Randomized trial of dose-dense versus conventionally scheduled and sequential versus concurrent combination chemotherapy as postoperative adjuvant treatment of node-positive primary breast cancer: first report of Intergroup Trial C9741/Cancer and Leukemia Group B Trial 9741. J Clin Oncol 2003;21(8):1431–9.

[21] Sparano JA, Wang M, Martino S, Jones V, Perez EA, Saphner T, et al. Weekly paclitaxel in the adjuvant treatment of breast cancer. N Engl J Med 2008;358(16):1663–71.

[22] Sparano JA, Zhao F, Martino S, Ligibel JA, Perez EA, Saphner T, et al. Long-term follow-up of the E1199 phase III trial evaluating the role of taxane and schedule in operable breast cancer. J Clin Oncol 2015;33(21):2353–60.

[23] Pinto AC, de Azambuja E. Improving quality of life after breast cancer: dealing with symptoms. Maturitas 2011;70(4):343–8.

[24] Jain M, Mukherjee K. Economic burden of breast cancer to the households in Punjab, India. Int J Med Public Health 2016;6(1):13–8.

[25] Sun L, Legood R, Dos-Santos-Silva I, Gaiha SM, Sadique Z. Global treatment costs of breast cancer by stage: a systematic review. PLoS One 2018;13(11). e0207993-e.

[26] Capri S, Russo A. Cost of breast cancer based on real-world data: a cancer registry study in Italy. BMC Health Serv Res 2017;17(1):84.

[27] Mittmann N, Porter JM, Rangrej J, Seung SJ, Liu N, Saskin R, et al. Health system costs for stage-specific breast cancer: a population-based approach. Curr Oncol 2014;21(6):281–93.

[28] Radhakrishna S. Cost-effective breast cancer care in India. Int J Adv Med Health Res 2018;5(1):1.

[29] Parmar V, Hawaldar R, Nadkarni MS, Badwe RA. Low axillary sampling in clinically node-negative operable breast cancer. Natl Med J India 2009;22(5):234–6.

[30] Lairson DR, Parikh RC, Cormier JN, Chan W, Du XL. Cost-effectiveness of chemotherapy for breast cancer and age effect in older women. Value Health 2015;18(8):1070–8.

[31] Earl HM, Hiller L, Vallier A-L, Loi S, Howe D, Higgins HB, et al. PERSEPHONE: 6 versus 12 months (m) of adjuvant trastuzumab in patients (pts) with HER2 positive (+) early breast cancer (EBC): randomised phase 3 non-inferiority trial with definitive 4-year (yr) disease-free survival (DFS) results. J Clin Oncol 2018;36(15_suppl):506.

[32] Lopez-Vivanco G, Salvador J, Diez R, Lopez D, De Salas-Cansado M, Navarro B, et al. Cost minimization analysis of treatment with intravenous or subcutaneous trastuzumab in patients with HER2-positive breast cancer in Spain. Clin Transl Oncol 2017;19(12):1454–61.

[33] Badwe R, Hawaldar R, Parmar V, Nadkarni M, Shet T, Desai S, et al. Single-injection depot progesterone before surgery and survival in women with operable breast cancer: a randomized controlled trial. J Clin Oncol 2011;29(21):2845–51.

[34] Mamiya H, Tahara RK, Tolaney SM, Choudhry NK, Najafzadeh M. Cost-effectiveness of palbociclib in hormone receptor-positive advanced breast cancer. Ann Oncol 2017;28(8):1825–31.

[35] Simsek C, Esin E, Yalcin S. Metronomic chemotherapy: a systematic review of the literature and clinical experience. J Oncol 2019;2019:5483791.

[36] Eaker S, Adami H-O, Sparén P. Reasons women do not attend screening for cervical cancer: a population-based study in Sweden. Prev Med 2001;32(6):482–91.

[37] Šarenac T, Mikov M. Cervical cancer, different treatments and importance of bile acids as therapeutic agents in this disease. Front Pharmacol 2019;10(484):1–29.

[38] Liu N, Mittmann N, Coyte PC, Hancock-Howard R, Seung SJ, Earle CC. Phase-specific healthcare costs of cervical cancer: estimates from a population-based study. Am J Obstet Gynecol 2016;214(5): e1–e11. 615.

[39] Helms LJ, Melnikow J. Determining costs of health care services for cost-effectiveness analyses: the case of cervical cancer prevention and treatment. Med Care 1999;37(7):652–61.

[40] McCrory DC, Matchar DB, Bastian L, Datta S, Hasselblad V, Hickey J, et al. Evaluation of cervical cytology. Evid Rep Technol Assess (Summ) 1999;(5):1–6.

[41] Insinga RP, Ye X, Singhal PK, Carides GW. Healthcare resource use and costs associated with cervical, vaginal and vulvar cancers in a large U.S. health plan. Gynecol Oncol 2008;111(2):188–96.

[42] Cromwell I, Ferreira Z, Smith L, van der Hoek K, Ogilvie G, Coldman A, et al. Cost and resource utilization in cervical cancer management: a real-world retrospective cost analysis. Curr Oncol 2016;23(Suppl 1):S14–22.

[43] Pendrith C, Thind A, Zaric GS, Sarma S. Costs of cervical cancer treatment: population-based estimates from Ontario. Curr Oncol 2016;23(2):e109–15.

[44] de Oliveira C, Bremner KE, Pataky R, Gunraj N, Chan K, Peacock S, et al. Understanding the costs of cancer care before and after diagnosis for the 21 most common cancers in Ontario: a population-based descriptive study. CMAJ Open 2013;1(1):E1–8.

[45] Subramanian S, Trogdon J, Ekwueme DU, Gardner JG, Whitmire JT, Rao C. Cost of cervical cancer treatment: implications for providing coverage to low-income women under the Medicaid expansion for cancer care. Womens Health Issues 2010;20(6):400–5.

[46] Newall AT, Jit M, Hutubessy R. Are current cost-effectiveness thresholds for low- and middle-income countries useful? Examples from the world of vaccines. Pharmacoeconomics 2014;32(6):525–31.

[47] Van Minh H, My NTT, Jit M. Cervical cancer treatment costs and cost-effectiveness analysis of human papillomavirus vaccination in Vietnam: a PRIME modeling study. BMC Health Serv Res 2017;17(1):353.

[48] Minion LE, Bai J, Monk BJ, Robin Keller L, Ramez EN, Forde GK, et al. A Markov model to evaluate cost-effectiveness of antiangiogenesis therapy using bevacizumab in advanced cervical cancer. Gynecol Oncol 2015;137(3):490–6.

[49] Ralaidovy AH, Gopalappa C, Ilbawi A, Pretorius C, Lauer JA. Cost-effective interventions for breast cancer, cervical cancer, and colorectal cancer: new results from WHO-CHOICE. Cost Eff Resour Alloc 2018;16:38.

[50] Dastidar GA, Gupta P, Basu B, Basu A, Shah JK, Seal SL. Is neo-adjuvant chemotherapy a better option for management of cervical cancer patients of rural India? Indian J Cancer 2016;53(1):56–9.

CHAPTER 14

Precision medicine for female-specific cancers: Strategies, challenges, and solutions

Rama Rao Malla[a] and Ganji Purnachandra Nagaraju[b]

[a]Cancer Biology Lab, Department of Biochemistry and Bioinformatics, Institute of Science, GITAM (Deemed to be University), Visakhapatnam, Andhra Pradesh, India
[b]Department of Hematology and Medical Oncology, Winship Cancer Institute, Emory University, Atlanta, GA, United States

Abstract

Female-specific cancers (FSCs) are a group of gynecological neoplasms that contribute to cancer mortality and morbidity worldwide. The etiology of FSCs is principally associated with lifestyle aspects. Moreover, morphological malformations and genetic and epigenetic alterations are also causing FSCs. Surgeries including robotic-assisted laparoscopic surgeries, resection of tumor organ, and radio and chemotherapy are effective therapies encouraged in the treatment of cancer. However, these therapies result in improved 5-year survival rates and are not curative; tumors tend to reoccur and/or metastasize. This remains a challenge for researchers. Targeted therapies that target the tumor site and reduce toxic effects on healthy cells are found to be effective. Precision medicine replaces the single medicine approach to overcome the hurdle of resistance developed by chemo drugs. Personalized medicine as per the stage and type of cancer has evolved and is found to be effective. Mutagenic therapy could be one among the novel therapeutic strategies for cancer. Additionally, regenerative medicine technologies that include engineering cancer stem cells and the extracellular matrix in the tumor microenvironment that plays a crucial role in metastasis are considered as novel therapeutic options to treat FSCs. Human papilloma virus vaccination and oral contraceptives are recommended as they tend to help prevent the formation of FSCs. This chapter focuses on novel therapeutic strategies developed by researchers to treat FSCs.

Keywords: Female-specific cancer, Tumor microenvironment, Resistance, Metastasis, Precision medicine.

Abbreviations

BRCA	breast cancer type 1 susceptibility protein
FOX	Fork head box
FSC	female-specific cancer
HER	human epidermal growth factor receptor
KRAS	Kirsten rat sarcoma
PM	precision medicine
PTEN	phosphatase and tensin homolog
TME	tumor microenvironment

G.P. Nagaraju, R.R. Malla (eds.)
A Theranostic and Precision Medicine Approach for Female-Specific Cancers
ISBN 978-0-12-822009-2
https://doi.org/10.1016/B978-0-12-822009-2.00014-5

Globally, incidence of female-specific cancers (FSCs) such as breast, cervical, ovarian, and endometrial cancers is rising despite advancements in theranostics. Several attempts have been made in the past decade to improve the outcome of FSCs by targeted cytotoxic agents. In addition, lack of effective and specific early detection methods makes treating FSCs more challenging. Additionally, remodeling of tumor microenvironments (TMEs), development of chemo- and radio-resistance, and lack of an effective immune system to counteract tumor cells leads to increased lethality and fatality in FSC patients. Although there has been improvement in treatments that do not affect normal tissues, the ultimate benefits have been so far disappointing. Therefore, in order to improve outcomes in patients with FSCs, a novel approach that modulates the intrinsic and acquired mechanisms of resistance is necessary. This chapter provides a deep understanding of personalized road-blocks for patients diagnosed with metastatic subtypes of FSCs. It also provides an in-depth understanding of the personalized treatment options currently available as well as prospective options. Finally, it explores the role of stakeholders and their role in successful implementation of precision medicine (PM) through integrated accuracy and personalized medicine approaches that have the potential to enhance patient care.

The era of single medicine approaches for cancers with phenotypic and genotypic similarities is rapidly declining and progressively being replaced with a PM approach [1]. PM for cancer is an approach for diagnosis, treatment, follow-up care, and prevention under survivorship [2]. Today, PM is recognized as a powerful solution for selective targeting of tumor cells with no or minimum damage to normal cells [3]. The goal of PM is to provide the most efficient treatment to each patient with specificity, to avoid treatment risks and side effects, and to develop therapies to target specific cellular and molecular pathways that are vital for tumor growth [4].

Personalized medicine for breast cancer [5] and gynecologic cancers is an evolving field [6]. PM-based treatment is a treatment for a group of specific cancers rather than a specific cancer [7]. For example, a single treatment can be prepared for breast cancer patients with different diagnoses, those who are positive for HER2, those with cancer in the lymph nodes, and women who are at the premenopausal stage with HER2-targeted therapy using trastuzumab [8]. However, scientific as well as logistical challenges hamper implementation at the clinical level [9]. For instance, the low frequency of genomic changes hinders randomized trials, since a large number of patients need to undergo screening [10]. However, scaling-up of patients with genomic changes for screening and the clustering of genomic alterations into pathways will resolve this limitation.

The research on PM for gynecologic cancers is still in progress. Some mutational drivers have been identified. For example, BRCA mutations, NOTCH, P13 K, BRAS/MEK, FOX 1, p53 in ovarian cancer, TP53, PTEN, P1K3CA, and KRAS in endometrial cancer, and P1K3CA, TP53, RB1 in cervical cancer [11]. Therapeutics that target mutational drivers are developing effectively by exploiting tumor-stromal interactions, vasculature, and deviant signaling axis [12].

Recently, a PM tumor board of ovarian cancer patients recommended targetable abnormalities and targeted therapies for implementation in clinics [13]. Thus, information of PM tumor boards is gaining importance in comparing patients with genotype-matched clinical trials [14]. However, additional research is a prerequisite in recognizing key pathways that are susceptible to specific treatments. Cancer and precancer proteomic, epigenomic, and genomic information defined the molecular mechanism of human papilloma virus-mediated carcinogenesis and provided personalized biomarkers for diagnosis [15]. Further, precision research is required for the subtyping of cervical cancers and development of PM to diagnose at an early stage and for effective treatment.

Epigenetic modifications are key factors of tumor initiation and progression. Epigenetics is an innovative discipline in PM that uncovers novel epigenetic processes and imparts promising biomarkers, targets, and therapeutics with promising applications in clinical settings [16]. Recently, TME is recognized as a multifaceted system with reflective heterogeneity, dynamic nature, and complexity in intercellular communication. Quantitative profiling of TME by immunohistochemistry, flow and mass spectrophotometry [17], high-throughput RNA sequencing [18], molecular imaging [3], and multiomic profiling paves the way to develop precision immune therapeutic for FSCs [19]. Machine learning approaches using supervised and unsupervised pattern analysis of datasets from cancer-derived omic profiles also support the development of PM for FSCs [20]. Various molecular in vitro diagnostic-based tests also support the development of PM, which offers feasible substitutions for cancer theranostics [16].

There are definitely numerous challenges ahead in implementing PM, but none of them are impossible to overcome. The expectations should be rational as PM will not be developed automatically nor overnight. However, healthcare stakeholders, including government, research and pharmaceutical industries, biomedical communities, patient communities, and regulatory bodies, have the ability to resolve these issues by making stringent laws, providing socioeconomic equality, developing early diagnostic tools, and providing regulatory frameworks to safeguard patients.

References

[1] Dalby M, Cree IA, Challoner BR, Ghosh S, Thurston DE. The precision medicine approach to cancer therapy: part 1—solid tumours. Acute Pain 2019;15:44.

[2] Harris EE. Precision medicine for breast cancer: the paths to truly individualized diagnosis and treatment. Int J Breast Cancer 2018;2018:1–8.

[3] Penet M-F, Krishnamachary B, Chen Z, Jin J, Bhujwalla ZM. Molecular imaging of the tumor microenvironment for precision medicine and theranostics. Adv Cancer Res 2014;124:235–56.

[4] Lumachi F, Chiara GB, Foltran L, Basso SM. Proteomics as a guide for personalized adjuvant chemotherapy in patients with early breast cancer. Cancer Genomics Proteomics 2015;12(6):385–90.

[5] (a) Odle TG. Precision medicine in breast cancer. Radiol Technol 2017;88(4):401m–421m. (b) Cheng L, Majumdar A, Stover D, Wu S, Lu Y, Li L. Computational cancer cell models to guide precision breast cancer medicine. Genes 2020;11(3):263.

[6] Corey L, Valente A, Wade K. Personalized medicine in gynecologic cancer: fact or fiction? Obstet Gynecol Clin North Am 2019;46(1):155–63.

[7] (a) Montemurro F, Valabrega G, Aglietta M. Trastuzumab treatment in breast cancer. N Engl J Med 2006;354(20):2186. author reply 2186. (b) Xuhong JC, Qi XW, Zhang Y, Jiang J. Mechanism, safety and efficacy of three tyrosine kinase inhibitors lapatinib, neratinib and pyrotinib in HER2-positive breast cancer. Am J Cancer Res 2019;9(10):2103–19.

[8] (a)Adamczyk A, Niemiec J, Janecka A, Harazin-Lechowska A, Ambicka A, Grela-Wojewoda A, Domagala-Haduch M, Cedrych I, Majchrzyk K, Kruczak A, Rys J, Jakubowicz J. Prognostic value of PIK3CA mutation status, PTEN and androgen receptor expression for metastasis-free survival in HER2-positive breast cancer patients treated with trastuzumab in adjuvant setting. Pol J Pathol 2015;66(2):133–41. (b)Harbeck N. Insights into biology of luminal HER2 vs. enriched HER2 subtypes: therapeutic implications. Breast 2015;24(Suppl 2):S44–8.

[9] Sachdev JC, Sandoval AC, Jahanzeb M. Update on precision medicine in breast cancer. Cancer Treat Res 2019;178:45–80.

[10] (a)Arnedos M, Vicier C, Loi S, Lefebvre C, Michiels S, Bonnefoi H, Andre F. Precision medicine for metastatic breast cancer—limitations and solutions. Nat Rev Clin Oncol 2015;12(12):693–704. (b)Low SK, Zembutsu H, Nakamura Y. Breast cancer: the translation of big genomic data to cancer precision medicine. Cancer Sci 2018;109(3):497–506.

[11] Corey L, Valente A, Wade K. Personalized medicine in gynecologic cancer: fact or fiction? Surg Oncol Clin 2020;29(1):105–13.

[12] Horowitz N, Matulonis UA. New biologic agents for the treatment of gynecologic cancers. Hematol Oncol Clin North Am 2012;26(1):133–56.

[13] Sanchez NS, Mills GB, Mills Shaw KR. Precision oncology: neither a silver bullet nor a dream. Pharmacogenomics 2017;18(16):1525–39.

[14] Ray-Coquard I, Pujade Lauraine E, Le Cesne A, Pautier P, Vacher Lavenue MC, Trama A, Casali P, Coindre JM, Blay JY. Improving treatment results with reference centres for rare cancers: where do we stand? Eur J Cancer 2017;77:90–8.

[15] (a)Wilting SM, Steenbergen RDM. Molecular events leading to HPV-induced high grade neoplasia. Papillomavirus Res 2016;2:85–8. (b)Mirnezami R, Nicholson J, Darzi A. Preparing for precision medicine. N Engl J Med 2012;366(6):489–91.

[16] Beltran-Garcia J, Osca-Verdegal R, Mena-Molla S, Garcia-Gimenez JL. Epigenetic IVD tests for personalized precision medicine in cancer. Front Genet 2019;10:621.

[17] Petitprez F, Sun CM, Lacroix L, Sautes-Fridman C, de Reynies A, Fridman WH. Quantitative analyses of the tumor microenvironment composition and orientation in the era of precision medicine. Front Oncol 2018;8:390.

[18] Lau D, Bobe AM, Khan AA. RNA sequencing of the tumor microenvironment in precision cancer immunotherapy. Trends Cancer 2019;5(3):149–56.

[19] Finotello F, Eduati F. Multi-omics profiling of the tumor microenvironment: paving the way to precision immuno-oncology. Front Oncol 2018;8:430.

[20] Azuaje F. Artificial intelligence for precision oncology: beyond patient stratification. NPJ Precis Oncol 2019;3:6.

CHAPTER 15

Environmental carcinogens and their impact on female-specific cancers

N. Srinivas[a], Rama Rao Malla[b], K. Suresh Kumar[a], and A. Ram Sailesh[a]
[a]Department of Environmental Science, Institute of Science, GITAM (Deemed to be University), Visakhapatnam, India
[b]Cancer Biology Lab, Department of Biochemistry and Bioinformatics, Institute of Science, GITAM (Deemed to be University), Visakhapatnam, Andhra Pradesh, India

Abstract

Environmentally derived substances or chemicals that may enter into the human body through food, air, or materials are called environmental carcinogens. Various natural and anthropogenic substances present in the environment are responsible for two thirds of cancers. Although differential exposures are associated with specific cancers types, environmental factors can be categorized into physical, chemical, and biological agents. Genotoxic agents are more likely to cause damage to the genetic material of a cell, which also leads to mutations that cause cancer. There are a few cancers that are usually found only in women, including breast cancer, endometrial cancer, ovarian cancer, and cervical cancer.

Keywords: Breast cancer, Environment, Genotoxic, Mutation, POPs.

Abbreviations

AhR	aryl hydrocarbon receptor
DDT	dichlorodiphenyltrichloroethane
DDE	dichlorodiphenyldichloroethylene
DNA	deoxyribo nucleic acid
EDC	endocrine disrupting chemicals
ER	estrogen receptor
FDA	Food and Drug Administration
GLOBOCAN	Global Cancer Incidence, Mortality, and Prevalence
HPV	human papilloma virus
OCP	organochlorine pesticide
PAH	polycyclic aromatic hydrocarbon
PCB	polychlorinated biphenyl
PFC	perfluorinated compound
POP	persistent organic pollutants
PR	progesterone receptor
UV	ultra violet
VOC	volatile organic compounds

G.P. Nagaraju, R.R. Malla (eds.)
A Theranostic and Precision Medicine Approach for Female-Specific Cancers
ISBN 978-0-12-822009-2
https://doi.org/10.1016/B978-0-12-822009-2.00015-7

1. Introduction

Over the years, the incidence of cancer has been increasing due to factors generated within the body and those received from outside the body or the environment. Current concepts of molecular mechanisms of malignancy propose that tumors are originating either directly or indirectly from environmental factors. These include chemical substances that are dangerous to human health at higher concentrations, otherwise known as carcinogens [1]. As per GLOBOCAN 2018, one out of three women and one out of two men may have cancer in their lifetimes.

2. Factors causing cancers

Cancer is a disease that causes damage to the cellular DNA or causes it to undergo mutation. It may also cause uncontrolled growth of altered cells and increase their ability to migrate from the original site to other parts of the body, which leads to mutation. Mutation is a permanent change that alters the amount of genetic material present in the cell or even alters the cell's genetic structure. The agents responsible for these mutations are called mutagens. Mutagenic substances are mostly carcinogenic, but not all substances cause mutations [2, 3].

2.1 Environmental carcinogens

Chemical substances, physical factors (such as radioactivity and sunlight), and prolonged exposures can lead to development of cancers in the human body; these substances are known as general carcinogens. Environmentally derived substances or chemicals that may enter into the human body through food, air, or materials are known as environmental carcinogens. Experimental evidences show that exposure to all carcinogens may not cause cancer in all cases. Different levels of exposure and their potential of causing cancer and many other factors like the intensity and length of exposure and an individual's genetic makeup may influence the risk of developing cancer [4]. In medical literature, environment is a concept that is used in a broader perspective that includes many of the nongenetic factors (lifestyle, diet, and infectious agents). It is also implicated in causing major types of cancers in humans [5].

Mechanisms of current biological trends reveal that the majority of the cancers originate from genetic disorders and the environment. Interference of cellular mechanisms and signal processing may help in the prevention of cancers caused by external factors [6].

3. Carcinogenic agents that induce cancer

Cancers can be induced due to several agents that are present both outside and inside the human body. Thus, scientists refer to everything that interacts outside the human body as

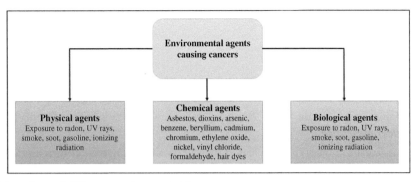

Fig. 1 Flow chart of different environmental factors.

the "environment." There are a variety of factors that can lead to increased risk of developing cancers. These include aging, family history, smoking, consumption of alcohol, organic and inorganic chemicals, ionizing radiations, sunlight, bacteria and viruses, and air and water pollution. These agents are further classified into physical, chemical, and biological agents based on the source of origin, as shown in Fig. 1. We discuss some of the factors that are present inside and outside the body in the sections that follow.

3.1 Agents inside the body

There are some agents responsible for causing cancer that can be found inside the body. Abnormal levels of hormones in the blood stream, gene alterations, or a weak immune system may lead to cancer [7]. People may differ in their ability to eliminate cancer cells or cancer-causing agents in their body due to exposure or damage to DNA [8]. These alterations may pass through individuals in families or be related to diet or exposure to carcinogens [9].

3.2 Agents outside the body

Various natural and anthropogenic substances present in the environment are responsible for two thirds of cancers [10]. These may be related to personal choices such as lifestyle activities like drinking alcohol, smoking, eating an unhealthy diet, not exercising, and so on [11]. Unhealthy diet, being overweight, and lack of exercise are major factors that contribute to breast as well as prostate cancers. Exposure to asbestos causes lung cancer, while exposure to benzidine causes bladder cancer [12]. Smoking is directly associated with oral, lung, kidney, and esophageal cancers, and indirectly associated with stomach, breast, cervix, pancreas, and liver cancers [13]. Environmental factors that can cause cancer can be categorized into physical, chemical, and biological agents [14].

3.3 Physical agents

3.3.1 Ultraviolet radiation

Melanoma arises due to exposure to UV radiation from the sun, tanning beds, and sun lamps, all of which can result in DNA damage as well as advanced aging of the skin [15].

3.3.2 Radon exposure

Radon is an inert radioactive gas released mainly by the decay of uranium as well as natural deposits underground. Traces of radon are also produced from lead with associated airborne particles [16]. It can easily dissipate into water in contact with the air and increase the risk of lung cancer [17].

3.3.3 Smoke

Smoke emission from vehicles and industries contributes to colon, lung, [18] breast, [19] and ovarian cancers [20].

Factors responsible for causing cancers	Name of carcinogen	Originating source	Affected sites	References
Physical agents	Radon exposure	Mine decay	Lungs	[1]
	Smoke	Emissions from vehicles, cigarette smoking, and air pollution	Colon and lungs	[21]
	Soot	Chimneys	Skin	[22]
	Gasoline	Petroleum products and oil	Blood and lungs	[23]
	Ionizing radiation	Exposure to X-rays	Bone marrow	[24]

3.4 Chemical agents

3.4.1 Metals

Water contaminated with arsenic is reported to be associated with liver, lung, bladder, kidney, and skin cancers. However, lung cancer is mainly noticed in individuals working with industries producing beryllium, cadmium compounds, and cadmium metals [25], particularly the defense, nuclear, and aerospace industries. Chromium is chiefly used in steel industries and is related to lung cancer in industry employees. Lead phosphate and acetates, chief components of cotton dyes and metal coatings, are known to cause kidney and brain cancers [26]. Nasal and lung cancers are also linked to nickel exposure, which is used in batteries, glazes, dental fillings, and steel manufacturing [27]. Diesel exhausts from rail and road works, garage works, car mechanics, and mines are also potential carcinogens and increase the risk of lung cancer [28].

Factors responsible for causing cancers	Name of carcinogen	Originating source	Affected sites
Chemical agents	Asbestos	Tiles (floor and roof)	Lungs (mesothelioma)
	Dioxins		
	Arsenic	Smelting and electrical activities, medication	Skin and lungs
	Benzene	Detergents, paints, rubber, and petroleum	Lymph nodes and blood
	Beryllium	Nuclear reactors, missile fuels	Lungs
	Cadmium	Coatings, paints, batteries	Prostate
	Chromium	Paints, preservatives, and pigments	Lungs
	Ethylene dioxide	Ripening agents, gases	Blood stream
	Nickel	Ferrous alloys, ceramics, and batteries	Lungs and nose
	Vinyl chloride	Glue and refrigerants	Liver
	Formaldehyde	Laboratories and hospitals	Pharynx and nose
	Hair dyes	Barbers and hair dressers	Bladder

3.4.2 Pesticides

Common pesticides like DDT, hexachlorobenzene, lead acetate, amitrole, toxaphene, and ethylene oxide are banned because they are carcinogenic. Studies over decades have demonstrated that farmers, manufacturers, applicators, and crop duster pilots have high risk of lymphatic, blood, stomach, brain, prostate, and melanoma cancers due to pesticide exposure [29].

3.4.3 Dioxins

Dioxins are byproducts of chemical processing, which deals with hydrocarbons and chlorine. They are released via pulp and paper bleaching as well as hospital and municipal waste, and are known to cause cancer [30].

3.4.4 Polycyclic aromatic hydrocarbons

The burning of carbon-containing materials like wood and fuel, as well as vehicle emissions and cigarette smoke, releases polyaromatic hydrocarbons (PAHs) into the air. Exposure to PAHs causes skin, lung, and urinary cancers [31].

3.4.5 Medical drugs

Chlorambucil, cyclophosphamide, and melphalan are cancer-treating drugs that are related to risk of secondary cancers like leukemia and some others [32]. Azathioprine and cyclosporine, which are used as immunosuppressants during organ transplantation, are also associated with risk of lymphoma [33]. These lifesaving drugs have a greater impact on the risk of additional cancers after years as determined by the Food and Drug Administration (FDA).

3.5 Biological agents

3.5.1 Viruses and bacteria

Bacteria and viruses are linked to various types of cancers. For example, human papilloma virus (HPV) is a sexually transmitted virus that is a primary cause of anal and cervical cancer [34]. Although very common, HPV is found in greater incidence among women who had their first sexual intercourse at an early age or who have had more than one sexual partner [35].

3.5.2 Toxins from fungi

Certain fungi that grow on common food products like peanuts and grains produce cancer causing-substances called aflatoxins [36]. Food contaminated by aflatoxin exposure can be considered as sources of exposure.

Factors responsible for causing cancers	Name of carcinogen	Originating source	Affected sites	References
Biological agents	Hepatic virus (B, C)	Drugs consumption, hospital workers	Liver	[1, 37]
	Burkitt's lymphoma	South African people	Lymph nodes	[38]
	Helicobacter pylori	Severe bacterial infection	Stomach	[39]
	HPV	More than one sexual partner	Skin, cervix, and head/neck	[40]

4. Other factors responsible for causing cancer

4.1 Tobacco

Tobacco-containing products such as cigars, cigarettes, and snuff, as well as exposure to tobacco smoke, account for up to one third of cancer deaths every year in the United States. Different tobacco products are linked to lung, kidney, mouth, colon, throat, esophageal, stomach, and lip cancers [41].

4.2 Diet/weight/physical inactivity

Obesity and being overweight are considered major causes of cancer [42]. People with lesser physical activity are prone to developing breast and colon cancer, while older, obese women in particular are prone to endometrial, esophageal, kidney, and colon cancers [43].

4.3 Alcoholic drinks

Alcohol consumption is directly associated with increased cancer risk of esophagus, throat, mouth, and voice box cancers in people who drink very often (usually two drinks per day) compared to people who only smoke [44].

Out of all the physical, chemical, and biological agents, persistent organic pollutants are found to be prominent in causing various types of cancers, especially in women [45].

5. Female-specific cancers

In normal conditions, the human body needs a certain number of cells. Old cells disintegrate and new cells regenerate in an orderly process. The process of cancer starts inside a cell [46]. When signs of cancer develop, new cells stop generating and existing cells start dividing, even when new cells are not needed. This abnormal growth results in tumors. With time, these tumor cells interfere with the process of normal and healthy living cells and invade them. Tumor development and metastasis takes several years and may not show symptoms initially [47]. Tumors can be classified into two types: benign and malignant. Benign tumors are site-specific and they do not spread easily to the other parts of body, whereas malignant tumors can escape from their places of origin and enter into the lymphatic system or blood flow and spread to other parts of the body. Over time, some benign tumors may become malignant.

Cancers are named after the specific type of organ or cell in which they grow [48]. Some of the common cancers include colon, breast, lung, and stomach cancers. Cancer that occurs in the skin cells, tissues, and eyes is termed melanoma; melanocytes are responsible for skin pigment. Lymphomas are cancer cells that develop in the lymphatic system and cause leukemia (blood cell cancer). Cancers that develop in the epithelial tissues of the liver, lung, breast, or skin are called carcinomas. Sarcoma is cancer that arises from bone cells, fat, muscles, connective tissues, and cartilage.

There are a few cancers that are usually found only in women, including breast, endometrial, ovarian, and cervical cancers [49].

5.1 Breast cancer

Breast cancer is one of the most common and primarily observed cancers in women. Its worldwide incidence is increasing at an alarming rate [50]. It might begin at any age, but elderly people are more prone to breast cancer. Due to certain factors, some women may have a greater chance of breast cancer when compared with others. Still, every woman should know about the risks for breast cancer and what can be done to help lower their risk. According to several research studies, certain chemicals present in the environment play an important role in causing breast cancer. Thus, study of chemicals in causing breast cancer is a major area for research [51]. Genotoxic agents are more likely to cause damage to the genetic material of a cell, which leads to mutations that cause cancer [52]. Ionizing and medical radiation exposure at a very young age can significantly increase the risk of breast cancer [53].

5.2 Endometrial cancer

Endometrial cancer is a cancer that affects the inner lining of the uterus called the endometrium. Increase in the number of endometrial cancers is directly proportional to increase in age of women. Treatment with tamoxifen may lower the risk of breast cancer, but it can rapidly affect the distribution of hormone levels [54]. Early onset of menstrual cycle, late menopause, family history of infertility history, or parity and nulliparity can increase the risk of cancer [55]. Women with a personal or family history of hereditary nonpolyposis colorectal cancer, polycystic ovary syndrome, or obesity are also at risk for endometrial cancer [56]. Prior history of breast or ovarian cancer may also increase risk of endometrial cancer [57].

5.3 Cervical cancer

Chronic infection with HPV is identified as one of most critical risk factors for cervical cancer [58]. This virus transmits through skin-to-skin contact via vaginal, anal, or oral sex [59]. Other risk factors for cervical cancer include smoking, being immunocompromised due to infections, obesity, and environmental pollutants [60].

6. Exposure to POPs and impact on women's health

Persistent organic pollutants (POPs) are organic compounds that contain a number of artificially produced chemical compounds like polychlorinated biphenyls (PCBs), dioxins and organochlorine (OC) pesticides, perfluorinated compounds (PFCs), and polyaromatic hydrocarbons (PAHs) that remain in the environment due to their resistance to photolytic and biochemical processes [61]. They accumulate in the adipose tissues of living organisms and ultimately transfer to the higher levels of the food chain [62]. Technological advancements in industrial activities have resulted in the release of different chemicals into the environment on a daily basis. These chemicals and pollutants pose threats to both humans and the ecosystem. Epidemiological studies over decades have reported that accidental or occupational exposures to metals as well as organic compounds lead to cancers [63].

7. Mechanism of action of POPs

The cellular and molecular pathways of POP exposure are linked with multiple receptor pathways. The toxic effects of POPs in different organs reflect the diverse mechanistic facets that influence disease outcome. However, activation of the aryl hydrocarbon receptor (AhR) pathway is the basic mechanism for mediating undesirable effects of POPs [64]. POPs mimic the activity of human endocrine hormones and disrupt endocrine homeostasis [65]. Dichlorodiphenyltrichloroethane (DDT), dichlorodiphenyldichloroethylene (DDE), PCBs, and dioxins are major endocrine-disrupting chemicals (EDCs) [66].

Exposure to POPs involves chemicals that may interact with more than one biological target, resulting in susceptibility to cancer as well as endocrine and cardio-metabolic disorders [67]. In addition, women, infants, children, and the elderly are more vulnerable to the effects of POPs. For instance, thyroid disorders are more recognized in women, and dementia typically affects older or elderly women [68].

8. Effects of POPs on cancer

Various researches and studies on exposure of POPs in women focus on breast cancer. Women with breast cancer were found to have high levels of PCBs. There is a significant association between breast cancers, perfluoroalkyl acids, and PCBs [69].

A study of Indian women reported that there is an interdependency between breast cancer and organochlorine pesticides (OCPs) that are obtained through consumption of contaminated food. Also, higher levels of pesticides were found in young women as compared to older women [70]. Clearance of POPs only occurs through child birth and lactation [71]. Similar study on Alaska native women with estrogen receptor (ER) or progesterone receptor (PR) tumor types found that the women had high concentrations of pesticides in their bodies [72]. A study from Spain that examined POPs in serum as well as adipose tissue of breast cancer verified that certain POPs may contribute to breast cancer aggressiveness [73].

A study of Japanese women reported an increase in the serum levels of OCPs and a few PCBs were found to be interlinked with hypomethylation of leukocyte DNA. Another study of the Inuit population of Greenland showed that high serum levels of POPs were inversely interdependent, indicating that POPs can mediate epigeneticsm, which may reflect carcinogenesis [74].

One research group identified that residential areas close to waste sites of hazardous chemicals that include POPs and VOCs were significantly related to breast cancer and hospitalization of women [75]. Consumption of fishes in the Great Lakes area enriches the concentration of serum total PCB and is positively linked to the occurrence of fibroids in women [76]. Uterine fibroids, also known as leiomyomas of the uterus, are benign tumors that develop from the smooth muscular tissue of the uterus. Women diagnosed with fibroids have shown higher mean concentrations of PCBs, POPs, and their metabolites in subcutaneous fat. Premenopausal women undergoing hysterectomy for fibroids also showed increased concentrations in the endometrium [76, 77].

9. Conclusion

Research has reported a positive relationship between OCPs and the occurrence of female-specific cancers including cervical, ovarian, endometrial, and breast cancers. Women suffering from uterine cancer have been recognized to have greater levels of OCPs like dieldrin in their blood.

Acknowledgment

The authors thank the GITAM for providing infrastructure facilities.

Conflict of interest

The authors declared that there is no conflict of interest.

Funding details

One of the authors, RamaRao Malla thank the DST-EMR (EMR/2016/002694), New Delhi, India for providing financial support.

References

[1] Parsa N. Environmental factors inducing human cancers. Iran J Public Health 2012;41(11):1–9.

[2] Soto AM, Sonnenschein C. Environmental causes of cancer: endocrine disruptors as carcinogens. Nat Rev Endocrinol 2010;6(7):363–70.

[3] (a) Baccarelli A, Bollati V. Epigenetics and environmental chemicals. Curr Opin Pediatr 2009;21 (2):243–51. (b) Ohshima H, Tatemichi M, Sawa T. Chemical basis of inflammation-induced carcinogenesis. Arch Biochem Biophys 2003;417(1):3–11.

[4] Valavanidis A. Environmental carcinogenic substances, exposure and risk assessment for carcinogenci potential. Classifications and Regulations by International and National Institutions. Scientific reviews, Athens, Greece: Department of Chemistry, National and Kapodistrian University of Athens, University Campus Zografou; 2017.

[5] Tomatis L. Cancer: causes occurrence and control. IARC Scientific Publications; 1990.

[6] Sonnenschein C, Soto AM. Theories of carcinogenesis: an emerging perspective. Semin Cancer Biol 2008;18(5):372–7.

[7] Bardeesy N, DePinho RA. Pancreatic cancer biology and genetics. Nat Rev Cancer 2002;2(12): 897–909.

[8] Ladiges W. Mouse models of XRCC1 DNA repair polymorphisms and cancer. Oncogene 2006; 25(11):1612–9.

[9] Salnikow K, Zhitkovich A. Genetic and epigenetic mechanisms in metal carcinogenesis and cocarcinogenesis: nickel, arsenic, and chromium. Chem Res Toxicol 2008;21(1):28–44.

[10] Wild CP. Environmental exposure measurement in cancer epidemiology. Mutagenesis 2009;24(2): 117–25.

[11] Weiderpass E. Lifestyle and cancer risk. J Prev Med Public Health 2010;43(6):459–71.

[12] Calle EE, Kaaks R. Overweight, obesity and cancer: epidemiological evidence and proposed mechanisms. Nat Rev Cancer 2004;4(8):579–91.

[13] Boyle P. Cancer, cigarette smoking and premature death in Europe: a review including the recommendations of European cancer experts consensus meeting, Helsinki, October 1996. Lung Cancer 1997; 17(1):1–60.

[14] Trosko JE, Chang C-C. Environmental carcinogenesis: an integrative model. Q Rev Biol 1978; 53(2):115–41.

[15] (a) Mead MN. Benefits of sunlight: a bright spot for human health. National Institute of Environmental Health Sciences; 2008. (b) Gasparro FP. Sunscreens, skin photobiology, and skin cancer: the need for UVA protection and evaluation of efficacy. Environ Health Perspect 2000;108 (Suppl 1):71–8.

[16] Varshney R. Natural radioactivity and Radon/Thoron measurement in environment & quantification of heavy elements. Aligarh Muslim University; 2013.

[17] Darby S, Hill D, Auvinen A, Barros-Dios J, Baysson H, Bochicchio F, Deo H, Falk R, Forastiere F, Hakama M. Radon in homes and risk of lung cancer: collaborative analysis of individual data from 13 European case-control studies. BMJ 2005;330(7485):223.

[18] Kloog I, Haim A, Stevens RG, Portnov BA. Global co-distribution of light at night (LAN) and cancers of prostate, colon, and lung in men. Chronobiol Int 2009;26(1):108–25.

[19] Matsumoto H, Adachi S, Suzuki Y. Bisphenol A in ambient air particulates responsible for the proliferation of MCF-7 human breast cancer cells and its concentration changes over 6 months. Arch Environ Contam Toxicol 2005;48(4):459–66.

[20] Hung L-J, Chan T-F, Wu C-H, Chiu H-F, Yang C-Y. Traffic air pollution and risk of death from ovarian cancer in Taiwan: fine particulate matter (PM2. 5) as a proxy marker. J Toxicol Environ Health A 2012;75(3):174–82.

[21] Møller P, Folkmann JK, Forchhammer L, Bräuner EV, Danielsen PH, Risom L, Loft S. Air pollution, oxidative damage to DNA, and carcinogenesis. Cancer Lett 2008;266(1):84–97.

[22] Hogstedt C, Jansson C, Hugosson M, Tinnerberg H, Gustavsson P. Cancer incidence in a cohort of Swedish chimney sweeps, 1958–2006. Am J Public Health 2013;103(9):1708–14.

[23] Okoro A, Ani E, Ibu J, Akpogomeh B. Effect of petroleum products inhalation on some haematological indices of fuel attendants in Calabar metropolis, Nigeria. Niger J Physiol Sci 2006;21:1–2.

[24] Madani I, De Neve W, Mareel M. Does ionizing radiation stimulate cancer invasion and metastasis? Bull Cancer 2008;95(3):292–300.

[25] (a) Huff J, Lunn RM, Waalkes MP, Tomatis L, Infante PF. Cadmium-induced cancers in animals and in humans. Int J Occup Environ Health 2007;13(2):202–12. (b) Tchounwou PB, Yedjou CG, Patlolla AK, Sutton DJ. Heavy metal toxicity and the environment. In: Molecular, clinical and environmental toxicology. Springer; 2012. p. 133–64.

[26] (a) Alghazal M, Šutiaková I, Kovalkovičová N, Legath J, Falis M, Pistl J, Sabo R, Beňová K, Sabova L, Váczi P. Induction of micronuclei in rat bone marrow after chronic exposure to lead acetate trihydrate. Toxicol Ind Health 2008;24(9):587–93. (b) Naja GM, Volesky B. Metals in the environment: toxicity and sources. In: Wang LK, Chen JP, Hung YT, Shammas NK, editors. Handbook on heavy metals in the environment. Boca Raton, FL: Taylor & Francis and CRC Press; 2009. p. 13–61 [chapter 2].

[27] (a) Martin S, Griswold W. Human health effects of heavy metals. Environ Sci Technol Briefs Citizens 2009;15:1–6. (b) Goyer RA, Clarkson TW. Toxic effects of metals. In: Casarett and Doull's toxicology: the basic science of poisons. 5:Center for Hazardous Substance Research (CHSR), Kansas State University; 1996. p. 696–8.

[28] Boffetta P, Harris RE, Wynder EL. Case-control study on occupational exposure to diesel exhaust and lung cancer risk. Am J Ind Med 1990;17(5):577–91.

[29] (a) Blair A, Zahm SH. Agricultural exposures and cancer. Environ Health Perspect 1995;103(Suppl 8):205–8. (b) Alavanja MC, Hoppin JA, Kamel F. Health effects of chronic pesticide exposure: cancer and neurotoxicity. Annu Rev Public Health 2004;25:155–97.

[30] (a)Kogevinas M. Human health effects of dioxins: cancer, reproductive and endocrine system effects. APMIS 2001;109(S103):S223–32; (b)Cole P, Trichopoulos D, Pastides H, Starr T, Mandel JS. Dioxin and cancer: a critical review. Regul Toxicol Pharmacol 2003;38(3):378–88.

[31] (a) Boffetta P, Jourenkova N, Gustavsson P. Cancer risk from occupational and environmental exposure to polycyclic aromatic hydrocarbons. Cancer Causes Control 1997;8(3):444–72. (b) Mastrangelo G, Fadda E, Marzia V. Polycyclic aromatic hydrocarbons and cancer in man. Environ Health Perspect 1996;104(11):1166–70.

[32] Cuzick J, Erskine S, Edelman D, Galton D. A comparison of the incidence of the myelodysplastic syndrome and acute myeloid leukaemia following melphalan and cyclophosphamide treatment for myelomatosis. Br J Cancer 1987;55(5):523–9.

[33] Opelz G, Döhler B. Lymphomas after solid organ transplantation: a collaborative transplant study report. Am J Transplant 2004;4(2):222–30.

[34] (a) Burd EM. Human papillomavirus and cervical cancer. Clin Microbiol Rev 2003;16(1):1–17. (b) Krzowska-Firych J, Lucas G, Lucas C, Lucas N, Pietrzyk Ł. An overview of human papillomavirus (HPV) as an etiological factor of the anal cancer. J Infect Public Health 2019;12(1):1–6.

[35] Castellsagué X. Natural history and epidemiology of HPV infection and cervical cancer. Gynecol Oncol 2008;110(3 Suppl 2):S4–7.

[36] (a) Wogan GN. Aflatoxins as risk factors for hepatocellular carcinoma in humans. Cancer Res 1992;52 (7 Suppl):2114s–2118s. (b) Dvorackova I. Aflatoxin inhalation and alveolar cell carcinoma. Br Med J 1976;1(6011):691. (c) Pitt J. Toxigenic fungi and mycotoxins. Br Med Bull 2000;56(1):184–92.

[37] (a) Buendia M. Mammalian hepatitis B viruses and primary liver cancer. Semin Cancer Biol 1992; 3(5):309–20. (b) Perz JF, Armstrong GL, Farrington LA, Hutin YJ, Bell BP. The contributions of hepatitis B virus and hepatitis C virus infections to cirrhosis and primary liver cancer worldwide. J Hepatol 2006;45(4):529–38.

[38] Bornkamm GW, Hausen HZ, Stein H, Lennert K, Rüggeberg F, Bartels H. Attempts to demonstrate virus-specific sequences in human tumors. IV. EB viral DNA in European Burkitt lymphoma and immunoblastic lymphadenopathy with excessive plasmacytosis. Int J Cancer 1976;17(2):177–81.

[39] Song ZQ, Zhou LY. Helicobacter pylori and gastric cancer: clinical aspects. Chin Med J (Engl) 2015;128(22):3101–5.

[40] (a) Rettig EM, D'Souza G. Epidemiology of head and neck cancer. Surg Oncol Clin N Am 2015; 24(3):379–96. (b) Yete S, D'Souza W, Saranath D. High-risk human papillomavirus in oral cancer: clinical implications. Oncology 2018;94(3):133–41.

[41] Shiels MS, Gibson T, Sampson J, Albanes D, Andreotti G, Beane Freeman L, Berrington de Gonzalez A, Caporaso N, Curtis RE, Elena J, Freedman ND, Robien K, Black A, Morton LM. Cigarette smoking prior to first cancer and risk of second smoking-associated cancers among survivors of bladder, kidney, head and neck, and stage I lung cancers. J Clin Oncol Off J Am Soc Clin Oncol 2014;32(35):3989–95.

[42] Bianchini F, Kaaks R, Vainio H. Overweight, obesity, and cancer risk. Lancet Oncol 2002;3(9): 565–74.

[43] Calle EE, Thun MJ. Obesity and cancer. Oncogene 2004;23(38):6365–78.

[44] (a) Pelucchi C, Gallus S, Garavello W, Bosetti C, La Vecchia C. Cancer risk associated with alcohol and tobacco use: focus on upper aero-digestive tract and liver. Alcohol Res Health 2006;29(3):193–8. (b) Elwood JM, Pearson JC, Skippen DH, Jackson SM. Alcohol, smoking, social and occupational factors in the aetiology of cancer of the oral cavity, pharynx and larynx. Int J Cancer 1984;34(5):603–12.

[45] Wikoff D, Fitzgerald L, Birnbaum L. Persistent organic pollutants: an overview. In: Dioxins and health, vol. 3; John Wiley & Sons, Inc.; 2012. p. 1–36

[46] Ewing J. Neoplastic diseases, a text-book on tumors: James Ewing … with 479 illustrations. WB Saunders Company; 1919.

[47] Nguyen DX, Bos PD, Massagué J. Metastasis: from dissemination to organ-specific colonization. Nat Rev Cancer 2009;9(4):274–84.

[48] Lengauer C, Kinzler KW, Vogelstein B. Genetic instabilities in human cancers. Nature 1998; 396(6712):643–9.

[49] (a) Weiderpass E, Labrèche F. Malignant tumors of the female reproductive system. In: Occupational cancers. Springer; 2014. p. 409–22. (b) Pike MC, Pearce CL, Wu AH. Prevention of cancers of the breast, endometrium and ovary. Oncogene 2004;23(38):6379–91.

[50] Forouzanfar MH, Foreman KJ, Delossantos AM, Lozano R, Lopez AD, Murray CJ, Naghavi M. Breast and cervical cancer in 187 countries between 1980 and 2010: a systematic analysis. Lancet 2011; 378(9801):1461–84.

[51] Cancer IB, Committee ERC. Breast cancer and the environment: prioritizing prevention. National Institute of Environmental Health Sciences; 2013.

[52] Lee SJ, Yum YN, Kim SC, Kim Y, Lim J, Lee WJ, Koo KH, Kim JH, Kim JE, Lee WS, Sohn S, Park SN, Park JH, Lee J, Kwon SW. Distinguishing between genotoxic and non-genotoxic hepato-carcinogens by gene expression profiling and bioinformatic pathway analysis. Sci Rep 2013;3:2783.

[53] (a) Land CE, Tokunaga M, Koyama K, Soda M, Preston DL, Nishimori I, Tokuoka S. Incidence of female breast cancer among atomic bomb survivors, Hiroshima and Nagasaki, 1950–1990. Radiat Res 2003;160(6):707–17. (b) Henderson TO, Amsterdam A, Bhatia S, Hudson MM, Meadows AT, Neglia JP, Diller LR, Constine LS, Smith RA, Mahoney MC, Morris EA, Montgomery LL, Landier W, Smith SM, Robison LL, Oeffinger KC. Systematic review: surveillance for breast cancer in women treated with chest radiation for childhood, adolescent, or young adult cancer. Ann Intern Med 2010;152(7):444–55 w144-54.

[54] van Leeuwen FE, Van den Belt-Dusebout A, Benraadt J, Diepenhorst F, Van Tinteren H, Coebergh J, Kiemeney L, Gimbrère C, Otter R, Schouten L. Risk of endometrial cancer after tamoxifen treatment of breast cancer. Lancet 1994;343(8895):448–52.

[55] (a) Ali AT. Reproductive factors and the risk of endometrial cancer. Int J Gynecol Cancer 2014; 24(3):384–93. (b) Modan B, Ron E, Lerner-Geva L, Blumstein T, Menczer J, Rabinovici J, Oelsner G, Freedman L, Mashiach S, Lunenfeld B. Cancer incidence in a cohort of infertile woman. Am J Epidemiol 1998;147(11):1038–42.

[56] (a) Bharati R, Jenkins MA, Lindor NM, Le Marchand L, Gallinger S, Haile RW, Newcomb PA, Hopper JL, Win AK. Does risk of endometrial cancer for women without a germline mutation in a DNA mismatch repair gene depend on family history of endometrial cancer or colorectal cancer? Gynecol Oncol 2014;133(2):287–92. (b) Schmeler KM, Soliman PT, Sun CC, Slomovitz BM, Gershenson DM, Lu KH. Endometrial cancer in young, normal-weight women. Gynecol Oncol 2005;99(2):388–92.

[57] Tulinius H, Egilsson V, Olafsdottir GH, Sigvaldason H. Risk of prostate, ovarian, and endometrial cancer among relatives of women with breast cancer. Br Med J 1992;305(6858):855–7.

[58] Bosch F, Munoz N, De Sanjosé S. Human papillomavirus and other risk factors for cervical cancer. Biomed Pharmacother 1997;51(6–7):268–75.

[59] Dietz CA, Nyberg CR. Genital, oral, and anal human papillomavirus infection in men who have sex with men. J Am Osteopath Assoc 2011;111(3_Suppl_2):S19–25.

[60] (a) Baay M, Verhoeven V, Avonts D, Vermorken J. Risk factors for cervical cancer development: what do women think? Sex Health 2004;1(3):145–9. (b) Wang LD-L, Lam WWT, Wu J, Fielding R. Hong Kong Chinese women's lay beliefs about cervical cancer causation and prevention. Asian Pac J Cancer Prev 2014;15(18):7679–86. (c) Scheurer ME, Danysh HE, Follen M, Lupo PJ. Association of traffic-related hazardous air pollutants and cervical dysplasia in an urban multiethnic population: a cross-sectional study. Environ Health 2014;13(1):52.

[61] (a) Ritter L, Solomon K, Forget J, Stemeroff M, O'Leary C. Persistent organic pollutants: an assessment report on: DDT, aldrin, dieldrin, endrin, chlordane, heptachlor, hexachlorobenzene, mirex, toxa-phene, polychlorinated biphenyls, dioxins and furans. International Programme on Chemical Safety (IPCS); 1995. (b) Ritter L, Solomon K, Forget J, Stemeroff M, O'Leary C. In: Persistent organic pol-lutants. An assessment report on: DDT–aldrin–dieldrin–chlordane–heptachlor–hexachlorobenzene–mirex–toxaphene, polychlorinated biphenyls, dioxins and furans. Second Meeting of ISG, Canberra, Australia; 1996. p. 5–8. (c) El-Shahawi MS, Hamza A, Bashammakh AS, Al-Saggaf WT. An overview on the accumulation, distribution, transformations, toxicity and analytical methods for the monitoring of persistent organic pollutants. Talanta 2010;80(5):1587–97.

[62] Vallack HW, Bakker DJ, Brandt I, Broström-Lundén E, Brouwer A, Bull KR, Gough C, Guardans R, Holoubek I, Jansson B, Koch R, Kuylenstierna J, Lecloux A, Mackay D, McCutcheon P, Mocarelli P, Taalman RD. Controlling persistent organic pollutants—what next? Environ Toxicol Pharmacol 1998;6(3):143–75.

[63] Chow W-H, Dong LM, Devesa SS. Epidemiology and risk factors for kidney cancer. Nat Rev Urol 2010;7(5):245.

[64] Sorg O. AhR signalling and dioxin toxicity. Toxicol Lett 2014;230(2):225–33.

[65] (a) Geyer HJ, Rimkus GG, Scheunert I, Kaune A, Schramm K-W, Kettrup A, Zeeman M, Muir DC, Hansen LG, Mackay D. Bioaccumulation and occurrence of endocrine-disrupting chemicals (EDCs), persistent organic pollutants (POPs), and other organic compounds in fish and other organisms including humans. In: Bioaccumulation—new aspects and developments. Springer; 2000. p. 1–166. (b) De Coster S, Van Larebeke N. Endocrine-disrupting chemicals: associated disorders and mechanisms of action. J Environ Public Health 2012;2012:713696. https://doi.org/10.1155/2012/713696.

[66] (a) Aoki Y. Polychlorinated biphenyls, polychlorinated dibenzo-p-dioxins, and polychlorinated dibenzofurans as endocrine disrupters—what we have learned from Yusho disease. Environ Res 2001; 86(1):2–11. (b) De Coster S, van Larebeke N. Endocrine-disrupting chemicals: associated disorders and mechanisms of action. J Environ Public Health 2012;2012:713696. (c) Boverhof DR, Kwekel JC, Humes DG, Burgoon LD, Zacharewski TR. Dioxin induces an estrogen-like, estrogen receptor-dependent gene expression response in the murine uterus. Mol Pharmacol 2006;69(5): 1599–606.

[67] Khalil N, Chen A, Lee M. Endocrine disruptive compounds and cardio-metabolic risk factors in children. Curr Opin Pharmacol 2014;19:120–4.

[68] (a) Fantini F, Porta D, Fano V, De Felip E, Senofonte O, Abballe A, D'Ilio S, Ingelido AM, Mataloni F, Narduzzi S, Blasetti F, Forastiere F. Epidemiologic studies on the health status of the population living in the Sacco River Valley. Epidemiol Prev 2012;36(5 Suppl 4):44–52. (b) Lee DH, Lind PM, Jacobs Jr. DR, Salihovic S, van Bavel B, Lind L. Association between background exposure to organochlorine pesticides and the risk of cognitive impairment: a prospective study that accounts for weight change. Environ Int 2016;89–90:179–84.

[69] Wielsøe M, Kern P, Bonefeld-Jørgensen EC. Serum levels of environmental pollutants is a risk factor for breast cancer in Inuit: a case control study. Environ Health 2017;16(1):56.

[70] Mathur V, Bhatnagar P, Sharma RG, Acharya V, Sexana R. Breast cancer incidence and exposure to pesticides among women originating from Jaipur. Environ Int 2002;28(5):331–6.

[71] Fernández-Rodríguez M, Arrebola JP, Artacho-Cordón F, Amaya E, Aragones N, Llorca J, Perez-Gomez B, Ardanaz E, Kogevinas M, Castano-Vinyals G, Pollan M, Olea N. Levels and predictors of persistent organic pollutants in an adult population from four Spanish regions. Sci Total Environ 2015;538:152–61.

[72] (a) Holmes AK, Koller KR, Kieszak SM, Sjodin A, Calafat AM, Sacco FD, Varner DW, Lanier AP, Rubin CH. Case-control study of breast cancer and exposure to synthetic environmental chemicals among Alaska Native women. Int J Circumpolar Health 2014;73:25760. (b) Holmes AK, Koller KR, Kieszak SM, Sjodin A, Calafat AM, Sacco FD, Varner DW, Lanier AP, Rubin CH. Case-control study of breast cancer and exposure to synthetic environmental chemicals among Alaska Native women. Int J Circumpolar Health 2014;73(1):25760.

[73] Arrebola JP, Fernández-Rodríguez M, Artacho-Cordón F, Garde C, Perez-Carrascosa F, Linares I, Tovar I, González-Alzaga B, Expósito J, Torne P, Fernández MF, Olea N. Associations of persistent organic pollutants in serum and adipose tissue with breast cancer prognostic markers. Sci Total Environ 2016;566–567:41–9.

[74] (a) Itoh H, Iwasaki M, Kasuga Y, Yokoyama S, Onuma H, Nishimura H, Kusama R, Yoshida T, Yokoyama K, Tsugane S. Association between serum organochlorines and global methylation level of leukocyte DNA among Japanese women: a cross-sectional study. Sci Total Environ 2014;490:603–9. (b) Rusiecki JA, Baccarelli A, Bollati V, Tarantini L, Moore LE, Bonefeld-Jorgensen EC. Global DNA hypomethylation is associated with high serum-persistent organic pollutants in Greenlandic Inuit. Environ Health Perspect 2008;116(11):1547–52.

[75] Lu X, Lessner L, Carpenter DO. Association between hospital discharge rate for female breast cancer and residence in a zip code containing hazardous waste sites. Environ Res 2014;134:375–81.

[76] Lambertino A, Turyk M, Anderson H, Freels S, Persky V. Uterine leiomyomata in a cohort of Great Lakes sport fish consumers. Environ Res 2011;111(4):565–72.

[77] Schaefer WR, Hermann T, Meinhold-Heerlein I, Deppert WR, Zahradnik HP. Exposure of human endometrium to environmental estrogens, antiandrogens, and organochlorine compounds. Fertil Steril 2000;74(3):558–63.

CHAPTER 16

CYP1B1 rs1056836 polymorphism and endometrial cancer risk: A meta-analysis

Samrat Rakshit and L.V.K.S. Bhaskar
Department of Zoology, Guru Ghasidas Vishwavidyalaya, Bilaspur, Chhattisgarh, India

Abstract

Endometrial cancer (EMC) is the most common gynecological cancer, accounting for more than 2% of cancer deaths in women worldwide. The present meta-analysis deals with the association between the CYP1B1 rs1056836 variant and risk of EMC. For the collection of data, we rigorously searched Google Scholar, PubMed, and Embase for relevant published articles. The association between CYP1B1 rs1056836 and EMC was assessed by calculating odds ratios (ORs) with 95% confidence intervals (CIs). Values of I^2 statistics were calculated with the help of the Cochrane Q test to determine heterogeneity of the study. To understand between-study heterogeneity, subgroup analysis and sensitivity analysis were performed. Funnel plots as well as Egger's tests were performed to determine publication bias. A total of 4804 cases of EMC and 7185 control patients involving the CYP1B1 rs1056836 polymorphism and EMC risk were investigated from 17 independent studies. Pooled analysis in a dominant genetic model showed that there is significant association between the CYP1B1 rs1056836 polymorphism and development of EMC (OR = 1.31; 95% CI 1.08–1.59; $P = 0.005$). There is significant heterogeneity between ethnic groups with no publication bias observed. In summary, this meta-analysis revealed that rs1056836 is a major risk factor for developing EMC. Further research is still needed to investigate the clinical and biological implications of these associations.

Keywords: Endometrial cancer (EMC), Single nucleotide polymorphisms (SNPs), Meta-analysis, Polymorphism, Confidence intervals (CIs), Odds ratios (ORs).

Abbreviations

CI	confidence interval
CYP1B1	cytochrome P-450 1B1
EMC	endometrial cancer
OR	odds ratio
SNPs	single nucleotide polymorphisms

1. Introduction

Endometrial cancer (EMC) is the most common gynecological cancer and accounts for more than 2% of cancer deaths in women worldwide. EMC is a very common female malignancy, second only to breast cancer [1, 2]. In comparison to other parts of the world, North America and parts of Europe have greater numbers of EMC patients, which may be

G.P. Nagaraju, R.R. Malla (eds.)
A Theranostic and Precision Medicine Approach for Female-Specific Cancers
ISBN 978-0-12-822009-2
https://doi.org/10.1016/B978-0-12-822009-2.00016-9

due to greater obesity and metabolic syndromes in these regions [1, 3]. Recent reports suggest that the incidence of EMC and associated mortality rates have increased and are likely to continue to increase during the next 10 years [1]. In 2012 alone, 319,605 new EMC cases were reported worldwide, of which 76,160 resulted in death [2]. Apart from genetic risk factors, some nongenetic risk factors such as obesity, physical inactivity, excess exogenous estrogen, insulin resistance, and tamoxifen use after breast cancer have been associated with an increased risk of EMC [2]. EMC was seen in up to 81% of obese women, 19%–36% being morbidly obese. Apart from current standard treatment for EMC, including removal of the uterus (hysterectomy) and removal of both fallopian tubes and ovaries, other treatments like adjuvant radiotherapy and chemotherapy can also be considered [4].

The CYP1B1 gene, which is generally overexpressed in human malignancies and activates a variety of carcinogens, is located on chromosome 2 p22-p21. This enzyme plays a major role in the metabolism of polycyclic aromatic hydrocarbons, procarcinogens, and certain anticancer agents that lead to activation of carcinogens and finally tumorigenesis [5]. CYP1B1, by hyperactivating protein and altering the tissue response to hormones and anticancer agents, can initiate the progression of cancer. In humans, more than 50 single nucleotide polymorphisms (SNPs) of CYP1B1 have been reported to date [6–8]. CYP1B1 expression is relatively high in the human endometrium, although its role is still unclear. The four most common polymorphisms of CYP1B1 (Arg by Gly at codon 48 (rs10012)—Ala by Ser at codon 119 (rs1056827), Leu by Val at codon 432 (rs1056836), and Asn by Ser at codon 453 (rs1800440)—have been characterized in different cancers including EMC. Catalytic activity of CYP1B1 is influenced by these four polymorphisms [7, 8]. The CYP1B1 gene codes a hydroxylase enzyme cytochrome P450 1B1 that converts estrogens to catecholestrogens or 2-hydroxy estrogens [9]. Polymorphisms at rs1056836 of the gene CYP1B1 confer a greater risk for EMC [10]. Previously, some recent meta-analyses were performed to study the association between CYP1B1 polymorphisms (rs1056836 and many others) and EMC. However, these meta-analyses did not include all published studies, and some original studies with larger sample sizes have been published since then. In order to obtain more conclusive evidence for the association between CYP1B1 Leu432Val and risk of EMC, we performed another meta-analysis.

2. Materials and method

2.1 Selection of data

We performed a thorough search in PubMed, Google Scholar, and EMBASE databases for association studies on EMC and CYP1B1 rs1056836 polymorphisms up to January 24, 2020. Before considering any report for the meta-analysis, references of eligible publications were checked manually. The search was continued until there were no such relevant studies left. Only studies published in English were considered [9–25].

We set the following criteria for eligible studies to conduct the current meta-analysis: (1) case-control studies with CYP1B1 genotypic polymorphism for rs1056836, (2) studies of the association between rs1056836 and EMC, and (3) studies with enough data to calculate odds ratio (OR) with a 95% confidence interval (CI) and P-value. Studies with (1) overlapping data, (2) case-only studies, (3) studies with no CYP1B1 rs1056836 polymorphism, and (4) articles published in languages other than English were excluded. For each study, we collected the lead author's name, publication year, country of origin, and the ethnicities and genotypes of the case and control groups used.

2.2 Statistical analyses

To evaluate the strength of relationship between CYP1B1 rs1056836 and risk of EMC, crude ORs with 95% CIs were deployed using a dominant model. Between-study heterogeneity was assessed with a Q test and I^2 statistic. As there is significant heterogeneity ($I^2 > 77\%$; $P < 0.001$), a random effects model was used to assess the pooled ORs. Publication bias was assessed using Egger's linear regression test and Begg's funnel plot. The leave-one-out method was used to perform sensitivity analysis. A subgroup analysis was conducted by ethnicity. We considered P-values less than 0.05 as statistically significant. Finally, we used the MetaGenyo web tool to perform analyses of the data [26].

3. Results

As depicted in Fig. 1, based on search strategy, 57 potentially relevant publications were identified through rigorous searching in online databases. We found and removed 14 duplicate studies. When the articles were checked for authenticity, one article was

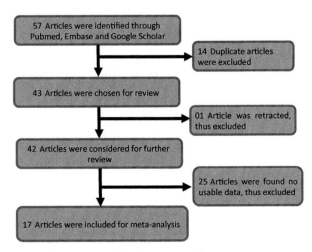

Fig. 1 Flow chart of study characteristics for the meta-analysis.

found to be retracted and thus we excluded it. After a close look at the titles and abstracts, 25 articles were found to be associated with either other SNPs or had no usable data and thus they were also excluded. Finally, 17 studies were found to meet the selection criteria and were included for meta-analysis. The current meta-analysis includes CYP1B1 rs1056836 polymorphism involving 4804 cases and 7185 controls. Genotype frequencies of CYP1B1 rs1056836 polymorphism from all studies are presented in Table 1. Hardy-Weinberg equilibrium deviation was found in some studies.

Between-study heterogeneity obtained by employing Q statistics indicated substantial between-study heterogeneity ($I^2 > 77\%$; $P < 0.001$). Fig. 2 represents a forest plot of each individual study as well as pooled studies ($n = 17$) involving association of rs1056836 and

Table 1 The distribution of rs1056836 SNP genotypes in EMC and control subjects.

References	Country	Ethnicity	EMC			Control			HW P-value
			CC	CG	GG	CC	CG	GG	
Sasaki et al. [11]	China	Asian	59	39	17	69	24	7	0.028
McGrath et al. [12]	America	Caucasian	61	113	45	193	316	146	0.441
Zimarina et al. [13]	Russia/ Norway	Caucasian	25	62	34	37	73	22	0.166
Rylander-Rudqvist et al. [14]	Sweden	Caucasian	195	336	134	425	676	279	0.733
Doherty et al. [15]	America	Caucasian	115	170	86	145	194	81	0.266
Rebbeck et al. [16]	America	Caucasian	119	371[a]		376	877[a]		NC
Tao et al. [17]	China	Asian	792	232	13	806	206	22	0.044
Cho et al. [18]	Korea	Asian	160	25	3	178	41	2	0.831
Ye et al. [19]	China	Asian	71	29	0	70	40	0	0.020
Hirata et al. [20]	America	Caucasian	53	64	33	55	72	38	0.130
Ashton et al. [21]	Australia	Asian	32	88	71	50	139	101	0.854
Sliwinski et al. [10]	Poland	Caucasian	29	37	34	33	31	36	< 0.001
Rebbeck et al. [22]	America	Caucasian	77	107[a]		154	37[a]		NC
El-Shennawy et al. [23]	Egypt	Caucasian	96	59	5	74	19	7	0.002
Lundin et al. [9]	Sweden/ Italy/ America	Caucasian	126	189	76	221	352	131	0.659
Li et al. [24]	China	Asian	165	67	18	188	54	8	0.103
Zhou et al. [25]	China	Asian	34	26	12	55	19	6	0.032

NC, not calculated.
[a]CG + GG.

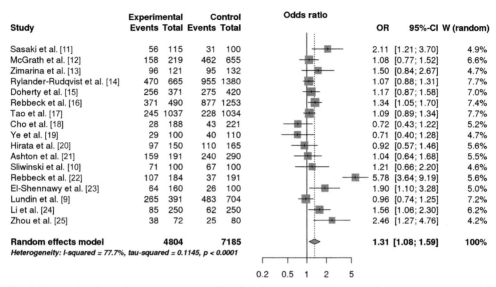

Fig. 2 Forest plot from the meta-analysis of EMC and rs1056836 SNP using dominant genetic model.

EMC. Pooled analysis in the dominant genetic model showed that there is significant association between CYP1B1 rs1056836 polymorphism and development of EMC (OR = 1.31; 95% CI 1.08–1.59; P = 0.005), as shown in Fig. 2. Subgroup analysis showed significant association between rs1056836 polymorphism and risk of EMC in Caucasian populations (OR = 1.38; 95% CI 1.05–1.80; P = 0.019), but not in Asian populations (OR = 1.22; 95% CI 0.91–1.64; P = 0.174). No significant asymmetry in the shape was found in Begg's funnel plot, indicative of no publication bias (Fig. 3). Absence of publication bias was further confirmed by Egger's test (dominant model, P = 0.211). To confirm robustness of our study, we performed sensitivity analysis by omitting each independent study at a time, which did not reveal any substantial difference in the pooled ORs (Fig. 4).

4. Discussion

We analyzed 4804 cases of EMC patients and 7185 controls collected from 17 independent studies that investigated the association between CYP1B1 rs1056836 polymorphism and risk of EMC. A positive association between CYP1B1 rs1056836 and risk of EMC was found in dominant genetic model. Significant between-study heterogeneity was observed. Subgroup analysis showed association between rs1056836 and EMC in Caucasians, but not in Asians. No publication bias was observed. Results of our meta-analysis are consistent with some of the previous meta-analyses that report increased risk of EMC

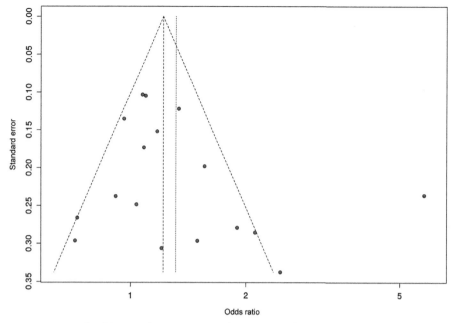

Fig. 3 Assessment of publication bias in meta-analysis using a funnel plot.

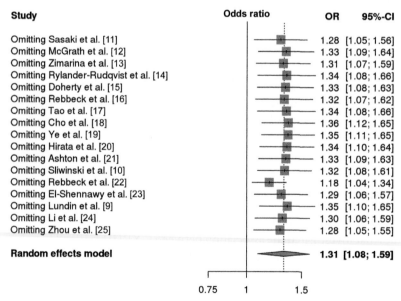

Fig. 4 Sensitivity analysis of this meta-analysis.

with the CYP1B1 rs1056836 polymorphism [6, 27, 28]. In contrast to this, other meta-analyses revealed no association between the CYP1B1 rs1056836 polymorphism and EMC risk [29].

The cytochrome *p450* enzyme system (CYP450) facilitates steroid hormone biosynthesis as well as helps in metabolic activation of carcinogens. CYP1B1, a major enzyme of this system, helps to convert estrogens into 4-hydroxy estrogens [23]. Generally, CYP1B1 is found to be overexpressed in human malignancies activating a variety of carcinogens. CYP1B1 catalyzes both the formation and oxidation of dihydrodiols of specific PAHs to carcinogenic dihydrodiol epoxides. Earlier studies have reported the role of CYP1B1 Leu432Val in risk of all cancers, but the results remain controversial. Zhou et al. reported higher frequency of CYP1B1 rs1056836 polymorphism genotypes G/G and C/G in an EMC group than in a control group [25]. In contrast to this, some other studies found no significant association between CYP1B1 rs1056836 polymorphism and EMC risk in population [21].

In summary, this meta-analysis revealed that rs1056836 is a major risk factor for developing EMC. Further research is still needed to investigate the clinical and biological implications of these associations.

References

[1] Clarke MA, Long BJ, Morillo ADM, Arbyn M, Bakkum-Gamez JN, Wentzensen N. Association of endometrial cancer risk with postmenopausal bleeding in women: a systematic review and meta-analysis. JAMA Intern Med 2018;178(9):1210–22.

[2] Raglan O, Kalliala I, Markozannes G, Cividini S, Gunter MJ, Nautiyal J, et al. Risk factors for endometrial cancer: an umbrella review of the literature. Int J Cancer 2019;145(7):1719–30.

[3] Morice P, Leary A, Creutzberg C, Abu-Rustum N, Darai E. Endometrial cancer. Lancet 2016;387 (10023):1094–108.

[4] Galaal K, Donkers H, Bryant A, Lopes AD. Laparoscopy versus laparotomy for the management of early stage endometrial cancer. Cochrane Database Syst Rev 2018;10: CD006655.

[5] Li C, Long B, Qin X, Li W, Zhou Y. Cytochrome P1B1 (CYP1B1) polymorphisms and cancer risk: a meta-analysis of 52 studies. Toxicology 2015;327:77–86.

[6] Liu J-Y, Yang Y, Liu Z-Z, Xie J-J, Du Y-P, Wang W, et al. Association between the CYP1B1 polymorphisms and risk of cancer: a meta-analysis. Mol Genet Genomics 2015;290(2):739–65.

[7] Falero-Perez J, Song Y-S, Sorenson CM, Sheibani N. CYP1B1: a key regulator of redox homeostasis. Trends Cell Mol Biol 2018;13:27.

[8] van den Berg M, van Duursen MB. Mechanistic considerations for reduced endometrial cancer risk by smoking. Curr Opin Toxicol 2019;14:52–9.

[9] Lundin E, Wirgin I, Lukanova A, Afanasyeva Y, Krogh V, Axelsson T, et al. Selected polymorphisms in sex hormone-related genes, circulating sex hormones and risk of endometrial cancer. Cancer Epidemiol 2012;36(5):445–52.

[10] Sliwinski T, Sitarek P, Stetkiewicz T, Sobczuk A, Blasiak J. Polymorphism of the ERα and CYP1B1 genes in endometrial cancer in a Polish subpopulation. J Obstet Gynaecol Res 2010;36(2):311–7.

[11] Sasaki M, Tanaka Y, Kaneuchi M, Sakuragi N, Dahiya R. Alleles of polymorphic sites that correspond to hyperactive variants of CYP1B1 protein are significantly less frequent in Japanese as compared to American and German populations. Hum Mutat 2003;21(6):652.

[12] McGrath M, Hankinson SE, Arbeitman L, Colditz GA, Hunter DJ, De Vivo I. Cytochrome P450 1B1 and catechol-O-methyltransferase polymorphisms and endometrial cancer susceptibility. Carcinogenesis 2004;25(4):559–65.

[13] Zimarina T, Kristensen V, Imyanitov E, Berstein L. Polymorphisms of CYP1B1 and COMT in breast and endometrial cancer. Mol Biol 2004;38(3):322–8.

[14] Rylander-Rudqvist T, Wedrén S, Jonasdottir G, Ahlberg S, Weiderpass E, Persson I, et al. Cytochrome P450 1B1 gene polymorphisms and postmenopausal endometrial cancer risk. Cancer Epidemiol Biomarkers Prev 2004;13(9):1515–20.

[15] Doherty JA, Weiss NS, Freeman RJ, Dightman DA, Thornton PJ, Houck JR, et al. Genetic factors in catechol estrogen metabolism in relation to the risk of endometrial cancer. Cancer Epidemiol Biomarkers Prev 2005;14(2):357–66.

[16] Rebbeck TR, Troxel AB, Wang Y, Walker AH, Panossian S, Gallagher S, et al. Estrogen sulfation genes, hormone replacement therapy, and endometrial cancer risk. J Natl Cancer Inst 2006;98(18):1311–20.

[17] Tao MH, Cai Q, Xu WH, Kataoka N, Wen W, Zheng W, et al. Cytochrome P450 1B1 and catechol-O-methyltransferase genetic polymorphisms and endometrial cancer risk in Chinese women. Cancer Epidemiol Biomarkers Prev 2006;15(12):2570–3.

[18] Cho YJ, Hur SE, Lee JY, Song IO, Moon H-S, Koong MK, et al. Single nucleotide polymorphisms and haplotypes of the genes encoding the CYP1B1 in Korean women: no association with advanced endometriosis. J Assist Reprod Genet 2007;24(7):271–7.

[19] Ye Y, Cheng X, Luo H-B, Liu L, Li Y-B, Hou Y-P, et al. CYP1A1 and CYP1B1 genetic polymorphisms and uterine leiomyoma risk in Chinese women. J Assist Reprod Genet 2008;25(8):389–94.

[20] Hirata H, Hinoda Y, Okayama N, Suehiro Y, Kawamoto K, Kikuno N, et al. CYP1A1, SULT1A1, and SULT1E1 polymorphisms are risk factors for endometrial cancer susceptibility. Cancer 2008;112(9):1964–73.

[21] Ashton KA, Proietto A, Otton G, Symonds I, McEvoy M, Attia J, et al. Polymorphisms in genes of the steroid hormone biosynthesis and metabolism pathways and endometrial cancer risk. Cancer Epidemiol 2010;34(3):328–37.

[22] Rebbeck TR, Su HI, Sammel MD, Lin H, Tran TV, Gracia CR, et al. Effect of hormone metabolism genotypes on steroid hormone levels and menopausal symptoms in a prospective population-based cohort of women experiencing the menopausal transition. Menopause 2010;17(5):1026.

[23] El-Shennawy GA, Elbialy A-AA, Isamil AE, El Behery MM. Is genetic polymorphism of ER-α, CYP1A1, and CYP1B1 a risk factor for uterine leiomyoma? Arch Gynecol Obstet 2011;283(6):1313–8.

[24] Li Y, Tan S-Q, Ma Q-H, Li L, Huang Z-Y, Wang Y, et al. CYP1B1 C4326G polymorphism and susceptibility to cervical cancer in Chinese Han women. Tumour Biol 2013;34(6):3561–7.

[25] Zhou J, Zhang L, Wei L, Wang J. Endometrial carcinoma-related genetic factors: application to research and clinical practice in China. BJOG 2016;123:90–6.

[26] Martorell-Marugan J, Toro-Dominguez D, Alarcon-Riquelme ME, Carmona-Saez P. MetaGenyo: a web tool for meta-analysis of genetic association studies. BMC Bioinformatics 2017;18(1):563.

[27] Teng Y, He C, Zuo X, Li X. Catechol-O-methyltransferase and cytochrome P-450 1B1 polymorphisms and endometrial cancer risk: a meta-analysis. Int J Gynecol Cancer 2013;23(3):422–30.

[28] Wang F, Zou Y-F, Sun G-P, Su H, Huang F. Association of CYP1B1 gene polymorphisms with susceptibility to endometrial cancer: a meta-analysis. Eur J Cancer Prev 2011;20(2):112–20.

[29] Wang X-W, Chen Y-L, Luo Y-L, Liu Q-Y. No association between the CYP1B1 C4326G polymorphism and endometrial cancer risk: a meta-analysis. Asian Pac J Cancer Prev 2011;12:2343–8.

CHAPTER 17

Nanotechnology advances in breast cancer

Kiranmayi Patnala[a], Soumya Vishwas[a], and Rama Rao Malla[b]

[a]Department of Biotechnology, Institute of Science, GITAM (Deemed to be University), Visakhapatnam, Andhra Pradesh, India
[b]Cancer Biology Lab, Department of Biochemistry and Bioinformatics, Institute of Science, GITAM (Deemed to be University), Visakhapatnam, Andhra Pradesh, India

Abstract

Cancer, as a major cause of mortality, is a worldwide concern. In the past decade, astounding progress has been made towards comprehending cancer progression and its treatment. However, with ever-increasing incidence, managing cancer clinically is still a challenge. The most common female cancer in developed as well as developing countries is breast cancer. It is the main cause of death among women aged 20–59 years and the second leading cause of death in women older than 60 years after lung cancer. Hence, more efficient treatment for breast cancer at late stages is important for the survival of patients. Nanotechnology, an integrative field of research combining chemistry, biology, engineering, and medicine is an emerging field that meets the demands of detecting and treating cancers. Nanotechnology is being used to develop novel therapeutic strategies for female-specific cancers. In particular, nanoparticle-based drug delivery systems provide potential benefits.

Keywords: Breast cancer, Cancer management, Chemotherapeutics, Nanotechnology, Nanotherapy.

Abbreviations

AAV2	adeno-associated virus type 2
ART	artemisinin
CDK4	cyclin-dependent kinase 4
CNTs	carbon nanotubes
CPSNP	calcium phosphosilicate nanoparticle
CPSNPs	calcium phosphosilicate nanoparticle based drug delivery system
DOX	toxorubicin
DTX	docetaxel
GNPs	gold NPs
H&E	hematoxylin and eosin
HER	human epidermal growth factor receptor
ICG	indocyanine green
LTNPs	lipid nanoparticles
MRI	magnetic resonance imaging
NDDS	nanotechnology-based drug-delivery
NIR	near infrared
NPs	nanoparticles
PBS	phosphate buffered saline
PDT	photodynamic therapy

G.P. Nagaraju, R.R. Malla (eds.)
A Theranostic and Precision Medicine Approach for Female-Specific Cancers
ISBN 978-0-12-822009-2
https://doi.org/10.1016/B978-0-12-822009-2.00017-0

PEG	polyethylene glycol
PINT	photoimmunonano therapy
PTA	photothermal ablation
QDs	quantum dots
SERM	selective estrogen receptor modulator
TAM	tamoxifen
Tf	transferrin
TNBC	triple-negative breast cancer
ZnMCPPc	zinc mono carboxy phenoxyphthalocyanine

1. Introduction

Cancer is the leading cause of human death worldwide [1, 2]. It is defined as the unbridled progression and proliferation of cells and usually occurs over a span of years during which the cells become mutated [3, 4]. In cancerous cells, the growth-promoting genes duplicate several times and become unstable by acquiring fatal characteristics as they divide [5]. Female-specific cancers start in a woman's primary or secondary generative organs. Main categories of cancer in women include ovarian cancer, cervical cancer, and breast cancer. These are also termed as gynecological cancers. In the past, significant progress was made towards understanding cancer progression and its treatment. However, with ever-increasing incidence, managing cancer clinically continues to be a great challenge. A large number of studies indicated that several biological processes like angiogenesis, cell death, binding of growth factors, transcription, or signal transduction become disturbed in cancer [6]. These studies then elicited the search for rational anti-cancer drugs thereby producing a number of new compounds for cancer treatment such as lapatinib, gefitinib, rituximab, trastuzumab, cetuximab, bevacizumab, and imatinib, all of which have been approved for routine clinical trials. Current therapeutic approaches depend on either destroying the cancer cells, terminating their blood supply, or altering the impaired mechanism of the genes [7]. Conventional treatment alternatives like excising cancerous parts surgically, chemotherapy, and radiation therapy have their own constraints [8]. A surgical approach is not beneficial in all cancer cases as it might lead to complete removal of the affected organ as well as cancer reoccurrence. Radiation, although successful in destroying cancer cells, is detrimental to surrounding healthy cells as well [7]. Chemotherapy, yet another technique to kill cancer cells by drug toxicity, depends either on inhibition of cell division or hindering nutrient uptake [9]. However, chemotherapy is seldom successful for treatment of cancers at advanced stages. Currently available chemotherapeutic agents can confer disease-free existence for a short time only. One of the reasons for chemotherapy failure is poor specificity of the anticancer drugs to reach the tumor site coupled with dose-limiting toxicity [10]. However, there is a possibility to develop advanced chemotherapeutics or site-specific drug delivery systems that can overcome the hurdles of drug toxicity and resistance [11]. These systems can potentially be used to destroy cancer cells without damaging healthy tissues and deliver drug molecules effectively and specifically to the site of action for maximum efficacy of treatment. At this juncture, approaches based on nanotechnology can be effective for treatment of cancer.

2. Nanotechnology in cancer management

Nanotechnology is an integrative arena of research combining chemistry, biology, engineering, and medicine. This emerging field meets the demands for innovative approaches in detecting and treating cancers [12]. Nanotechnology can be defined as fabrication and/or utilization of materials at a nanoscale level. This can be achieved either by extending a single group of atoms or by diminishing immense materials into particles of nanoscale. Owing to their adjustable physiochemical characteristics, nanoparticles (NPs) have gained prominence in technological advancements. Different biological therapies like cancer diagnosis and treatment, imaging, and delivery of drugs use a variety of biological nanomaterials [13]. These NPs vary in size up to several hundred nanometers and readily interact with the biomolecules present on or inside the cells [14]. NPs are devised as nanoplatforms for diagnosis, imaging, and treatment of cancers. They are effective in targeted drug delivery by overcoming numerous physical, biomedical, or biological barriers. Cancer nanodevices are made up of gold NPs (GNPs), paramagnetic NPs, carbon nanotubes (CNTs), quantum dots (QDs), liposomes, and magnetic resonance imaging (MRI) agents that are used for imaging during the course of surgery. NP-based methods are used for detecting DNA and protein with high specificity and offer obvious advantages over traditional methods [12, 15–17]. Development of bioaffinity NP probes have led to easier methods of molecular and cellular imaging, drug targeting, and designing nanodevices for early detection of cancer. These advances provide plausible chances for personalized cancer treatment in which protein and genetic biomarkers can be used depending on the individual patient profile [18]. Literature on drug payload and stability show that NP formulations are well made with elevated drug-carrying capacity, which are apt for dispensing both hydrophobic and hydrophilic substances by various paths [19]. These formulations have potential to transport active drugs to cancerous cells by using the distinctive physiology of tumors selectively. The nanotechnology-based drug-delivery systems (NDDS) targeting explicitly the cancer cells show longer shelf life with better distribution and administration of drugs via various routes when compared to conventional therapies (Fig. 1).

Development in material properties of NP systems improve their distribution directly to the affected site [20]. Cancer therapeutic NPs currently in use include micelles, liposomes dendrimers, lipid NPs, protein NPs, metallic NPs, viral NPs, ceramic NPs, polymeric NPs (PNPs), and CNTs [21]. Liposomes are spherical vesicles within the size range of 50–450 nm that are composed of steroids and phospholipids. The structure of

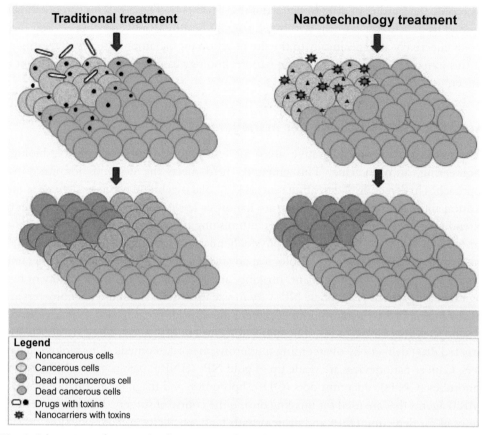

Fig. 1 Advantages of nanotechnology over traditional treatment of cancer. The nanoformulations have the potential to carry active drugs to cancerous cells by using the unique pathophysiology of tumor selectively. These nanotechnology-based drug delivery systems or NDDS target explicitly the cancer cells, leaving the adjacent non-cancerous cells unaffected. The system exhibits longer self-life with better distribution of drugs when compared to traditional chemotherapies.

liposomes is comparable to the structure of a cell membrane. This property thereby facilitates the incorporation of drugs into cells effectively [22]. Liposomes can be used with hydrophobic or hydrophilic drugs. These liposomes are capable of stabilizing the therapeutic compounds and improving their distribution. Moreover, they are biocompatible and biodegradable. Conventional liposomes are filled with both materials. Incorporation of polyethylene glycol (PEG) on to the surface of PEGylated liposomes achieves steric equilibrium. Ligands like carbohydrates, peptides, and antibodies are linked to the exterior of the ligand-targeted liposomes or tailed to the previously attached PEG chains. Theranostic liposomes are an amalgamation of conventional, PEGylated, and ligand-targeted liposomes [23].

Polymeric micelles (smaller than 100 nm) are structures made of amphiphilic copolymers that aggregate on their own in aqueous solution forming a core shell structure. Hydrophobic drugs like paclitaxel, docetaxel, or camptothecin can be loaded into the hydrophobic core, and the hydrophilic shell can solubilize the system in water thus stabilizing core particles. Due to their narrow distribution and slow renal excretion, they easily accumulate in affected tissues through the Enhanced Permeability and Retention (EPR) effect. Additionally, their polymeric shell prevents vague interactions with biological elements [24, 25]. Dendrimers are highly branched, monodisperse, distinct globular structures. These structures can be considered as outstanding agents for drug delivery as their surface can be functionalized easily [26]. Their structure allows the addition and exhibition of antigen molecules at their margin, making them multifunctional. Drugs are generally loaded into core cavities by hydrogen bonds, hydrophobic interactions, chemical linkages, or conjugation to the polymer scaffold [27].

Particles of silica, iron oxide, gold, and silver are examples of inorganic NPs. While the majority of NPs are undergoing clinical trials, only a few have been accepted for clinical use. Silver and gold NPs, unlike liposomes, dendrimers, and micelles, have a particular property of surface plasmon resonance (SPR). They have good biocompatibility and versatility in surface functionalization [28]. Nanocrystals are unalloyed particles of a drug within 1000 nm in size. These drugs do not need an attached carrier molecule and can be stabilized using surfactants or polymeric steric stabilizers. Addition of nanosuspension will alleviate nanocrystal suspension in any marginal liquid medium like water or liquid PEG and oils [29]. Specific characteristics of nanocrystals allow them to overcome pitfalls like increased stickiness to membranes, increased dissolution velocity, and increased saturation solubility.

Quantum dots (QDs) are semiconductor nanocrystals ranging from 2 to 10 nm in size. Depending upon their size, they exhibit optical properties like photoluminescence and absorbance. They emit light at <650 nm, which is a crucial characteristic for biomedical imaging as biological tissues exhibit reduced light scattering and low absorption. QDs therefore overpower the conventionally used organic dyes in the field of nanomedicine [30]. Additionally, QDs of diverse compositions and sizes can be excited to a wide range of spectra using the same light source [31, 32]. QDs are therefore very attractive for collective imaging. QDs, as contrasting agents, have also been extensively studied for imaging and sensoring in drug delivery [30, 33, 34]. Biopolymer NPs include polysaccharides and proteins originating from biological sources [35]. Being protein-based NPs, they can easily decompose and attach to particular drugs or intended ligands.

3. Breast cancer

Breast cancer is the most common cancer in females worldwide. It is a leading cause of death in women of all ages. A known cause of breast cancer mortality is metastasis to lymph nodes, lungs, liver, bones, and brain. Therefore, more effective treatment at late stages is crucial for the survival of patients. Current treatment includes surgery, radiotherapy, hormone therapy, and chemotherapy [36]. Lack of specificity is the noteworthy limitation associated with these therapies. In these cases, NP-based drug-delivery systems provide potential benefits.

4. Nanotechnology in chemotherapeutics

Taxanes and anthracyclines are the key chemotherapeutics used for late stages of breast cancer [37]. Tamoxifen (TAM) NPs, fulvestrant, and aromatase inhibitors are other well-known medications. Tamoxifen is an "anti-estrogen" drug that blocks estrogen's action by binding to its receptors in breast cancer cells. At present, most breast cancers are treated with TAM [38].

Doxil, PEGylated liposomes encapsulating doxorubicin, is an FDA-approved drug that intercalates between DNA base pairs and inhibits its synthesis and transcription. It is used to treat sarcoma, breast, and ovarian cancers [39]. Studies show a monthly dose of liposomal doxorubicin of $50\,mg/m^2$ is as efficient as the same dosage of conventional doxorubicin given every 21 days. More importantly, the risk of developing cardiotoxicity associated with peak concentration of free doxorubicin is significantly reduced with liposomal doxorubicin [40].

Genexol-PM is a nanomedicine currently undergoing clinical trials. It is paclitaxel-doped poly(D,L–lactic acid) that was demonstrated clinically to deliver a higher dose of paclitaxel than conventional chemotherapy [41]. In Phase II clinical trials, Genexol-PM was used to treat metastatic breast cancer and the overall rate of response was 43.5%–73.7%. The systemic toxicity and side effects could be contributed to the absence of polyoxyethylated castor oil, which was used to increase paclitaxel solubility and was associated with hypersensitivity reactions [42].

Paclitaxel binds to microtubules and promotes polymerization of tubulin thereby stabilizing microtubule polymers and inhibiting polymer disassembly during cell division [43]. Thus motility, mitosis, and transportation within cancer cells are blocked, leading to apoptosis [44]. Abraxane is an FDA-approved drug with particle size of ~130 nm and is prepared by homogenization of paclitaxel and human serum albumin under high pressure [45, 46]. In NP albumin–bound paclitaxel (nab-paclitaxel), albumin and paclitaxel are connected through reversible non-covalent binding. This complex binds to the gp60 receptor on the cell surface, and is later transcytosed to the extravascular space through caveolae [47].

In normal saline solution, the nab-paclitaxel can be reconstituted to <10 mg/mL, which is much greater than Cremophor EL paclitaxel. Therefore, nab-paclitaxel allows safe administration of greater doses and a significant shorter injection time without tubing or premedications for hypersensitivity reactions [45, 48]. In a Phase III clinical trial of metastatic breast cancer patients, an injection dose nearly 50% greater than the tolerated dose of Taxol was tested and the rate of clearance and distribution volume were found to be greater for nab-paclitaxel as was the response rate (34% vs 19%).

Nanomedical studies show that drugs like docetaxel (DTX), paclitaxel (Taxol), and doxorubicin (DOX), among others can be effectively targeted or delivered in a controlled manner to the affected site via graphene, fullerenes, and CNTs [49–54]. The bioavailability of DTX increases up to four times, whereas clearance decreases up to 50% when fullerene C60 is used in combination with DTX (C60/DTX) for system conformation [51]. In addition, this C60/DTX system is compatible with erythrocytes and releases DTX in a controlled manner with efficacy of 84.32% within 2 h. Moreover, the C60/DTX system showed greater cytotoxicity on MCF-7 and MDA-MB-231 cell lines when compared to using free DTX. HA-MWCNTs/Tf@ART, a drug delivery system that targets tumor in MCF-7 cells, was developed by Zhang et al. [55] to treat these cells in vitro. In this system MWCNTs functionalized with hyaluronic acid act as carriers of artemisinin (ART) drug targeting transferrin (Tf) ligand. ART showed inhibitory effect against tumors by its enhanced accumulation in cancer cells even under radiation.

According to studies, functionalized MWCNTs can act as effective cell probes in MRI because CNTs can carry imaging [56, 57]. Studies also showed that gadolinium (Gd) CNTs could act as high-quality agents for in vivo MRI done on mice [58]. CNTs functionalized with phospholipids were more biocompatible and stable during circulation in the reticuloendothelial system [59]. Zinc mono carboxy phenoxy phthalocyanine in conjugation with spermidine and adsorbed onto SWCNT (i.e., ZnMCPPc-spermine-SWCNT) showed activity of photodynamic therapy (PDT) on MCF-7 cells [60]. When ZnMCPPc-spermine-SWCNT was tested on these MCF-7 cell lines without exposing to light, it showed nontoxicity on the tested breast cancer cells. Also, a test with 40 mM of spermine showed improved PDT effect. A multifunctional system for diagnosis of cancer, magnetic targeting, PDT, and radiofrequency-assisted thermal therapy was developed [61]. This system was established using folic acid, PEG, iron oxide NPs, and fullerene C60. Furthermore, toxicity studies did not show noted results either in vivo or in vitro. Graphene nanodots can be used for cancer imaging of internal organs or deeply embedded tissues effectively. When tested on MDA-MB231 cell lines, these carboxyl functionalized nanodots of graphene could efficiently kill more than 70% of breast cancer cells, thereby proving their potential use in photodynamic therapies [62].

Hwang et al. developed a system to identify breast cancer at early stages. The system used a graphene sensor that aided in detecting not only early stage cancer but also any mutations that may have occurred [63]. Researchers at the University of Manchester stated that graphene could be used to target cancer stem cells selectively, as graphene exhibited nontoxicity to normal cells [64]. A graphene sensor proficient in identifying mutations and early-stage breast cancer was developed by Hwang et al. [63]. It was also supported by other researchers stating that graphene oxide is not toxic to healthy cells and can act as an anticancer agent specifically targeting cancer stem cells [64]. In spite of these techniques being new, they show promising results in treating or preventing breast cancer. Clinical trials of various nanomedicines for breast cancer are listed in Table 1.

Table 1 Clinical trials of nanomedicines for breast cancer.

Drugs	US manufacturer	Platform	Drug status in US
Abraxane	Abraxis Bioscience	Nanoparticle albumin-bound paclitaxel	Approved in 2005
Doxil	Janssen Products	PEGylated liposomal/doxorubicin hydrochloride	Approved in 1995
NK-105 Clinical Trials Database: NCT01644890	Nippon Kayaku	PEG-polyaspartate/paclitaxel	Phase III
Genexol-PM Clinical Trials Database: NCT00876486	Samyang Biopharmaceuticals	PEG-poly(D,L-lactide)/paclitaxel	Phase III
Myocet Clinical Trials Database: NCT00294996	Sopherion Therapeutics	Non-PEGylated liposomal/doxorubicin	Phase III
NK-012 Clinical Trials Database: NCT00951054	Nippon Kayaku	PEG-polyglutamic acid/SN-38	Phase II
LEP-ETU Xyotax Clinical Trials Database: NCT00148707	INSYS Therapeutics Dana-Farber Cancer Institute	Liposome/paclitaxel Paclitaxel poliglumex	Phase II Phase II
Liposomal annamycin Clinical Trials Database: NCT00012129	New York University, School of Medicine	Liposome/semi-synthetic doxorubicin analogue anamycin	Phase I/II
ThermoDox Clinical Trials Database: NCT00826085	Celsion	Heat-activated liposomal/doxorubicin	Phase I/II

Table 1 Clinical trials of nanomedicines for breast cancer—cont'd

Drugs	US manufacturer	Platform	Drug status in US
Rexin-G Clinical Trials Database: NCT00505271	Epeius Biotechnologies	Targeting protein tagged phospholipid/ microRNA-122	Phase I/II
SPI-077 Clinical Trials Database: NCT01861496	LiPlasome Pharma	Stealth liposomal cisplatin	Phase I
BIND-014 Clinical Trials Database: NCT01300533	BIND	PEG-polylactic-*co*-glycolic acid/docetaxel	Phase I
Nanoxel Clinical Trials Database: NCT00915369	Fresenius Kabi Oncology	PEG-poly(D,L-lactide)/ docetaxel	Phase I
S-CKD602 Clinical Trials Database: NCT00177281	Alza	PEGylated liposomal/ CKD602	Phase I

5. Nanotechnology in novel therapies

HER2 is a receptor tyrosine kinase that plays a significant role in regulation of genes, growth of cells and their mobility, propagation, apoptosis, and other responses for cancerous cells [65]. HER2/HER3 is the most vigorous and tumor-stimulating dimerized combination [66] that is overexpressed in most breast tumors [65, 67]. The overexpression of HER2 can be used in the development of targeted nanomedicine to treat breast cancer. The first clinical HER2 targeting agent is trastuzumab (Herceptin) [68], which has two antigen-specific sites that can stick to HER2, thereby preventing the activation of tyrosine kinase and HER2 dimerization [69]. Trastuzumab activates the immune cells for cell-mediated cytotoxicity to promote death of cancer cells [70]. The presence of trastuzumab also enhances the therapeutic effects of other chemotherapeutics [71]. Trastuzumab is extensively used as a single agent or combined with other chemotherapeutics in adjuvant therapy to reduce cancer recurrence and recover the complete survival of HER2-positive patients clinically [72].

The recognition and binding of the two antigen-specific sites in trastuzumab allows it to target the ligand directly to cancer cells overexpressing HER2 [69]. Anhorn et al.

developed doxorubicin-doped albumin NPs (DOX–NP) surface functionalized with trastuzumab (DOX–NP–trastuzumab). This DOX–NP–trastuzumab can bind to 73% of HER2 in breast cancer cells (SK-Br-3). This increase in cellular binding was dependent on the concentration of trastuzumab and not observed with MCF-7 cells, a cell line that feebly expresses HER2 [73]. Unfortunately, several patients acquire trastuzumab resistance within a year after the initial response to trastuzumab formulations [65]. Therefore, other HER2-targeting strategies have been developed. Pertuzumab is another monoclonal antibody that binds to HER2 and blocks the homo and hetero dimerization of HER receptors. Current trials on pertuzumab concentrate on evaluating the efficacy of pertuzumab with other drugs to treat breast cancer [74–76]. Another type of HER2 inhibitor is the tyrosine kinase antagonist lapatinib (Tykerb). A recent in vivo study suggests that lipid nanoparticles (LTNPs) were approximately three- to fivefold more effective then free lapatinib and lapatinib suspension, respectively [77]. Furthermore, the uptake of LTNPs by breast cancer cells was effective and led to apoptosis [78].

An important subset of breast cancer is triple-negative breast cancer (TNBC). It is categorized either by weak expression of progesterone receptors, estrogen receptors, and HER2, or their complete absence. Nearly 15% of female patients with breast cancer are a victim to TNBC, which is aggressive and highly metastatic [79]. The management of TNBC is a clinical challenge due to the high proliferative rates of cells, early recurrence, and poor survival rate. Chemotherapeutics for treating TNBC involve cisplatin, taxanes, and anthracyclines [80], but the overall outcome for TNBC patients is not good [81]. The overall survival rate estimated in 1 year, 3 years, and 5 years for TNBC patients is 90%, 74%, and 64%, while that for non–TNBC patients is 97%, 89%, and 81%, respectively [82]. Novel therapies for TNBC are the need of the hour. Alam et al. reported a novel therapy for TNBC that induced necrotic death and inhibited TNBC tumor growth in vivo using adeno-associated virus type 2 (AAV2). Triple-negative MDA-MB-435 cells were implanted subcutaneously and a significant destruction of tumor growth was recorded after receiving AAV2 [83]. This novel virus infection has shown potential for the treatment of breast cancer.

An LNP loaded with siRNA against cyclin-dependent kinase 4 (CDK4) showed a 16-fold increase in intracellular uptake of siRNA by MDA–MB-468 breast cancer cells in in vitro conditions [84]. The CDK4 suppression and G1 cell cycle arrest (13.8%) in MDA-MB-468 cells was efficiently induced by LNP–siRNA, while no CDK4 mRNA or G1 cell cycle downregulation occurred after incubation with free CDK4 siRNA. This indicates the potential of LNP–siRNA to obstruct cancer cell growth without side effects [84]. Honma et al. prepared a stable RPN2 siRNA–atelocollagen complex and tested the effects on improving docetaxel sensitivity of the MCF-7-ADR breast cancer cell line that is resistant to multiple drugs [85]. In the MCF-7-ADR cells in vitro study, RPN2 siRNA–atelocollagen considerably inhibited cell growth and induced apoptosis when compared to free siRNA with docetaxel. MCF-7-ADR cells treated with RPN2 siRNA–atelocollagen were more sensitive to paclitaxel (2.6-fold) and docetaxel (3.5-fold) than cells transfected with non-targeting siRNAs. A week's treatment with this complex plus docetaxel effectively decreased breast tumor size in mice implanted with MCF-7-ADR cells. There were no clear improvements when mice were treated with docetaxel, RPN2 siRNA, or siRNA–atelocollagen complex individually. Li et al. also reported the enhanced chemosensitivities to doxorubicin, paclitaxel, and cisplatin facilitating delivery of target siRNAs [86].

In NP-mediated photothermal ablation (PTA), the NPs show photothermal effect. They transform the energy absorbed from near-infrared (NIR) radiations into heat energy [87]. Most of the nanomaterials used for this therapy are metallic NPs that encounter ineffective clearance in vivo [88]. Nanomaterials based on gold and silver particles have high absorption potential in the NIR region, thereby exhibiting high photothermal effect [89]. Carpin et al. reported silica-gold nanoshells conjugated with anti-HER2 antibody for photothermal ablation in breast cancer cells [90]. Significantly more anti-HER2 nanoshells were bound to SK-BR-3 (trastuzumab-sensitive) and BT474 AZ LR (trastuzumab-resistant) when compared with HER2-negative breast cancer cells MCF-10A.

A calcium phosphosilicate NP-based drug-delivery system (CPSNPs) is considered a prominent method to encapsulate various chemotherapeutics [91]. Calcium phosphate (CP) is relatively insoluble at physiological pH (pH 7.4), but more soluble in solid tumor microenvironment (pH 5.8–7.8) and the late-stage endo-lysosomes (pH 4.8–5.5) where the encapsulants are released after [92, 93]. Docetaxel-doped CPSNPs are yet another set of emerging nanomedicines for treating breast cancer. For passive growth through increased penetrability and retention, docetaxel that is surface functionalized with PEG can be used. Similarly, docetaxel with anti-CD71 could be used for targeting breast cancer cells actively [94, 95]. It was shown by Barth et al. that targeting of CD71-CPSNPs to breast cancer cells is effective based on an study with an athymic nude mice model [95].

CPSNPs doped with indocyanine green (ICG-CPSNPs) are used for NIR bioimaging that acts as a photosensitizer for novel Photo Immuno Nano Therapy (PINT) [96, 97]. Athymic nude mice with MDA-MB-231 breast cancer tumors (T-cell deficient) and BALB/cJ mice with murine 410.4 breast cancer cells (T-cell competent) were injected with PEGylated ICG-CPSNPs, free ICG, PEGylated CPNSPs without encapsulant, and citrate surface-functionalized ICG-CPSNPs followed by NIR laser treatment. The tumor growth in both models was effectively suppressed after PINT-mediated therapy by PEGylated ICG-CPSNPs. PINT was used for the first time to treat chronic myeloid leukemia, a nonsolid tumor, signifying its potential application in the treatment of metastatic breast cancer (Table 2).

Table 2 Summary of novel nanotherapies for breast cancer.

Active agents	Strategy	Nanotherapy
Trastuzumab, pertuzuma, and lapatinib	Targets overexpressed HER2 BCs by inhibiting tyrosine kinase activity and homo or heterodimerization of HER2	Targeted nanotherapy
Nano-encapsulated AAV2	Inhibits BC tumor growth by inducing necrosis	Gene therapy
siRNA–atelocollagen complex	Inhibits the growth and induce apoptosis of tumor cells by improving the uptake of siRNA–atelocollagen complex	Gene therapy
Metallic-based nanomaterials (AU or Ag) or nanomaterials with high photothermal conversion effects	Target BC by release heat energy above the thermal ablation threshold	Photothermal ablation (PTA)
CPSNPs-doped docetaxel, indocyanine green (photosensitizer)	Target BCs at late stage endo-lysosome formation by releasing active agents	Photodynamic therapy

6. Conclusion

Cancer nanotechnology is a quickly expanding field in cancer biomedicine. It has wide applications in the fields of diagnostics and treatment of breast cancer. Use of NPs is more efficacious and potentially safer when compared to other conventional treatments for breast cancer. There are several FDA-approved systems that depend on NPs for effective cancer treatment. However, there are still other NP-based systems that are undergoing clinical trials. The NPs used in cancer therapies are so adaptable that they can easily target cancer sites of different types and have the potential to deliver various active agents to the targeted sites. Although the unique properties of different NPs make nanomedical studies strenuous, they have great potential to improve the condition of the patient. Further human efforts are desirable for implementing techniques not only to treat breast cancer but also to control its recurrence and propagation.

Funding

The authors thank the DST-EMR (EMR/2016/002694, dt. 21st August 2017), New Delhi, India for supporting the project.

Conflict of interest

The authors declared that there is no conflict of interest.

Acknowledgment

The authors are grateful to the Department of Biotechnology, Institute of Science, Gandhi Institute of Technology and Management, Visakhapatnam, Andhra Pradesh, India.

References

[1] Siegel RL, Miller KD, Jemal A. Cancer statistics, 2015. CA Cancer J Clin 2015;65(1):5–29.
[2] Arnold R. Prospective cancer treatment found in the Cowpea Mosaic Virus. Microrev Cell Mol Biol 2016;1(1):1–3.
[3] Anand P, Kunnumakkara AB, Sundaram C, Harikumar KB, Tharakan ST, Lai OS, Sung B, Aggarwal BB. Cancer is a preventable disease that requires major lifestyle changes. Pharm Res 2008;25(9):2097–116.
[4] Duncan R. Polymer conjugates as anticancer nanomedicines. Nat Rev Cancer 2006;6(9):688–701.
[5] Couvreur P, Vauthier C. Nanotechnology: intelligent design to treat complex disease. Pharm Res 2006;23(7):1417–50.
[6] Sikora K. The impact of future technology on cancer care. Clin Med 2002;2(6):560–8.
[7] Singh OP, Nehru R. Nanotechnology and cancer treatment. Asian J Exp Sci 2008;22(2):6.
[8] Rajitha B, Malla RR, Vadde R, Kasa P, Prasad GLV, Farran B, Kumari S, Pavitra E, Kamal MA, Raju GSR, Peela S, Nagaraju GP. Horizons of nanotechnology applications in female specific cancers. Semin Cancer Biol 2019;**19**:30090–2.
[9] Chidambaram M, Manavalan R, Kathiresan K. Nanotherapeutics to overcome conventional cancer chemotherapy limitations. Int J Pharm Pharm Sci 2011;14(1):67–77.
[10] Sinha R, Kim GJ, Nie S, Shin DM. Nanotechnology in cancer therapeutics: bioconjugated nanoparticles for drug delivery. Mol Cancer Ther 2006;5(8):1909–17.
[11] Ranganathan R, Madanmohan S, Kesavan A, Baskar G, Krishnamoorthy YR, Santosham R, Ponraju D, Rayala SK, Venkatraman G. Nanomedicine: towards development of patient-friendly drug-delivery systems for oncological applications. Int J Nanomedicine 2012;7:1043–60.
[12] Cai W, Gao T, Hong H, Sun J. Applications of gold nanoparticles in cancer nanotechnology. Nanotechnol Sci Appl 2008;1:17–32.
[13] Ghanbari H, de Mel A, Seifalian AM. Cardiovascular application of polyhedral oligomeric silsesquioxane nanomaterials: a glimpse into prospective horizons. Int J Nanomedicine 2011;6:775–86.
[14] Elsersawi A. World of nanobioengineering: potential big ideas for the future. Author House.
[15] Kircher MF, Mahmood U, King RS, Weissleder R, Josephson L. A multimodal nanoparticle for preoperative magnetic resonance imaging and intraoperative optical brain tumor delineation. Cancer Res 2003;63(23):8122–5.
[16] Ferrari M. Cancer nanotechnology: opportunities and challenges. Nat Rev Cancer 2005;5(3):161–71.
[17] Jamieson T, Bakhshi R, Petrova D, Pocock R, Imani M, Seifalian AM. Biological applications of quantum dots. Biomaterials 2007;28(31):4717–32.
[18] Cai W, Chen X. Multimodality molecular imaging of tumor angiogenesis. J Nucl Med 2008;49(Suppl. 2):113s–128s.
[19] Gelperina S, Kisich K, Iseman MD, Heifets L. The potential advantages of nanoparticle drug delivery systems in chemotherapy of tuberculosis. Am J Respir Crit Care Med 2005;172(12):1487–90.
[20] Byrne JD, Betancourt T, Brannon-Peppas L. Active targeting schemes for nanoparticle systems in cancer therapeutics. Adv Drug Deliv Rev 2008;60(15):1615–26.
[21] Hahn MA, Singh AK, Sharma P, Brown SC, Moudgil BM. Nanoparticles as contrast agents for in-vivo bioimaging: current status and future perspectives. Anal Bioanal Chem 2011;399(1):3–27.
[22] Bozzuto G, Molinari A. Liposomes as nanomedical devices. Int J Nanomedicine 2015;10:975.
[23] Sercombe L, Veerati T, Moheimani F, Wu SY, Sood AK, Hua S. Advances and challenges of liposome assisted drug delivery. Front Pharmacol 2015;6:286.
[24] Xu W, Ling P, Zhang T. Polymeric micelles, a promising drug delivery system to enhance bioavailability of poorly water-soluble drugs. J Drug Deliv 2013;2013:340315.

[25] Miyata K, Christie RJ, Kataoka K. Polymeric micelles for nano-scale drug delivery. React Funct Polym 2011;71(3):227–34.

[26] Zhu J, Shi X. Dendrimer-based nanodevices for targeted drug delivery applications. J Mater Chem B 2013;1(34):4199–211.

[27] Ordikhani F, Erdem Arslan M, Marcelo R, Sahin I, Grigsby P, Schwarz JK, Azab AK. Drug delivery approaches for the treatment of cervical cancer. Pharmaceutics 2016;8(3):23.

[28] Choi S-J, Lee JK, Jeong J, Choy J-H. Toxicity evaluation of inorganic nanoparticles: considerations and challenges. Mol Cell Toxicol 2013;9(3):205–10.

[29] Du J, Li X, Zhao H, Zhou Y, Wang L, Tian S, Wang Y. Nanosuspensions of poorly water-soluble drugs prepared by bottom-up technologies. Int J Pharm 2015;495(2):738–49.

[30] Volkov Y. Quantum dots in nanomedicine: recent trends, advances and unresolved issues. Biochem Biophys Res Commun 2015;468(3):419–27.

[31] Liu J, Lau SK, Varma VA, Moffitt RA, Caldwell M, Liu T, Young AN, Petros JA, Osunkoya AO, Krogstad T, Leyland-Jones B, Wang MD, Nie S. Molecular mapping of tumor heterogeneity on clinical tissue specimens with multiplexed quantum dots. ACS Nano 2010;4(5):2755–65.

[32] Xu G, Zeng S, Zhang B, Swihart MT, Yong KT, Prasad PN. New generation cadmium-free quantum dots for biophotonics and nanomedicine. Chem Rev 2016;116(19):12234–327.

[33] Shi Y, Pramanik A, Tchounwou C, Pedraza F, Crouch RA, Chavva SR, Vangara A, Sinha SS, Jones S, Sardar D, Hawker C, Ray PC. Multifunctional biocompatible graphene oxide quantum dots decorated magnetic nanoplatform for efficient capture and two-photon imaging of rare tumor cells. ACS Appl Mater Interfaces 2015;7(20):10935–43.

[34] Han HS, Niemeyer E, Huang Y, Kamoun WS, Martin JD, Bhaumik J, Chen Y, Roberge S, Cui J, Martin MR, Fukumura D, Jain RK, Bawendi MG, Duda DG. Quantum dot/antibody conjugates for in vivo cytometric imaging in mice. Proc Natl Acad Sci U S A 2015;112(5):1350–5.

[35] Bassas-Galia M, Follonier S, Pusnik M, Zinn M. Natural polymers: a source of inspiration. In: Bioresorbable polymers for biomedical applications. Elsevier; 2017. p. 31–64.

[36] Hortobagyi GN. Treatment of breast cancer. N Engl J Med 1998;339(14):974–84.

[37] Peto R, Davies C, Godwin J, Gray R, Pan HC, Clarke M, Cutter D, Darby S, McGale P, Taylor C, Wang YC, Bergh J, Di Leo A, Albain K, Swain S, Piccart M, Pritchard K. Comparisons between different polychemotherapy regimens for early breast cancer: meta-analyses of long-term outcome among 100,000 women in 123 randomised trials. Lancet 2012;379(9814):432–44.

[38] Davies C, Godwin J, Gray R, Clarke M, Cutter D, Darby S, McGale P, Pan HC, Taylor C, Wang YC, Dowsett M, Ingle J, Peto R. Relevance of breast cancer hormone receptors and other factors to the efficacy of adjuvant tamoxifen: patient-level meta-analysis of randomised trials. Lancet 2011; 378(9793):771–84.

[39] Barenholz Y. Doxil®–the first FDA-approved nano-drug: lessons learned. J Control Release 2012; 160(2):117–34.

[40] O'Brien ME, Wigler N, Inbar M, Rosso R, Grischke E, Santoro A, Catane R, Kieback D, Tomczak P, Ackland S. Reduced cardiotoxicity and comparable efficacy in a phase III trial of pegylated liposomal doxorubicin HCl (CAELYX™/Doxil®) versus conventional doxorubicin for first-line treatment of metastatic breast cancer. Ann Oncol 2004;15(3):440–9.

[41] Tang X, Loc WS, Dong C, Matters GL, Butler PJ, Kester M, Meyers C, Jiang Y, Adair JH. The use of nanoparticulates to treat breast cancer. Nanomedicine 2017;12(19):2367–88.

[42] Gelderblom H, Verweij J, Nooter K, Sparreboom A. Cremophor EL: the drawbacks and advantages of vehicle selection for drug formulation. Eur J Cancer 2001;37(13):1590–8.

[43] Mackler NJ, Pienta KJ. Drug insight: use of docetaxel in prostate and urothelial cancers. Nat Clin Pract Urol 2005;2(2):92–100 quiz 1 p following 112.

[44] Jordan MA, Wilson L. Microtubules as a target for anticancer drugs. Nat Rev Cancer 2004;4 (4):253–65.

[45] Ibrahim NK, Desai N, Legha S, Soon-Shiong P, Theriault RL, Rivera E, Esmaeli B, Ring SE, Bedikian A, Hortobagyi GN, Ellerhorst JA. Phase I and pharmacokinetic study of ABI-007, a Cremophor-free, protein-stabilized, nanoparticle formulation of paclitaxel. Clin Cancer Res 2002; 8(5):1038–44.

[46] Miele E, Spinelli GP, Miele E, Tomao F, Tomao S. Albumin-bound formulation of paclitaxel (Abraxane ABI-007) in the treatment of breast cancer. Int J Nanomedicine 2009;4:99–105.

[47] Hawkins MJ, Soon-Shiong P, Desai N. Protein nanoparticles as drug carriers in clinical medicine. Adv Drug Deliv Rev 2008;60(8):876–85.

[48] Sparreboom A, Scripture CD, Trieu V, Williams PJ, De T, Yang A, Beals B, Figg WD, Hawkins M, Desai N. Comparative preclinical and clinical pharmacokinetics of a cremophor-free, nanoparticle albumin-bound paclitaxel (ABI-007) and paclitaxel formulated in Cremophor (Taxol). Clin Cancer Res 2005;11(11):4136–43.

[49] Hashemi M, Yadegari A, Yazdanpanah G, Omidi M, Jabbehdari S, Haghiralsadat F, Yazdian F, Tayebi L. Normalization of doxorubicin release from graphene oxide: new approach for optimization of effective parameters on drug loading. Biotechnol Appl Biochem 2017;64(3):433–42.

[50] Shi J, Wang B, Wang L, Lu T, Fu Y, Zhang H, Zhang Z. Fullerene (C60)-based tumor-targeting nanoparticles with "off-on" state for enhanced treatment of cancer. J Control Release 2016;235:245–58.

[51] Raza K, Thotakura N, Kumar P, Joshi M, Bhushan S, Bhatia A, Kumar V, Malik R, Sharma G, Guru SK. C60-fullerenes for delivery of docetaxel to breast cancer cells: a promising approach for enhanced efficacy and better pharmacokinetic profile. Int J Pharm 2015;495(1):551–9.

[52] Jiang T, Sun W, Zhu Q, Burns NA, Khan SA, Mo R, Gu Z. Furin-mediated sequential delivery of anticancer cytokine and small-molecule drug shuttled by graphene. Adv Mater 2015;27(6):1021–8.

[53] Xu Z, Zhu S, Wang M, Li Y, Shi P, Huang X. Delivery of paclitaxel using PEGylated graphene oxide as a nanocarrier. ACS Appl Mater Interfaces 2015;7(2):1355–63.

[54] Zhou T, Zhou X, Xing D. Controlled release of doxorubicin from graphene oxide based charge-reversal nanocarrier. Biomaterials 2014;35(13):4185–94.

[55] Zhang H, Ji Y, Chen Q, Jiao X, Hou L, Zhu X, Zhang Z. Enhancement of cytotoxicity of artemisinin toward cancer cells by transferrin-mediated carbon nanotubes nanoparticles. J Drug Target 2015;23(6):552–67.

[56] Hernández-Rivera M, Zaibaq NG, Wilson LJ. Toward carbon nanotube-based imaging agents for the clinic. Biomaterials 2016;101:229–40.

[57] Servant A, Jacobs I, Bussy C, Fabbro C, Da Ros T, Pach E, Ballesteros B, Prato M, Nicolay K, Kostarelos K. Gadolinium-functionalised multi-walled carbon nanotubes as a T1 contrast agent for MRI cell labelling and tracking. Carbon 2016;97:126–33.

[58] Marangon I, Ménard-Moyon C, Kolosnjaj-Tabi J, Béoutis ML, Lartigue L, Alloyeau D, Pach E, Ballesteros B, Autret G, Ninjbadgar T. Covalent functionalization of multi-walled carbon nanotubes with a gadolinium chelate for efficient T1-weighted magnetic resonance imaging. Adv Funct Mater 2014;24(45):7173–86.

[59] Mallick K, Strydom AM. Biophilic carbon nanotubes. Colloids Surf B Biointerfaces 2013;105:310–8.

[60] Ogbodu RO, Limson JL, Prinsloo E, Nyokong T. Photophysical properties and photodynamic therapy effect of zinc phthalocyanine-spermine-single walled carbon nanotube conjugate on MCF-7 breast cancer cell line. Synth Met 2015;204:122–32.

[61] Shi J, Wang L, Gao J, Liu Y, Zhang J, Ma R, Liu R, Zhang Z. A fullerene-based multi-functional nanoplatform for cancer theranostic applications. Biomaterials 2014;35(22):5771–84.

[62] Nurunnabi M, Khatun Z, Reeck GR, Lee DY, Lee YK. Photoluminescent graphene nanoparticles for cancer phototherapy and imaging. ACS Appl Mater Interfaces 2014;6(15):12413–21.

[63] Hwang MT, Landon PB, Lee J, Choi D, Mo AH, Glinsky G, Lal R. Highly specific SNP detection using 2D graphene electronics and DNA strand displacement. Proc Natl Acad Sci U S A 2016;113(26):7088–93.

[64] Fiorillo M, Verre AF, Iliut M, Peiris-Pagés M, Ozsvari B, Gandara R, Cappello AR, Sotgia F, Vijayaraghavan A, Lisanti MP. Graphene oxide selectively targets cancer stem cells, across multiple tumor types: implications for non-toxic cancer treatment, via "differentiation-based nano-therapy". Oncotarget 2015;6(6):3553–62.

[65] Meric-Bernstam F, Hung MC. Advances in targeting human epidermal growth factor receptor-2 signaling for cancer therapy. Clin Cancer Res 2006;12(21):6326–30.

[66] Pinkas-Kramarski R, Lenferink AE, Bacus SS, Lyass L, van de Poll ML, Klapper LN, Tzahar E, Sela M, van Zoelen EJ, Yarden Y. The oncogenic ErbB-2/ErbB-3 heterodimer is a surrogate receptor of the epidermal growth factor and betacellulin. Oncogene 1998;16(10):1249–58.

[67] Tan M, Yao J, Yu D. Overexpression of the c-erbB-2 gene enhanced intrinsic metastasis potential in human breast cancer cells without increasing their transformation abilities. Cancer Res 1997; 57(6):1199–205.

[68] Carter P, Presta L, Gorman CM, Ridgway JB, Henner D, Wong WL, Rowland AM, Kotts C, Carver ME, Shepard HM. Humanization of an anti-p185HER2 antibody for human cancer therapy. Proc Natl Acad Sci U S A 1992;89(10):4285–9.

[69] Hudis CA. Trastuzumab—mechanism of action and use in clinical practice. N Engl J Med 2007; 357(1):39–51.

[70] Park S, Jiang Z, Mortenson ED, Deng L, Radkevich-Brown O, Yang X, Sattar H, Wang Y, Brown NK, Greene M, Liu Y, Tang J, Wang S, Fu YX. The therapeutic effect of anti-HER2/neu antibody depends on both innate and adaptive immunity. Cancer Cell 2010;18(2):160–70.

[71] Pegram M, Hsu S, Lewis G, Pietras R, Beryt M, Sliwkowski M, Coombs D, Baly D, Kabbinavar F, Slamon D. Inhibitory effects of combinations of HER-2/neu antibody and chemotherapeutic agents used for treatment of human breast cancers. Oncogene 1999;18(13):2241–51.

[72] Romond EH, Perez EA, Bryant J, Suman VJ, Geyer Jr. CE, Davidson NE, Tan-Chiu E, Martino S, Paik S, Kaufman PA, Swain SM, Pisansky TM, Fehrenbacher L, Kutteh LA, Vogel VG, Visscher DW, Yothers G, Jenkins RB, Brown AM, Dakhil SR, Mamounas EP, Lingle WL, Klein PM, Ingle JN, Wolmark N. Trastuzumab plus adjuvant chemotherapy for operable HER2-positive breast cancer. N Engl J Med 2005;353(16):1673–84.

[73] Anhorn MG, Wagner S, Kreuter J, Langer K, von Briesen H. Specific targeting of HER2 overexpressing breast cancer cells with doxorubicin-loaded trastuzumab-modified human serum albumin nanoparticles. Bioconjug Chem 2008;19(12):2321–31.

[74] Baselga J, Cortés J, Kim SB, Im SA, Hegg R, Im YH, Roman L, Pedrini JL, Pienkowski T, Knott A, Clark E, Benyunes MC, Ross G, Swain SM. Pertuzumab plus trastuzumab plus docetaxel for metastatic breast cancer. N Engl J Med 2012;366(2):109–19.

[75] Gianni L, Pienkowski T, Im YH, Roman L, Tseng LM, Liu MC, Lluch A, Staroslawska E, de la Haba-Rodriguez J, Im SA, Pedrini JL, Poirier B, Morandi P, Semiglazov V, Srimuninnimit V, Bianchi G, Szado T, Ratnayake J, Ross G, Valagussa P. Efficacy and safety of neoadjuvant pertuzumab and trastuzumab in women with locally advanced, inflammatory, or early HER2-positive breast cancer (NeoSphere): a randomised multicentre, open-label, phase 2 trial. Lancet Oncol 2012;13(1):25–32.

[76] Baselga J, Gelmon KA, Verma S, Wardley A, Conte P, Miles D, Bianchi G, Cortes J, McNally VA, Ross GA, Fumoleau P, Gianni L. Phase II trial of pertuzumab and trastuzumab in patients with human epidermal growth factor receptor 2-positive metastatic breast cancer that progressed during prior trastuzumab therapy. J Clin Oncol Off J Am Soc Clin Oncol 2010;28(7):1138–44.

[77] Gao H, Chen C, Xi Z, Chen J, Zhang Q, Cao S, Jiang X. In vivo behavior and safety of lapatinib-incorporated lipid nanoparticles. Curr Pharm Biotechnol 2014;14(12):1062–71.

[78] Gao H, Cao S, Chen C, Cao S, Yang Z, Pang Z, Xi Z, Pan S, Zhang Q, Jiang X. Incorporation of lapatinib into lipoprotein-like nanoparticles with enhanced water solubility and anti-tumor effect in breast cancer. Nanomedicine 2013;8(9):1429–42.

[79] Foulkes WD, Smith IE, Reis-Filho JS. Triple-negative breast cancer. N Engl J Med 2010;363 (20):1938–48.

[80] Hudis CA, Gianni L. Triple-negative breast cancer: an unmet medical need. Oncologist 2011;16 (Suppl. 1):1–11.

[81] Tan DS, Marchió C, Jones RL, Savage K, Smith IE, Dowsett M, Reis-Filho JS. Triple negative breast cancer: molecular profiling and prognostic impact in adjuvant anthracycline-treated patients. Breast Cancer Res Treat 2008;111(1):27–44.

[82] Liedtke C, Mazouni C, Hess KR, André F, Tordai A, Mejia JA, Symmans WF, Gonzalez-Angulo AM, Hennessy B, Green M, Cristofanilli M, Hortobagyi GN, Pusztai L. Response to neoadjuvant therapy and long-term survival in patients with triple-negative breast cancer. J Clin Oncol Off J Am Soc Clin Oncol 2008;26(8):1275–81.

[83] Alam S, Bowser BS, Israr M, Conway MJ, Meyers C. Adeno-associated virus type 2 infection of nude mouse human breast cancer xenograft induces necrotic death and inhibits tumor growth. Cancer Biol Ther 2014;15(8):1013–28.

[84] Wang X, Yu B, Wu Y, Lee RJ, Lee LJ. Efficient down-regulation of CDK4 by novel lipid nanoparticle-mediated siRNA delivery. Anticancer Res 2011;31(5):1619–26.

[85] Honma K, Iwao-Koizumi K, Takeshita F, Yamamoto Y, Yoshida T, Nishio K, Nagahara S, Kato K, Ochiya T. RPN2 gene confers docetaxel resistance in breast cancer. Nat Med 2008;14(9):939–48.

[86] Li YT, Chua MJ, Kunnath AP, Chowdhury EH. Reversing multidrug resistance in breast cancer cells by silencing ABC transporter genes with nanoparticle-facilitated delivery of target siRNAs. Int J Nanomedicine 2012;7:2473–81.

[87] Guo Y, Zhang Z, Kim DH, Li W, Nicolai J, Procissi D, Huan Y, Han G, Omary RA, Larson AC. Photothermal ablation of pancreatic cancer cells with hybrid iron-oxide core gold-shell nanoparticles. Int J Nanomedicine 2013;8:3437–46.

[88] Longmire M, Choyke PL, Kobayashi H. Clearance properties of nano-sized particles and molecules as imaging agents: considerations and caveats. Nanomedicine 2008;3(5):703–17.

[89] Huang X, Jain PK, El-Sayed IH, El-Sayed MA. Gold nanoparticles: interesting optical properties and recent applications in cancer diagnostics and therapy. Nanomedicine 2007;2(5):681–93.

[90] Carpin LB, Bickford LR, Agollah G, Yu TK, Schiff R, Li Y, Drezek RA. Immunoconjugated gold nanoshell-mediated photothermal ablation of trastuzumab-resistant breast cancer cells. Breast Cancer Res Treat 2011;125(1):27–34.

[91] Altinoğlu EI, Adair JH. Near infrared imaging with nanoparticles. Wiley Interdiscip Rev Nanomed Nanobiotechnol 2010;2(5):461–77.

[92] Panyam J, Labhasetwar V. Biodegradable nanoparticles for drug and gene delivery to cells and tissue. Adv Drug Deliv Rev 2003;55(3):329–47.

[93] Pinto O, Tabaković A, Goff T, Liu Y, Adair J. Calcium phosphate and calcium phosphosilicate mediated drug delivery and imaging. In: Intracellular delivery. Springer; 2011. p. 713–44.

[94] Altinoğlu EI, Russin TJ, Kaiser JM, Barth BM, Eklund PC, Kester M, Adair JH. Near-infrared emitting fluorophore-doped calcium phosphate nanoparticles for in vivo imaging of human breast cancer. ACS Nano 2008;2(10):2075–84.

[95] Barth BM, Sharma R, Altinoğlu EI, Morgan TT, Shanmugavelandy SS, Kaiser JM, McGovern C, Matters GL, Smith JP, Kester M, Adair JH. Bioconjugation of calcium phosphosilicate composite nanoparticles for selective targeting of human breast and pancreatic cancers in vivo. ACS Nano 2010; 4(3):1279–87.

[96] Barth BM, Erhan IA, Shanmugavelandy SS, Kaiser JM, Crespo-Gonzalez D, DiVittore NA, McGovern C, Goff TM, Keasey NR, Adair JH, Loughran Jr. TP, Claxton DF, Kester M. Targeted indocyanine-green-loaded calcium phosphosilicate nanoparticles for in vivo photodynamic therapy of leukemia. ACS Nano 2011;5(7):5325–37.

[97] Barth BM, Shanmugavelandy SS, Kaiser JM, McGovern C, Altınoğlu E, Haakenson JK, Hengst JA, Gilius EL, Knupp SA, Fox TE, Smith JP, Ritty TM, Adair JH, Kester M. PhotoImmunoNanoTherapy reveals an anticancer role for sphingosine kinase 2 and dihydrosphingosine-1-phosphate. ACS Nano 2013;7(3):2132–44.

Index

Note: Page numbers followed by *f* indicate figures and *t* indicate tables.